Ion Exchange Technology

Ion Exchange Technology

Edited by Carla Mayfield

Larsen & Keller
www.larsen-keller.com

Ion Exchange Technology
Edited by Carla Mayfield
ISBN: 979-8-88836-011-8 (Hardback)

Larsen & Keller
Published by Larsen and Keller Education,
5 Penn Plaza,
19th Floor,
New York, NY 10001, USA

Cataloging-in-Publication Data

Ion exchange technology / edited by Carla Mayfield.
p. cm.
Includes bibliographical references and index.
ISBN 979-8-88836-011-8
1. Ion exchange. 2. Chemical engineering.
3. Ion exchange--Industrial applications.
4. Ion exchange resins. I. Mayfield, Carla.
TP156.I6 I56 2023
660.297 23--dc23

For more information regarding Larsen and Keller Education and its products, please visit the publisher's website www.larsen-keller.com

Contents

Preface

Ion exchange refers to a reversible chemical reaction in which an ion attached to an immobile solid particle is exchanged for a similarly charged ion from a wastewater solution. It is a method of purifying aqueous solutions utilizing solid polymeric ion-exchange resin. Apart from its use in purification of drinking water, this technology is broadly used for the separation of a wide range of medicinally and industrially significant chemicals. Various industries that are dependent on the ion exchange technology comprise power, pharmaceutical, chemical processing, nuclear mining, food and waste water processing industries. This book is a compilation of chapters that discuss the most vital concepts and emerging trends related to ion exchange technology. From theories to research to practical applications, studies related to all contemporary topics of relevance to this subject have been included herein. Those in search of information to further their knowledge will be greatly assisted by this book.

This book is a comprehensive compilation of works of different researchers from varied parts of the world. It includes valuable experiences of the researchers with the sole objective of providing the readers (learners) with a proper knowledge of the concerned field. This book will be beneficial in evoking inspiration and enhancing the knowledge of the interested readers.

In the end, I would like to extend my heartiest thanks to the authors who worked with great determination on their chapters. I also appreciate the publisher's support in the course of the book. I would also like to deeply acknowledge my family who stood by me as a source of inspiration during the project.

Editor

1

Producing the Gradient Changes in Glass Refraction by the Ion Exchange Method — Selected Aspects

Roman Rogoziński

Abstract

This chapter presents the chosen aspects of the ion exchange technological method. Results of research refer to the refractive index profiles of planar waveguides produced by the ion exchange method in various glasses. All refractive index profiles were determined by measuring the effective refractive indices of the waveguide modes. In the first part of this chapter the processes of electrodiffusion doping soda-lime glass with silver ions has been described. The second part contains a description of the two measurement methods allowing to estimate the equilibrium concentration basing on the example of the concentration of sodium in soda-lime glass. The third part describes the phenomenon of stress birefringence arising in borosilicate BK-7 glass doped with potassium ions K^+. The fourth section presents the results of research on the real-time control of ion exchange diffusion processes. The results confirming the validity of this idea with examples of soda-lime glass (Menzel-Gläser company), BK-7 (Schott company) and Pyrex (Borosilicate 33 of Corning company) doped with Ag^+ silver ions and K^+ potassium ions are also presented here.

Keywords: ion exchange, planar waveguides, electrodiffusion

1. Introduction

The development of the ion exchange technological method as applied to the preparation of gradient changes of the refraction in the surface area of the glass was initiated in the 1970s. The produced changes in refraction can be used to conduct the electromagnetic wave along the surface of the glass (then we talk about waveguide structures) or provide refractive areas for wave passing (perpendicular) through the glass. In each of these cases we have to deal with the gradient structures of planar topology (Fig.1). In the case of waveguide structures we can distinguish planar waveguides (Fig.1a) or waveguides of channel type (Fig.1b). In the first case the refractive index gradient has only one coordinate, while in the waveguides of channel type it has two. On the basis of such structures various elements of planar waveguide optics can be constructed, such as: splitters of Y type [1], splitters of 1xN type [2], multiplexers [3-7], directional couplers [8], ring resonators [9-10], planar polarizers [11], planar interferometers [12-13] or gradient MMI structures [14-16]. These elements can cooperate with the fiber waveguides. In addition to the passive components of planar optics, by the ion exchange method the structures of lasers and optical amplifiers can also be produced. For this purpose glasses doped with rare earth ions are used [17-23]. On the basis of gradient planar waveguide structures the optical sensors are also produced [24-28]. By the ion exchange method the refractive structures are produced as well [29-32]. In such structures (e.g., lenses or microlenses) the refraction gradient is generally three-dimensional (Fig.1c).

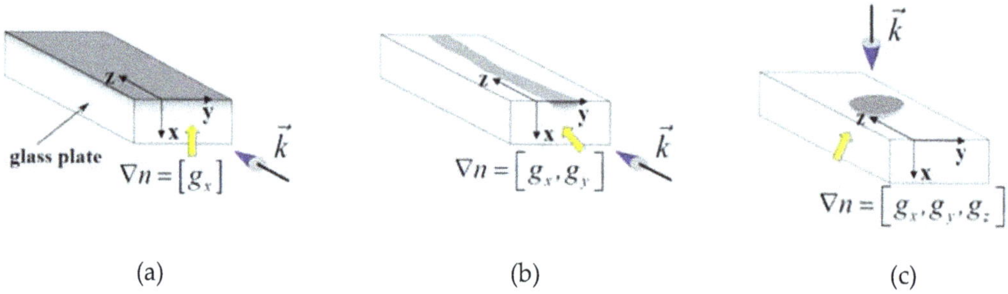

Figure 1. The classification of gradient structures in glass: (a) planar of slab type, (b) planar of channel type, (c) refractive.

The ion exchange technological method is relatively cheap both in terms of used materials, and the apparatus required for its implementation, which is undoubtedly its main advantage. It has been mainly applied to the composite oxide glasses. The physical basis of this method is the phenomenon of ionic conductivity of glasses. This phenomenon occurs in the glasses that contain modifier ions possessing low binding energy with glass network. In these glasses at elevated temperature the modifier ions obtain energy to enable them to migrate within the glass network [33]. This migration is manifested in the form of ionic current flowing through the glass in the presence of an electric field generated in the area. The migration of modifier ions at an elevated temperature enables also the realization of ion exchange in glass. This situation occurs when at a sufficiently high temperature the glass is in contact with the phase of the source of these ions. Most often it is a liquid phase in the form of molten salt [33]. The most commonly used liquid sources of admixture ions are molten nitrates, which have low melting points. Such sources can be considered as sources of constant productivity. At sufficiently high temperature ionized admixture obtains sufficient energy to overcome the potential barrier at the surface of the glass. At the same time one modifier ion has to "come out" from the glass to the source of admixture, so that the electrical neutrality is maintained throughout the volume of the glass. In this way, the concentration gradient of admixture introduced into the glass is initiated in the surface area of glass. In this area the concentration gradient of the glass modifier ions is also formed. It has a direction opposite to the gradient of concentration of the admixture. Consequently, in the glass there are two opposing streams of ions: a stream of admixture ions directed into the glass and a stream of modifier ions directed to the surface of the glass. This phenomenon has a diffusive character. Its effect is to create in the glass area the gradient changes of its optical properties (refraction) and mechanical properties (stress). The local changes in the refractive index of the glass, which are the results of ion exchange, are the effect of the difference of their electric polarizabilities, as well as differences in their ionic radii.

This chapter is based on the physical basis of this phenomenon described in detail in work [33]. Here the chosen aspects of the ion exchange technological method are going to be presented. Results of research refer to the refractive index profiles of planar waveguides produced by the ion exchange method in various glasses. All refractive index profiles were determined by measuring the effective refractive indices of the waveguide modes. The goniometric method of synchronous angles measurements with the use of a prism coupler has been used. A detailed description of the measurement method and the method for determining the refractive index profiles was given in the work [33] (pp.166-172).

In the first part of this chapter, the process of electrodiffusion doping soda-lime glass with silver ions has been described. The second part contains a description of the two measurement methods allowing to estimate the equilibrium concentration basing on the example of the concentration of sodium in soda-lime glass. The third part describes the phenomenon of stress birefringence arising in borosilicate BK-7 glass doped with potassium ions K^+. The fourth section presents the results of research on the real-time control of ion exchange diffusion processes. The idea of this method has been described in work [33]. Here are presented the results confirming the validity of this idea with examples of soda-lime glass (Menzel-Gläser company), BK-7 (Schott company) and Pyrex (Borosilicate 33 of Corning company) doped with Ag^+ silver ions and K^+ potassium ions.

2. Ion exchange in glasses in the presence of an electric field

In the two-component model of the ion exchange in the glass [33] there are two oppositely directed streams of ions: stream of admixture ions $\vec{\Phi}_A$ and stream of modifier ions $\vec{\Phi}_B$ (Fig.2). The stream of admixture ions diffusing in the glass in the ion exchange process is the sum of the two components [33] (p.174):

$$\vec{\Phi}_A = -D_A \nabla c_A + \mu_A c_A \vec{E}_0 \quad \left(m^{-2}s^{-1}\right) \tag{1}$$

The first one is related to the local concentration gradient of admixture in the glass. The second component expresses the electric drift of admixture ions in the local electric field. This electric field can be the effect of a difference in mobility of exchanged ions. In addition to the ion exchange processes based on a purely thermal diffusion of admixture ions, some processes are also carried out in the presence of an external electric field. In such processes an additional parameter occurs: the intensity of the external electric field $\vec{E}_e$ (Fig.2). These processes are briefly called electrodiffusion. They are characterized by a directed migration of ions introduced into the glass caused by the effects of an electric field.

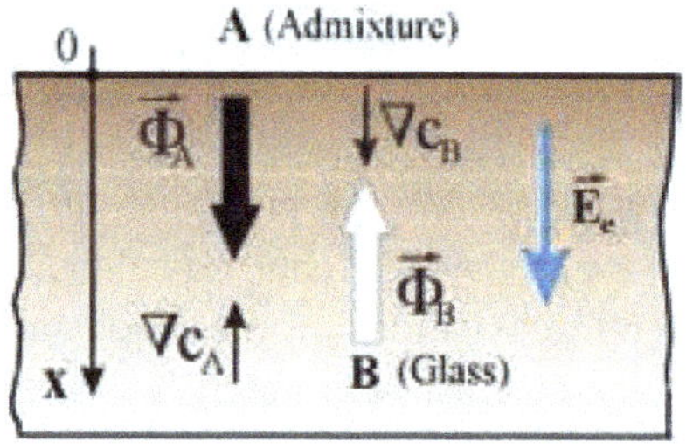

Figure 2. Ion exchange process in glass.

The local field $\vec{E}_d$ is then a superposition of the diffuse field which is the result of differential mobility of exchanged ions and external field $\vec{E}_e$:

$$\vec{E}_0 = \vec{E}_d + \vec{E}_e \tag{2}$$

The form of the function describing the distribution of the admixture concentration $c_A(x)$, introduced into the glass in the electrodiffusion process is dependent from the interaction between the diffusion component $D_A c_A$ and the electric migration component $\mu_A c_A \vec{E}_0$ of the admixture stream (D_A and μ_A denote, respectively, the diffusion coefficient of the admixture and its electric mobility). In the case of electrodiffusion processes the local electric field takes the form [33] (p.176):

$$\vec{E}_0 = \frac{\vec{E}_e}{1-\alpha u} + \frac{HkT}{e} \cdot \frac{\alpha}{1-\alpha u} \nabla u, \tag{3}$$

where $u(x) = c_A(x)/c_0$, $\alpha = 1 - \mu_A(u)/\mu_B(1 - u)$.

In the above equations, c_0 - equilibrium concentration [33] (p.175), H - the correlation coefficient ($0 < H < 1$) [34], k - the Boltzmann constant, T - the absolute temperature, μ_A and μ_B - electrochemical mobilities of admixture ions and modifier ions respectively.

In the electrodiffusion processes the external electric field may have polarization established during the process. Then we talk about the electrodiffusion processes with fixed electric polarization. The polarization of this field can also be changed during the implementation of the ion exchange process. Such processes are called processes with a change of electric polarization.

2.1. Electrodiffusion processes with fixed electric polarization

In the case of such electrodiffusion processes implemented with the use of a molten source of admixture, the molten salt ($AgNO_3$) contacts with one side of the glass plate. To the other side of the plate the solid electrode is applied. Between the molten salt and the solid electrode a potential difference is produced. The molten salt is at a positive potential in relation to the glass. With such polarization the silver ions Ag^+ in glass drift toward the electric field. At the same time the modifier ions (Na^+) in the volume of the glass drift toward the negative electrode. There their reduction to the atomic form occurs. As a result, between the glass surface and the negative electrode the metallic sodium is generated. This reduces the adhesion of the electrode to the glass surface.

Figure 3a presents the method of implementation of the electrodiffusion process with fixed polarization. The role of the solid electrode here is served by electrically conductive glue layer resistant to high temperatures. Figure 3b shows refractive index profiles of planar waveguides produced by the electrodiffusion processes for different values of electric field intensity. The durations of these processes are almost the same. Each process has been carried out at the temperature $T = 300°C$. For comparison the refractive index profile of the waveguide produced in the same conditions (time and temperature) in the diffusion process ($\vec{E}_e = 0$) is presented. The presented refractive index profiles have been determined for the wavelength $\lambda = 677$ nm.

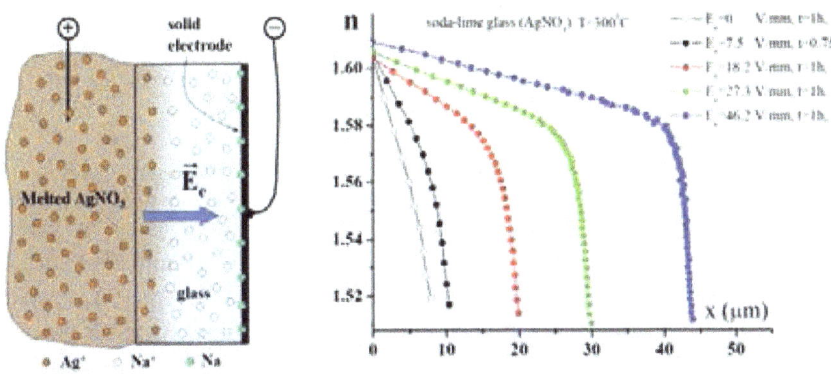

(a) (b)

Figure 3. (a) Implementation of electrodiffusion processes with fixed polarization. (b) Refractive index profiles of planar waveguides ($\lambda = 677$ nm) produced in electrodiffusion processes for different values of electric field intensity.

Due to the kinetics of diffusion assisted by an electric field, the electrodiffusion processes can be used as a tool for producing deep waveguide structures in a relatively short time. This is not the only argument in favor of the use of this type of technological processes. Much more important are there the opportunities to influence the shape of refractive index profiles of waveguides generated in the electrodiffusion processes through changes in polarization direction of the electric field.

2.2. Electrodiffusion processes with a change of the electric polarization

If a change in direction of the vector of external field $\vec{E}_e$ is made in the equation (3) describing the local electric field $\vec{E}_0$ in the electrodiffusion process, then this can obtain the value of the field $\vec{E}_0$ at which the change occurs in the direction of flow of admixture ions in the glass. This occurs only in situations where admixture ions are supplied on both sides of the glass substrate. It is then that the condition of continuity of ionic current flow through the glass can be satisfied. The electrodiffusion processes with the change in polarization of the external electric field are carried out under conditions of glass substrate being in contact with the liquid admixture sources on both sides [35-37].

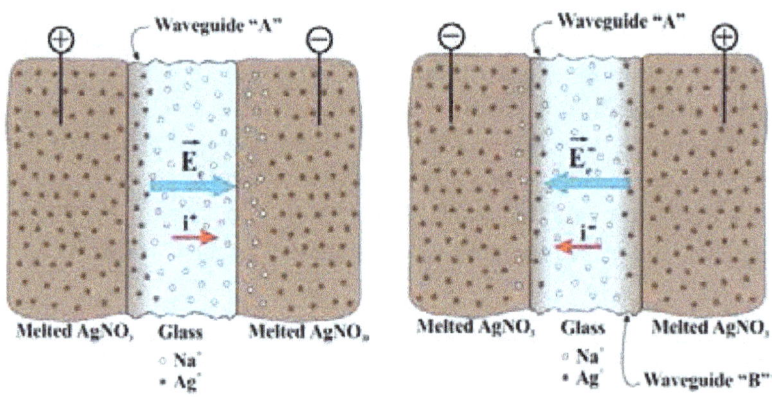

(a) (b)

Figure 4. The implementation of electrodiffusion processes with a change of the direction of the electric polarization.

The idea of using a cyclic change of polarization of the external electric field in the electrodiffusion processes has been described by Houde-Walter and Moore [35]. The authors have indicated there a theoretical ability to influence the final form of the refractive index profile of the waveguide in the electrodiffusion processes in which a multiple change of polarization of the electric field was applied. Such processes create major possibilities to influence the final form of the admixture distribution in the glass. The factors that determine here the shape of the resulting distribution of admixture in the glass are both the value and the polarization direction of the external electric field, as well as temporal relations between states with opposite polarizations. If a possibility of multiple repetitions of polarization changes of $\vec{E}_e$ field is also taken into consideration, the number of these factors will further increase.

In the electrodiffusion processes, where a change of the direction of polarization of the external electric field $\vec{E}_e$ is made, the admixture is introduced into the glass on both sides of the substrate.

Therefore the waveguide structures are formed on both sides of the glass plate. In order to distinguish between these two structures the following designations are used (Fig.4): waveguide "A" is formed on the side of the substrate, where the initial polarization of the electric field enhanced the process of migrating admixture into the glass, while the waveguide "B" is formed on the opposite side of the substrate, after the reversal of polarization. Figure 4 also indicates the conventionally adopted positive and negative polarization states.

Sample	Waveguide	Temperature of the process (°C)	Duration of the process (min)	Polarization and the electric field intensity (V/mm)	Electric charge (C)
1	A	272±1	30/30	+18.2/9.1	1.130
	B		30/30	18.2/+9.1	0.572
2	A	298 1	75/15	+18.2/9.1	5.680
	B		15/75	18.2/+9.1	0.572
3	A	298±1	60/30	+18.2/9.1	4.755
	B		30/60	18.2/+9.1	1.172
4	A	298±1	30/60	+18.2/9.1	3.280
	B		60/30	18.2/+9.1	2.951
5	A	298±1	45/45	+18.2/13.6	3.636
	B		45/45	18.2/+13.6	0.144
6	A	298±1	45/45	+18.2/9.1	3.719
	B		45/45	18.2/+9.1	1.774
7	A	298±1	45/45	+4.5/18.2	0.908
	B		45/45	4.5/+18.2	4.164
8	A	299±1	30/30	+18.2/9.1	2.554
	B		30/30	18.2/+9.1	1.252
9	A	299±1	45/45	+18.2/18.2	4.654
	B		45/45	18.2/+18.2	4.777
10	A	299±1	50/40	+18.2/9.1	3.688
	B		40/50	18.2/+9.1	1.803
11	A	300±1	45/45	+13.6/18.2	2.965
	B		45/45	13.6/+18.2	4.240
12	A	301±1	45/45	+9.1/18.2	1.628
	B		45/45	9.1/+18.2	3.791

Table 1. Electrodiffusion processes with a change of direction of the electric polarization.

Significant differences in the shapes of refractive index profiles that arise in the case of changes in the polarization direction of the electric field during the process can be explained by considering the situation where the polarization of the applied electric field is changed only once. During such a process, with a positive polarization (Fig.4a), the admixture ions enter the glass substrate form the waveguide "A". On the other side of the glass plate (despite its contact with the source of admixture) a waveguide is not formed, because the glass ions pass into the liquid phase due to the current i^+ flowing through the substrate.

After the change of polarization (Fig.4b), the situation is reversed. Through the glass substrate the current i^- now flows in the opposite direction. Waveguide "B" starts forming on the other side of the glass, while the opposite stream of admixture ions previously introduced into the glass with the glass modifiers ions changes the distribution of the introduced admixture in the positive polarization. Consequently, the shapes of the refractive index profile of waveguide "A" changes.

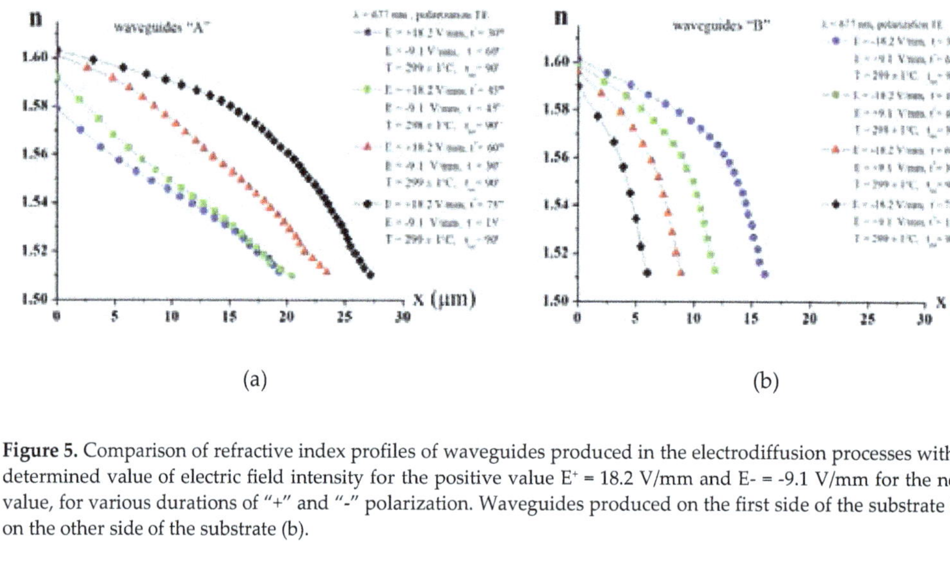

(a) (b)

Figure 5. Comparison of refractive index profiles of waveguides produced in the electrodiffusion processes with a predetermined value of electric field intensity for the positive value E^+ = 18.2 V/mm and E- = -9.1 V/mm for the negative value, for various durations of "+" and "-" polarization. Waveguides produced on the first side of the substrate (a) and on the other side of the substrate (b).

The parameters of chosen technological processes with the use of a single change of the direction of the electric field polarization are summarized in Table 1. The soda-lime glass was used as the substrate. As liquid source of admixture on both sides of the substrate the molten silver nitrate ($AgNO_3$) was used. The electrodiffusion processes were realized with the use of the laboratory stand described in [33] (p.164). The temperatures of the processes were within the range: 272÷301°C. The given duration of each process consists of two values relating respectively to the positive and negative polarization. In each glass plate two waveguides ("A" and "B") were produced on the opposite sides. The table shows the total electric charge of admixture ions that were introduced into the glass substrate from both sides. Figure 5a,b shows the refractive index profiles of waveguides produced in electrodiffusion processes, for which the field intensity for the "+" and "-" polarization have been established according to Fig.4.

Here the ratio of the duration of the polarization t^+ and t- was a variable. In each case, the duration of the whole process was the same and was $t_{tot} = 90'$. The waveguides of "A" and "B" were formed on opposite sides of the glass plate. The equivalent centric symbols in both the charts (a) and (b) denote the same glass substrate in which the adequate waveguides were formed.

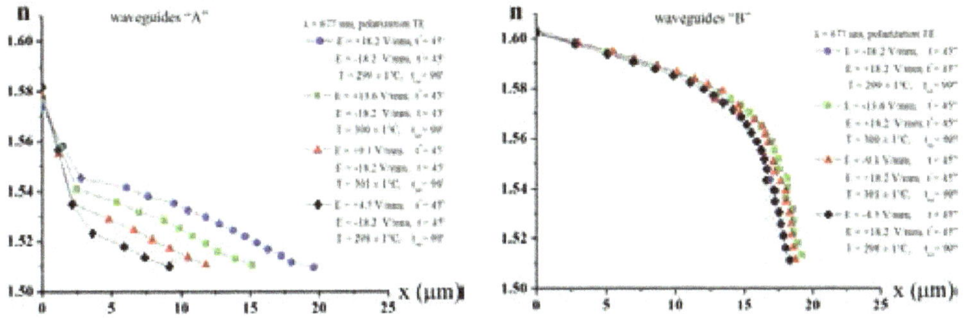

(a) (b)

Figure 6. Comparison of refractive index profiles of waveguides produced by electrodiffusion processes with fixed polarization durations $t^+ = 45'$ and t- = 45', for different values of electric field intensity for polarization "+" and "-". Waveguides produced on the first side of the substrate (a) and the other side of the substrate (b).

Figure 6a,b shows a comparison of refractive index profiles of waveguides produced by processes in which a division of the duration of each polarization t^+ = t- has been set, with the total duration of the process $t_{tot} = 90'$. The values of the electric field intensity were altered for the two polarization states. As mentioned earlier, the waveguides of types "A" and "B" were formed here on the opposite sides of the glass plate. Here again, the same centric symbols in both figures (a and b) indicate the same glass substrates, in which adequate waveguides have been formed. In the case of this sequence of processes, a strong influence of polarization field "-" on the shape of waveguide of type "A" can be seen (Fig.6a). In the extreme case, for the values of electric field intensity: E = +4.5 V/mm and E = -18.2 V/mm, the obtained admixture distribution gives the concave shape of the refractive index profile.

As mentioned earlier, by selecting the duration of a specific state of polarization and electric field intensity, a specified form of the refractive index profile of the waveguide can be deliberately produced. Figure 7 presents the refractive index profiles of two waveguides of type "A" produced in the electrodiffusion processes in which the duration of polarization was t^+ = t- = 30' and the values of the electric field intensity were E = +18.2 V/mm and E = -9.1 V/mm respectively. The dependencies n(x) are of almost linear nature. The temperatures of both processes were T = 299°C and 272°C respectively. The resulting refractive index profiles vary in depth, which is a result of electric mobility of ions depending on the temperature. In the case of a waveguide produced at a lower temperature, the effect of the diffusion component on the final form of the shape of the refractive index profile in comparison with the electric drift is much smaller.

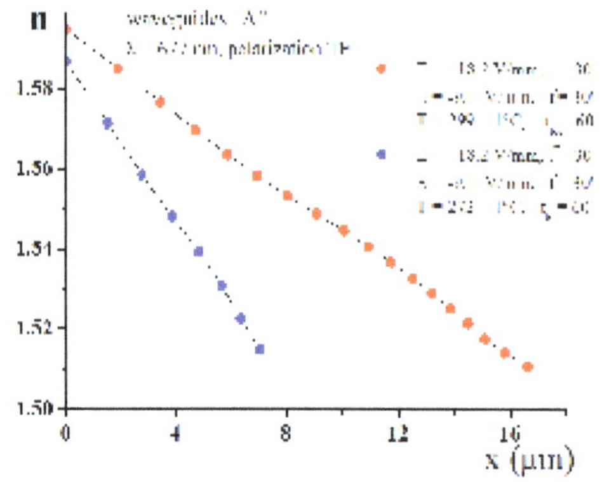

Figure 7. Refractive index profiles of produced waveguides with shapes close to linear.

As an illustration of the applicability of the sequence of polarization direction changes of the electric field, Fig.8 shows the refractive index profiles of waveguides produced on both sides of the glass substrate in the electrodiffusion process, in which six different polarization states were applied. The total duration of the process was t_{tot} = 60 min. The durations of each of the processes were equal and were t^{+} = t- = 10'. In the case of waveguide of type "A" (Fig.4a), a 12-modal structure was obtained (TE polarization, λ = 677 nm). The shapes of the refractive index profile of this waveguide are similar to that obtained in the electrodiffusion processes with fixed polarization direction of the electric field (compare Fig.3b). For the resulting waveguide of type "B" (10-modal structure) a monotonic course of the refractive index profile with characteristic inflection point at a depth corresponding to the position of the turning point of the mode of 3rd order was obtained (Fig.8b). In both figures the total values of the electric charge that has passed through the substrate in either direction are given.

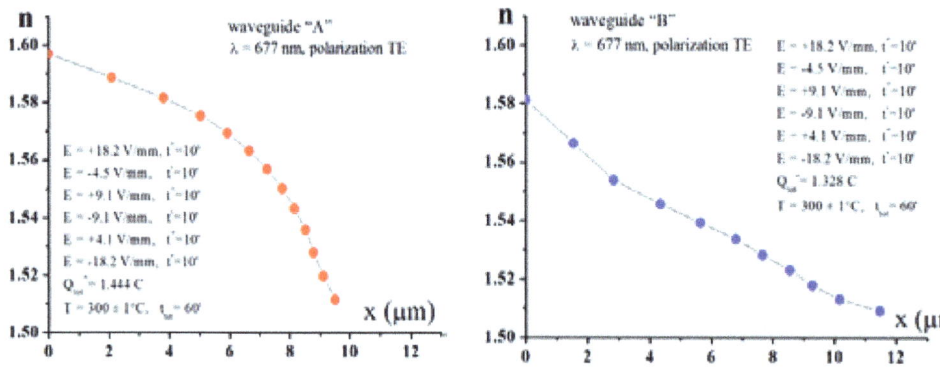

(a) (b)

Figure 8. The refractive index profiles of the waveguides produced by electrodiffusion processes with a multiple change of the direction of polarization and of the values of the electric field intensity. The waveguide produced on the first side of the substrate (a) and on the other side of the substrate (b).

The presented measurement results of refractive index profiles of waveguides produced in the electrodiffusion processes (in which there is a change of polarization direction of the applied electric field) indicate a high possibility of the use of such processes to the intended shaping of refractive index profiles of produced waveguide structures.

3. Determination of the equilibrium concentration of modifier ions in the glass

In the theoretical calculations describing the ion exchange processes in glasses a two-component model of this phenomenon is adopted [33]. The nonlinear diffusion equation (electrodiffusion) is solved in the field of normalized concentration of exchanged ions (admixture ions introduced into the glass and mobile ions of its modifier). This normalization is made in relation to the equilibrium concentration, which is the total concentration in the glass of both types of exchanged ions. The value of this concentration remains constant during the whole process of ion exchange. This section presents two methods of estimating the value of this concentration. One of them is based on a calculation of the electric charge flowing through the glass in the electrodiffusion processes. The second one is based on a measurement of the mass of the glass substrate before and after the diffusion process. The experimental results are presented in relation to soda-lime glass, in which the $Ag^+ \leftrightarrow Na^+$ ion exchange processes were carried out.

3.1. The dependence of changes of the refractive index profile of the waveguide on the electric charge flowing in the electrodiffusion process

In the electrodiffusion process carried out at time τ, the amount of the charge Q_τ, which has passed through the glass can be determined on the basis of the current dependency i(t):

$$Q_\tau = \int_0^\tau i(t)dt \tag{4}$$

Taking into account that $u(x) = c(x)/c_0$, where c_0 (m^{-3}) is the normalized concentration of mobile components of the glass, (which is a concentration of mobile modifier ions in the glass before the exchange process) the equation describing the relationship of the refractive index profile n(x) with the normalized concentration of the admixture u(x) takes the form:

$$n(x) = n_b + \Delta n_s \cdot \frac{c(x)}{c_0}, \tag{5}$$

where n_b - refractive index of the glass before the ion exchange process, Δn_s - increment of the refractive index of the glass (caused by the ion exchange process) at its surface, $c_A(x)$ - function describing the distribution of the absolute concentration of admixture ions introduced into the glass (m^{-3}).

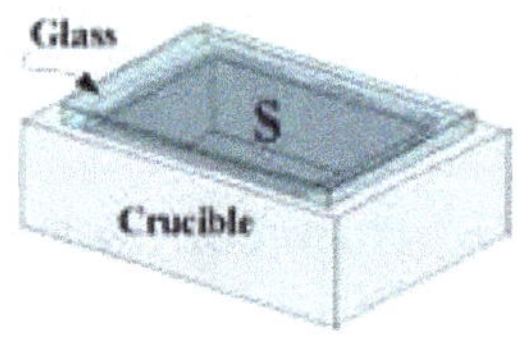

Figure 9. Determination of the cross section *S* of the exchange process.

By introducing the function, $\delta n(x) = n(x) - n_b$, from the equation (5) the following can be obtained.

$$c(x) = \delta n(x) \cdot \frac{c_0}{\Delta n_s} \quad (m^{-3}) \tag{6}$$

This equation describes the distribution of concentration of admixture introduced into the glass, expressed by the parameters of the refractive index profile n(x) and of the equilibrium concentration c_0. Integrating the expression (6) over the entire volume of the glass *V*, the total number of admixture ions N_τ introduced into the glass during the electrodiffusion process in the time τ is obtained as follows.

$$N_\tau = \int_V c(x) dV = \frac{c_0 \cdot S}{\Delta n} \int_0^\infty \delta n(x) dx \tag{7}$$

In this equation, *S* is the surface area of the glass substrate (cross section for the exchange - Fig.9) which comes in contact with a liquid admixture contained in the crucible. At the same time the total charge (4), which has flowed through the glass, allows to specify the number of admixture ions having a valence *w*, which have been introduced into the glass in the time τ:

$$N_\tau = \frac{1}{w \cdot e} \int_0^\tau i(t) d\tau = \frac{Q_\tau}{w \cdot e}, \tag{8}$$

where *e* - the elementary charge.

Comparing (7) and (8) the following equation is obtained.

$$\int_0^\infty \delta n(x) dx = \frac{\Delta n_s}{e \cdot w \cdot S \cdot c_0} \cdot Q_\tau \quad (m) \tag{9}$$

By calculating the above integral the product Sc_0 of the cross-section of the exchange and the equilibrium concentration can be determined.

Table 2 summarizes the results of calculation product Sc_0 based on equation (9) for several electrodiffusion processes carried out in the substrate of soda-lime glass, using a pure silver nitrate $AgNO_3$ as the source of admixture of Ag^+ ions. It also presents the maximum changes in refractive index profile Δn_s at the surface of the glass calculated on the basis of the determined refractive index profiles.

Q_τ (C)	$\int_0^\infty \delta n(x)dx$ (m)	Δn_s	E (V/mm)	T (°C)	τ (min)
0.572	$0.2967 \cdot 10^6$	0.0789	9.1	299 ± 1	15
1.172	$0.5064 \cdot 10^6$	0.0846	9.1	299 ± 1	30
1.774	$0.7264 \cdot 10^6$	0.0879	9.1	298 ± 1	45
2.951	$1.0289 \cdot 10^6$	0.0902	9.1	299 ± 1	60
3.791	$1.3337 \cdot 10^6$	0.0912	18.2	301 ± 1	45
4.164	$1.2699 \cdot 10^6$	0.0916	18.2	298 ± 1	45
4.240	$1.3527 \cdot 10^6$	0.0925	18.2	300 ± 1	45

Table 2. Electrodiffusion process parameters.

Figure10 shows the method of calculating the product of Sc_0 according to the equation (10) on the basis of data from Table 2.

The cross-section *S* for the exchange process shown in Fig.9 is defined by the geometry of the crucible. During the process it grows as a result of the gradual penetration of the molten salt into contact area of the glass and the crucible. The electric field is not uniform near the inner edge of the crucible. Based on the above facts, it is expected that the refractive index profile of the waveguide in these areas of the glass which are at the edge of the crucible is significantly different from the form that is obtained in the central portion of the glass plate. The integrals in Table 2 were calculated assuming a uniform shape of refractive index profile in the entire area of glass. With these reservations, on the basis of the data from Fig.10, the equilibrium concentration c_0 of mobile ions in the glass substrate that was used in the process can be estimated. Assuming the cross-section of the crucible $S = (3.6 \pm 0.7) \cdot 10^{-4}$ m^2, identical for all processes listed in Table 2, the mean value $c_0 = (5.7 \pm 1.5) \cdot 10^{27}$ m^{-3} was obtained.

3.2. The glass mass (weight) change due to the $Ag^+ \leftrightarrow Na^+$ ion exchange

Another phenomenon accompanying the process of ion exchange in the glass is a change in the weight of the glass. This phenomenon is the more perspicuous the larger the difference of the masses of exchanged ions. This is easy to observe in the case of heavy silver ions Ag^+, which replace the mobile sodium ions Na^+ in the glass. For this type of exchange ($Ag^+ \leftrightarrow Na^+$) this

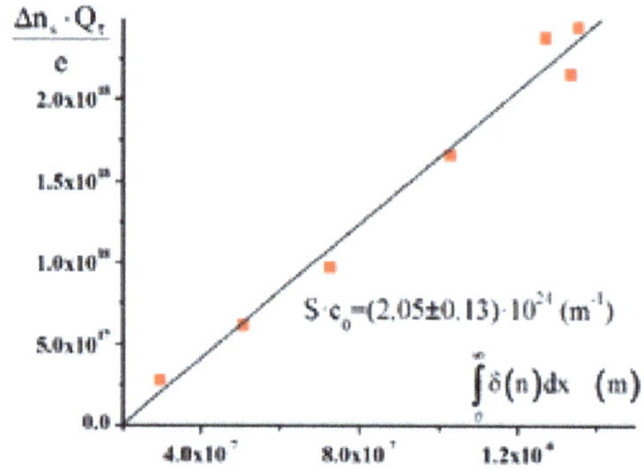

Figure 10. The calculation of the product Sc_0 according to (10).

effect will be greater if the sodium modifier is more in the glass. The difference of the masses of atoms (silver $\bar{m}_{Ag}$ and sodium$\bar{m}_{Na}$) can be expressed as:

$$\Delta M = \bar{m}_{Ag} - \bar{m}_{Na} = \frac{M_{Ag} - M_{Na}}{N_A}, \tag{10}$$

where M_{Ag} and M_{Na} - molar masses, N_A - Avogadro's number.

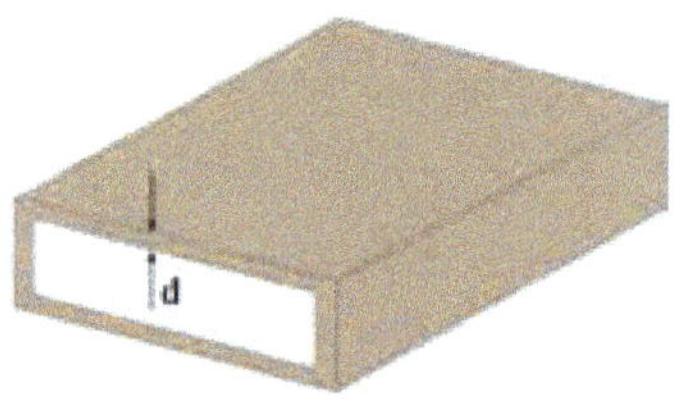

Figure 11. The cross section of a glass plate with a marked depth *d* of doped region.

The glass mass difference caused by the ion exchange can be calculated based on the amount of the ions N_e exchanged in the glass:

$$\Delta m = m_e - m_0 = N_e \cdot \Delta M, \tag{11}$$

where m_e, m_0 - glass masses before and after the ion exchange process respectively.

In the case of planar waveguides the amount of ions exchanged in the glass can be estimated by measuring their refractive index profile n(x). The *x* coordinate is calculated here into the glass (on the surface $x = 0$) to the direction perpendicular to its surface. This profile can also be described theoretically using the normalized concentration u(x) of admixture ions introduced into the glass - see equation (5).

Thus, appearing in equation (11), the quantity of ions exchanged in the glass N_e can be expressed as follows:

$$N_e = S \cdot \int_0^d c_A(x)dx = Sc_0 \cdot \int_0^d u(x)dx, \tag{12}$$

where S - the total glass surface through which the ion exchange process was conducted, d - the depth of the area of doped glass (Fig.11).

On the basis of (10-12), the weight change of the glass due to the ion exchange of $Ag^+ \leftrightarrow Na^+$ is expressed in the equation:

$$\Delta m = \frac{M_{Ag} - M_{Na}}{N_A} c_0 S \cdot \int_0^d u(x)dx \tag{13}$$

The above relationship shows the possibility of estimating the equilibrium concentration c_0 on the basis of experimentally designated values: Δm, S and the refractive index profile n(x) of produced waveguide.

κ	Δm (g)	S (mm^2)	$\int_0^d u(x)\,dx\,(\mu m)$	$S \cdot \int_0^d u(x)\,dx\,(mm^3)$	T_{ave}(°C)	t(h)
1	0.0147	2995.37	6.09	18.24	295 ± 1	2.12
1	0.0088	2515.12	4.71	11.84	290 ± 1	1.38
0.01	0.0183	2258.58	10.71	24.19	353 ± 1	3.24
0.001	0.0056	2184.85	3.09	6.75	354 ± 1	7.35
0.001	0.0057	2210.15	3.31	7.31	354 ± 1	7.52
0.0005	0.0058	2134.67	2.83	6.04	352 ± 1	23.33
0.00025	0.0029	2341.48	1.57	3.67	355 ± 1	26.02
0.00025	0.0036	2383.02	1.49	3.55	354 ± 1	26.00
0.00025	0.0030	2205.24	1.44	3.17	354 ± 1	26.00

Table 3. Mass change of glass substrates in the diffusion processes and the integrals of the normalized concentrations of the admixture.

Table 3 shows the results of an experiment involving the determination of weight increase of glass plates subjected to processes of diffusion doping with Ag^+ ions. The substrates were made of soda-lime glass (Menzel-Gläser Company). This glass contains a large amount of sodium oxide Na_2O. Thus in the case of $Ag^+ \leftrightarrow Na^+$, a considerable increase of the weight is obtained.

Chemical composition in % by weight of the glass [38]: 72.2% SiO_2, 14.3% Na_2O, 6.4% CaO, 4.3% MgO, 1.2% Al_2O_3, 1.2% K_2O, 0.3% SO_3, 0.03% Fe_2O_3. Diffusion processes were carried out with a liquid source of admixture. The silver nitrate $AgNO_3$ and its sodium nitrate $NaNO_3$ solutions have been used. These solutions are determined by the molar fraction κ [39] listed in Table 3.

Figure 12 shows the dependence of the mass increase of the glass substrates from the product of the total surface of the glass and the integral of the normalized concentration of the admixture (Ag^+ ions).

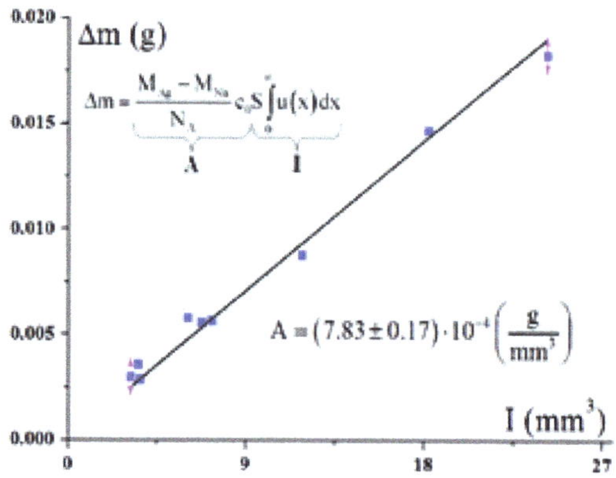

Figure 12. The dependence of the mass increase of the sample on the product of its surface and the integral of the normalized concentration of the admixture.

This dependency is based on the results shown in Table 3. For : M_{Ag} = 107.87 (g/mol), M_{Na} = 22.99 (g/mol), $N_A = 6.02 \cdot 10^{23}$ (mol^{-1}), the following equation is obtained: $A/c_0 = 1.41 \cdot 10^{-22}$ (g).

Determined by this method the value of the equilibrium concentration in the glass is: $c_0 = A/1.41\times10^{-22} = (5.6 \pm 0.2)\ 10^{27}(m^{-3})$.

The results of the equilibrium concentration c_0 obtained by integrating the electric charge (Section 2.1) and by method of weighing presented here, comply within the limits of uncertainty calculation.

4. Optical birefringence phenomenon of the gradient area as the effect of ion exchange

Ion exchange in glass processes with the use of liquid source of admixture ions, among which the most widely used are nitrates, are carried out at temperatures much lower than transition temperature of glasses T_g. For example, for the borosilicate glass BK-7 (Schott), which is widely used in these processes, the temperature T_g = 557°C [40] is much higher than the temperature

T_{diff} = 400°C of the implementation of the $K^+ \leftrightarrow Na^+$ ion exchange processes with the use of liquid potassium nitrate KNO_3 as the source of K^+ ions. In such cases, changing the refractive index of the glass in its surface area where the ion exchange occurs is due not only to the difference of their electric polarizability, but also the result of the elastooptic phenomenon generated by mechanical stresses resulting in this area [41]. These stresses are the result of changes in the volume of glass in the doping area, which results from the difference of ionic radii of components exchanged as well as the difference in thermal expansion between the doped region and the rest of the glass.

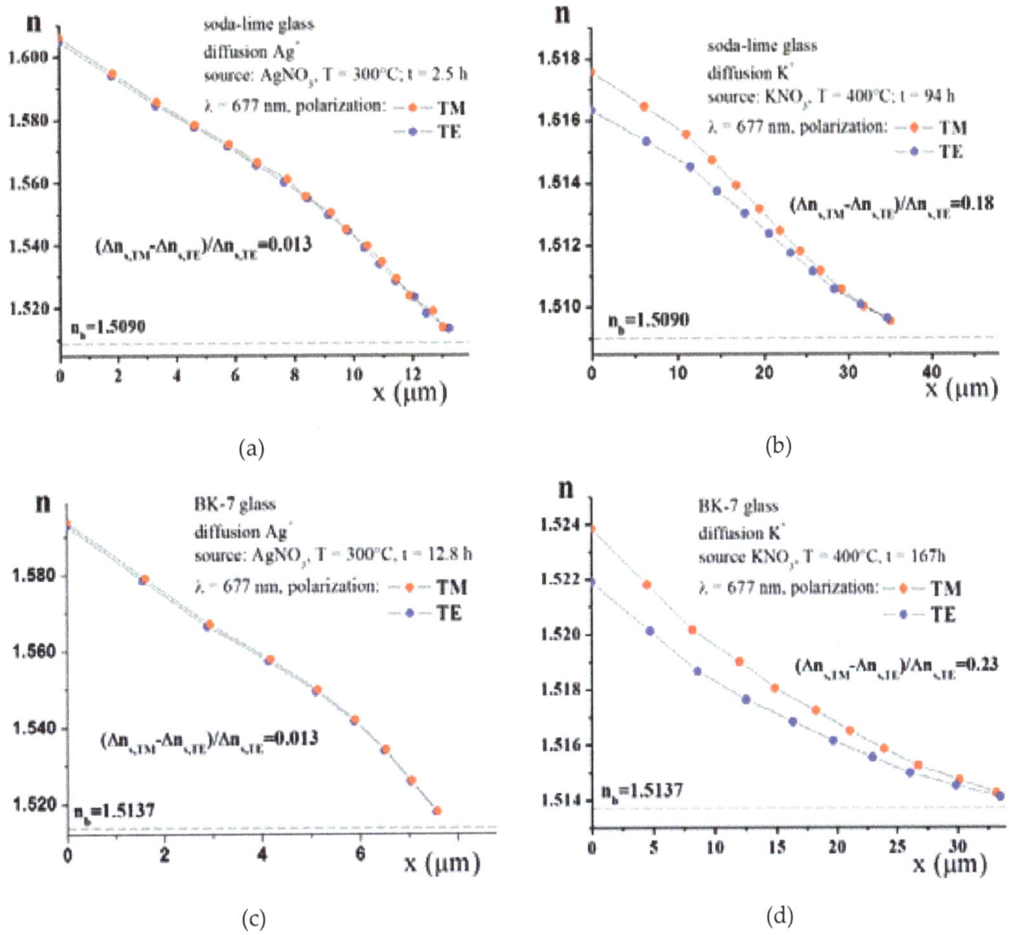

Figure 13. Comparison of refractive index profiles of planar waveguides produced in soda-lime glass doped with ions: Ag^+ (a) or K^+ (b) with the refractive index profiles produced in BK-7 glass doped with ions: Ag^+ (c) or K^+ (d).

The ionic radii of the most commonly used admixtures K^+ and Ag^+ are 1.33 Å and 1.26 Å [41] respectively. Thus, in relation to the ionic radius of sodium (0.95 Å [41]), which, due to one of the lowest binding energies with glass network, is its most easily replaceable component, the change in volume in the case of $K^+ \leftrightarrow Na^+$ exchange is much larger than in the $Ag^+ \leftrightarrow Na^+$ exchange. After completion of the process, when the glass is cooled to a low room temperature, a resulted difference generates stress in the doping area of the glass. Also, the difference in thermal expansion of the doping area, in relation to the rest of the glass, makes a significant contribution to the resulting stress. According to the principle of additivity [41], the coefficient of thermal expansion of glass is:

$$\alpha = \sum_i \alpha_i \cdot c_i \, , \tag{14}$$

where α_i and c_i represent, the coefficient of thermal expansion and the mole fraction of i-th component in glass respectively.

The coefficients of thermal expansion for sodium and potassium are $\alpha_{Na} = 39.5 \cdot 10^{-6}\ K^{-1}$ and $\alpha_K = 46.5 \cdot 10^{-6}\ K^{-1}$ [41] respectively. These values are related to the concentration of Na_2O and K_2O in the glass. These data indicate that ion exchange processes of $K^+ \leftrightarrow Na^+$ in the glasses will be accompanied by the generation of significant mechanical stresses. The presence of these stresses in the waveguide layer of the glass is apparent in the propagation of the electromagnetic wave. A difference in propagation constants of the modes of the same order occurs for a monochromatic wave depending on its state of polarization. This phenomenon is called stress birefringence. Its scale depends on the type of glass and the admixture into the glass. This is illustrated in Fig.13, in which the refractive index profiles of planar waveguides produced in two types of glass (soda-lime and BK-7) by doping them with silver ions Ag^+ and potassium ions K^+ are presented. These profiles were determined by measuring the propagation constants for the two polarization states: TE and TM. Measurement uncertainties of the effective refractive indices of the modes do not exceed $3 \cdot 10^{-4}$. In all the presented refractive index profiles, the modes with TM polarization have bigger values of effective refractive indices. As there is only a polarization mode dispersion it is therefore inversely than in the case of a waveguide without stresses.

Type of glass	**Type of admixture ions**			
	Ag^+		**K^+**	
soda-lime	$n_{s,TE}$=1.6049	$\Delta n_{s,TE}$=0.0959	$n_{s,TE}$=1.5163	$\Delta n_{s,TE}$=0.0073
n_b=1.5090	$n_{s,TM}$=1.6062	$\Delta n_{s,TM}$=0.0972	$n_{s,TM}$=1.5176	$\Delta n_{s,TM}$=0.0086
BK-7	$n_{s,TE}$=1.5929	$\Delta n_{s,TE}$=0.0792	$n_{s,TE}$=1.5219	$\Delta n_{s,TE}$=0.0082
n_b=1.5137	$n_{s,TM}$=1.5939	$\Delta n_{s,TM}$=0.0802	$n_{s,TM}$=1.5238	$\Delta n_{s,TM}$=0.0101

Table 4. Changes in the refractive index (for λ = 677nm) at the glass surface depending on the type of admixture ions (data provided by refraction profiles of Fig.13).

For each case of the refractive index profile shown in Fig.13, the relative change in the refractive index at the glass surface for both polarization states, in relation to the TE polarization, has been shown. These figures can provide a quantitative measure of the scale of the birefringence when comparing this phenomenon in various glass-admixture systems. A comparison of Fig. 13a,c shows that the size of the birefringence due to the doping of both types of glasses with the silver ions Ag^+ is practically the same, although the changes in refractive indices at the surface differ quite significantly (Table 4). In the case of doping both kinds of glass with potassium ions K^+ the measure of the birefringence defined above is significantly higher in the case of BK-7 glass.

4.1. Refractive birefringence and modal birefringence

In the description of the phenomenon of birefringence, which extends over the entire doping area of glass, we can use the concept of birefringence as a function whose domain is the depth of doping area of the glass, or the row of modes. The following definitions of stress birefringence refractive δn_σ and modal δN_σ are introduced.

Refractive birefringence is determined by the refractive index profiles for the TE and TM polarization of the waveguide structure and is a function of the depth x counted from the glass surface:

$$\delta n_\sigma(x) = n_{TM}(x) - n_{TE}(x) \tag{15}$$

The modal birefringence concerns the difference in the effective refractive indices of the modes of the same order for the TE and TM polarization. This value is a function of the mode order:

$$\delta N_\sigma(m) = N_{TM}(m) - N_{TE}(m) \tag{16}$$

The sense of the values incorporated herein is illustrated in Fig.14. Refractive birefringence describes the difference in the refractive indices of the refractive index profiles for the TE and TM polarization that occur at a depth x in the waveguide. It is related to the stresses arising in the glass, in relation to the concentration of admixture introduced into the volume of the glass during the ion exchange processes. Function $\delta n_\sigma(x)$ reaches a maximum at the glass surface ($x = 0$) and decreases monotonically to zero at the point of glass, which was not reached by the admixture. The modal birefringence can be determined only for those rows of modes for which simultaneously there exist modes of both TE and TM polarization.

4.2. The relationship between birefringence and stresses

Quantification of changes in the refractive index of glass, in which the mechanical stresses arise, requires knowledge of the nature of the stress and elastooptic constants of the glass. The resulting stresses in the doped glass are of compressive nature [41]. In the case of a planar waveguide structure, for which the doped region extends from the surface into the glass, there is a one-dimensional distribution of the admixture concentration. In a direction perpendicular to the surface of the glass the doped region is free to deform, as a result of which at this direction

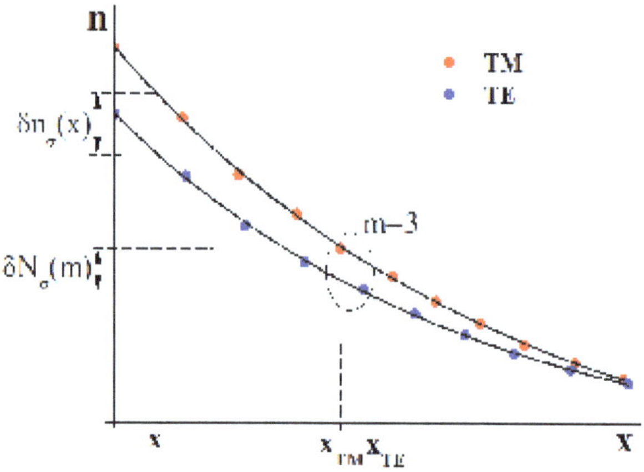

Figure 14. Determination of refractive birefringence and modal birefringence in the planar waveguide.

the stresses do not occur $\sigma_{xx}(x) = 0$. In contrast, at the directions parallel to the glass surface, the stresses generated in the doped area will be the same: $\sigma_{xy}(x) = \sigma_{xz}(x)$. The geometries of the directions of the stresses are shown in Fig.15. These stresses are the functions of a depth x only, the same as one-dimensional concentration distribution of the admixture introduced into the glass.

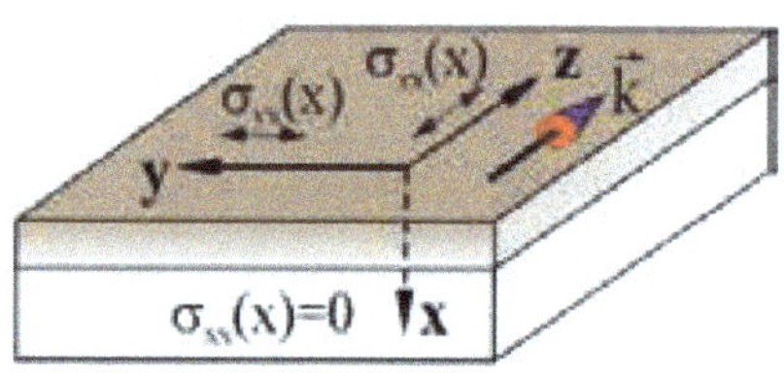

Figure 15. The geometries of the directions of the stresses in a planar waveguide.

Changes in the refractive index of the glass, which result from the generated stress, are determined by the elastooptic constants [42]:

$$C_1 = \frac{dn_{//}}{d\sigma} \left(\frac{m^2}{N}\right) \qquad C_2 = \frac{dn_{\perp}}{d\sigma} \left(\frac{m^2}{N}\right) \tag{17}$$

In the above equations, $dn_{//}$ and dn mean the differentials of change in refractive index for the wave with polarizations respectively parallel and perpendicular to the direction of stress σ.

The refractive index profiles $n_{TM}(x)$ and $n_{TE}(x)$ for the waveguides with TM and TE polarizations, propagating in the direction of the axis z (Fig.15), in a waveguide are (in the presence of stress) described by the Maxwell-Neumann equations [41,43]:

$$\begin{aligned} n_{TM}(x) &= n_0(x) + C_1\sigma_{xx}(x) + C_2\left[\sigma_{yx}(x) + \sigma_{zx}(x)\right] \\ n_{TE}(x) &= n_0(x) + C_1\sigma_{yx}(x) + C_2\left[\sigma_{xx}(x) + \sigma_{zx}(x)\right], \end{aligned} \tag{18}$$

where $n_0(x)$ is a refractive index profile of the waveguide in the absence of stress.

After taking into account the assumptions: $\sigma_{xx}(x) = 0$ and $\sigma_{yx}(x) = \sigma_{zx}(x) = \sigma(x)$, the equations (18) reduce to the following form:

$$\begin{aligned} n_{TM}(x) &= n_0(x) + 2C_2\sigma(x) \\ n_{TE}(x) &= n_0(x) + (C_1 + C_2)\sigma(x) \end{aligned} \tag{19}$$

These relationships enable us to align the birefringence of refraction (15) with the distribution of stress $\sigma(x)$:

$$\delta n_\sigma(x) = n_{TM}(x) - n_{TE}(x) = (C_2 - C_1)\sigma(x) \tag{20}$$

The elastooptic constants for the BK-7 glass for the wavelength of $\lambda = 677$ nm are [40]:

$$C_1 = \text{-}0.5 \cdot 10^{-6} \quad (\text{mm}^2/\text{N}) \qquad C_2 = \text{-}3.3 \cdot 10^{-6} \quad (\text{mm}^2/\text{N}) \tag{21}$$

They allow to specify the value of the stress generated in the doped area of the glass. As it stems from the equation (20):

$$\sigma(x) = \frac{n_{TM}(x) - n_{TE}(x)}{C_2 - C_1} \tag{22}$$

The maximum stresses which occur at the surface of the BK-7 glass ($x = 0$), by doping it with potassium ions K^+ and silver ions Ag^+, can be determined based on the data presented in Table 4. They are $|\sigma(0)_{K+}| = 678$ (N/mm^2) and $|\sigma(0)_{Ag+}| = 375$ (N/mm^2) respectively.

Figure 16 illustrates the distributions of stresses occurring in the BK-7 glass after the diffusion processes of potassium ions K^+, and silver ions Ag^+, depending on their normalized concentrations u. This presentation allows to compare these functions, defined on the basis of different refractive index profiles.

4.3. The dependence of stress birefringence on the duration of the diffusion processes

During the doping of the glass, the emerging stresses are also accompanied by the relaxation processes. They are reflected in the reduction of the role of stresses in the changes of the

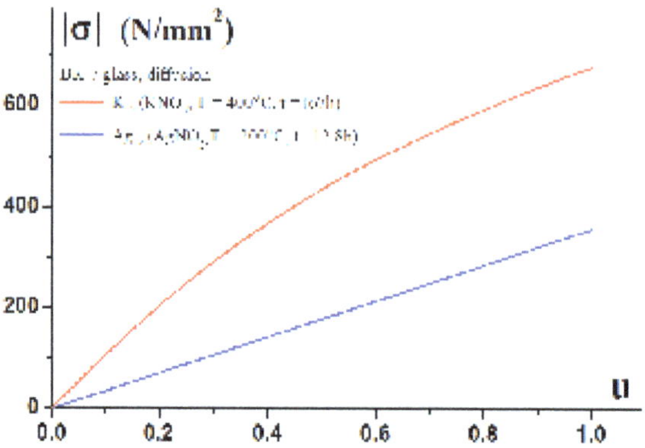

Figure 16. The dependence of stresses on normalized concentration of admixture in the BK-7 glass.

refractive index of the doped area of the glass. This phenomenon can be observed by comparing the refractive index profiles of waveguides produced in processes with different diffusion times.

The tests were carried out on BK-7 glass doped with potassium ions K^+ in the diffusion processes lasting from 24 to more than 500 h. The resulting waveguide structures were multimode. This guaranteed the fidelity of reconstruction of their refractive index profiles.

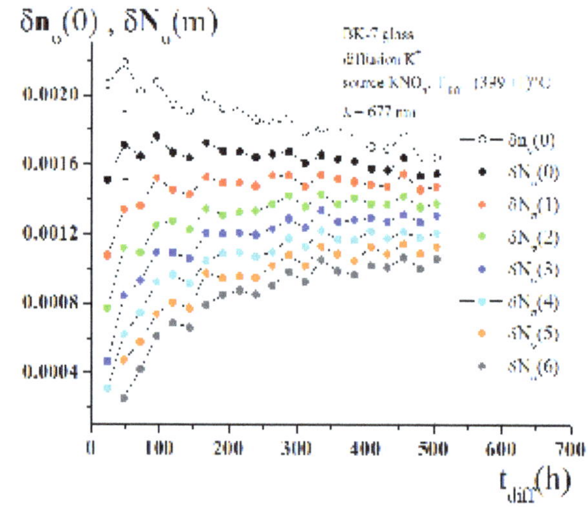

Figure 17. Refractive birefringence (at the glass surface) and modal birefringence, depending on the duration of the diffusion process.

The methodology of the experiment was as follows: 21 glass substrates made of BK-7 glass with the dimensions: 8 × 30 × 1.5 mm were prepared. They were polished on one side and then placed in the holders of silica glass tubes, allowing their individual extraction from the crucible containing the molten potassium nitrate. Due to the long duration of the process, the volume of salt (KNO_3) filling the crucible was large (~300 cm^3). During the entire process, the molten salt was continuously stirred. This ensured thermal and concentration uniformity for the entire bath. On alternate days an additional portion of salt was introduced into the crucible in order to supplement salt losses as a result of evaporation. From all the 21 substrates placed at the beginning of the process in the crucible, one was removed every 24 h. This substrate, after cooling and washing in distilled water, was subjected to further testing. The duration of the whole process was 504 h. The temperature of the salt in the crucible was measured with a thermocouple and recorded continuously. The averaged temperature was (399 ± 1)°C.

The measurements of synchronous angles of modes in produced waveguides were performed using a prism coupler [33] made of PSK-3 glass (of Schott company). They were carried out for a wavelength $\lambda = 677$ nm and the TE and TM polarization. The effective refractive indices of the modes N_m calculated on their basis have the uncertainty of measurement at the level of $\Delta N_m = 0.0002$.

Figure 17 shows the change in the refractive birefringence $\delta n_\sigma(0)$ at the surface of the waveguides and in modal birefringence $\delta N_\sigma(m)$ for the modes of 0-6 row, depending on the duration of the diffusion process. The course of the changes $\delta n_\sigma(0)$ indicates a decrease in the birefringence on the surface of the waveguide with increasing duration of the diffusion. In contrast, the courses of modal birefringence tend to increase with the diffusion time. However, the nature of this growth is a function of the order of the mode. In the case of a zero-order mode, even a downward trend can be seen. The strongest increase in birefringence occurs for the modes of higher-order. The modal birefringence (16) is not related to the location in the waveguide, where there is a defined concentration of the admixture. It can be however assumed, that it reflects the nature of the stresses associated with the depth of diffusion of admixture into the glass.

4.4. The changes of stress birefringence in the heating processes

The relaxation phenomenon of a waveguide layer, in which the doping of the glass resulted in the appearance of stresses, is clearly visible after the heating processes. In order to observe this phenomenon, in the BK-7 glass the long-term heating processes were implemented in relation to the waveguide produced by the diffusion process of KNO_3 in 72 hours [44]. The durations of these heating processes, after which the measurements of effective refractive indices of modes for both polarization states made were 24, 48, 96, 192, and 384 h respectively. The temperature of the heating processes was (399 ± 1)°C.

Based on the determined effective refractive indices, the refractive index profiles of the waveguides were reconstructed and their modal birefringence as well as the refractive birefringence at the surface of the glass was estimated. The nature of the changes of these values depending on the duration of the heating process is shown in Fig.18a.

Birefringences for the time $t = 0$ shown in the graphs represent the values of these terms after the diffusion processes. In the case of the refractive birefringence at the glass surface, the character of its changes is always a decreasing function of the duration of the heating process. Whereas, for the modal birefringence, the course of the dependence of $\delta N_\sigma(m)$ from the duration of the heating is dependent on the order of m mode, which can be clearly seen in the cases of the modes of higher order. For the modes of 5th and 6th order (Fig.18a), the increase of the heating duration in the first 25 h is accompanied by the increase of the birefringence. This effect can be explained by the generation of stresses in the deeper areas of the glass, which are reached by the admixture due to the diffusion widening of the doping area during the heating of the waveguide structure.

On the basis of equation (22) and elastooptic constants (21) it can be calculated how the stresses at the glass surface change as a result of the heating processes. A graph of this relationship is shown in Fig.18 b. There can be seen that after a heating time of about 50 h, the value of stresses at the glass surface decreases to half its value after the diffusion process.

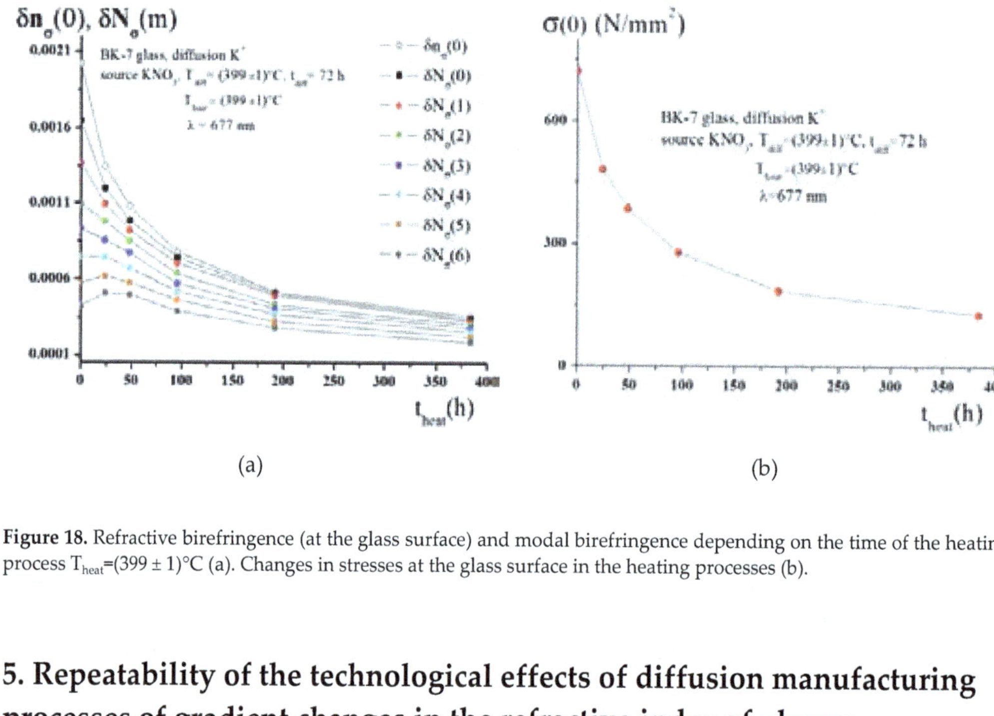

Figure 18. Refractive birefringence (at the glass surface) and modal birefringence depending on the time of the heating process T_{heat}=(399 ± 1)°C (a). Changes in stresses at the glass surface in the heating processes (b).

5. Repeatability of the technological effects of diffusion manufacturing processes of gradient changes in the refractive index of glasses

The ion exchange in glass method can be considered as a technological method. With its use, the predictable and repeatable results are obtained. Those effects are the changes in glass refraction, which are of a gradient character. They are described by the refractive index profiles. Determination of the refractive index profiles is easy to implement in the case of planar waveguides. For this purpose a waveguide method with the use of selective excitation of modes using a prism coupler is used [33]. In this method, the measurements of synchronous angles are used, for which excitations of waveguide modes are obtained. Such measurements are made using a goniometer. Precision of the measurements of these angles is at the level of several tens of seconds. Based on the measured synchronous angles, the effective refractive indices of the waveguide modes are calculated. The uncertainties of their determination are at the level of 10^{-4}. Based on the set of effective refractive indices of modes, a reconstruction of the refractive index profile of the waveguide is performed. For this purpose the procedure proposed in 1976 by White and Heidrich [45] is used. This procedure is based on the mode equation of the waveguide. Its specific application is shown in [33] (p.169).

The control of technological processes of diffusion doping of glasses in the ion exchange processes was proposed in the work [33] (p.181). The described method is used in producing one-dimensional planar waveguides. The basis of this method is to determine the temperature dependences of diffusion coefficients of ions exchanged in a glass-admixture system. Knowledge of these relationships allows to calculate the kinetics of diffusion process in the current process temperature. This is done by measuring the temperature of the process carried out by a thermocouple placed in the immediate vicinity of the glass plate. This measurement occurs at defined points in time t_p. Moments of the temperature measurement are a total multiples of Δt time step which is numerically integrated over one-dimensional diffusion equation [33] (p. 177):

$$\frac{\partial u}{\partial t}=\frac{D_{0A}e^{Au}}{1-\alpha u}\cdot\frac{\partial^2 u}{\partial x^2}+\frac{D_{0A}e^{Au}\left[\alpha+(1-u)A\right]-u(1-\alpha)^2 D_{0B}Be^{B(1-u)}}{(1-\alpha u)^2}\left(\frac{\partial u}{\partial x}\right)^2, \tag{23}$$

where

$$\alpha=1-\frac{D_{0A}}{D_{0B}}\cdot\exp\left[u(A+B)-B\right] \tag{24}$$

In the above equations, u(x) is a function describing the distribution of normalized concentration of admixture ions introduced into the glass. D_{0A},D_{0B}, A, and B are coefficients that describe the functional dependence of diffusion coefficients of admixture ions $D_A(u)$ and the modifier $D_B(u)$ on normalized concentration of admixture introduced into the glass. These functions have the form [33]:

$$D_A(u)=D_{0A}e^{Au},\quad D_B(u)=D_{0B}e^{B(1-u)} \tag{25}$$

The dependence of coefficients D_{0A} and D_{0B} on temperature is described by Arrhenius equations:

$$D_{0i}(T)=D_{0i}^{*}\cdot\exp\left(-\frac{\Delta Q_i}{RT}\right)\qquad(i=A,B), \tag{26}$$

where ΔQ_i - activation energy of the i-th ion type, R - universal gas constant.

The function u(x,t), which is a solution of the equation (23), transforms into one-dimensional refractive index profile n(x,t) according to the relation (5):

$$n(x,t) = n_b + \Delta n_s \cdot u(x,t) \tag{27}$$

Thus, solving the diffusion equation is carried out in parallel with the realization of the diffusion process. The diffusion coefficients appearing in this equation are calculated at points in time t_p based on the knowledge of their temperature dependencies. In this way the course of the process temperature $T_p = T(t_p)$ by the diffusion coefficients $D_{0A}(T_p)$ and $D_{0B}(T_p)$ affects the form of solution $u(x,t_p)$ of the diffusion equation (23). On the basis of equation (27) the dependence of function describing the refraction distribution in the glass $n(x,t_p)$ is obtained. In turn, with the use of equation (28), the effective refractive indices of the modes $N_m(t_p)$ corresponding with the refractive index profile $n(x,t_p)$ are calculated (for the assumed wavelength and state of polarization). The modal equation in the case of planar optical waveguides with a monotonic refractive index profile has the form:

$$k_0 \int_0^{x_m} \sqrt{n^2(x,t_p) - N_m^2(t_p)}\,dx = \pi\left(m + \frac{1}{4}\right) + \mathrm{arctg}\left(r\sqrt{\frac{N_m^2(t_p) - n_c^2}{n_s^2 - N_m^2(t_p)}}\right), \tag{28}$$

where k_0 - wave number of an electromagnetic wave in a free space, x_m - position of the turning point of the m-th mode, $n(x,t_p)$ - refractive index profile of the waveguide, $N_m(t_p)$ - effective refractive index of the m-th mode, n_c - refractive index of the ambient (coverage) of the waveguide, n_s - refractive index of the waveguide at the glass surface, $r = 1$ (for TE polarization), $r = (n_s/n_c)^2$ (for TM polarization).

Visualization of the refractive index profile $n(x,t_p)$ and the resulting set of the effective refractive indices of modes $N_m(t_p)$ provide a direct control of the diffusion process. After the process, the duration of which was t_{diff}, a registered final refractive index profile of the glass $n(x,t_{diff})$ and a set of effective refractive indices $\{N_m(t_{diff}): m = 0,1...M-1\}$ (M - number of modes) are obtained. Figure 19 schematically shows the principle of control of diffusion processes described earlier [46].

During the diffusion process control using the method mentioned earlier, the heat uniformity of the source of admixture has to be ensured. This is achieved by continuous mixing of the contents of the crucible. Thermocouple used to the measurements of the process temperature should be as close as possible to the glass substrate. The fulfillment of these conditions in the implemented technological processes was provided by a special handle made of silica glass (for details see [33] p.182).

In the continuation of this section the results being comparisons of effective refractive indices were presented. These comparisons relate to the effective refractive indices of modes calculated (N_{calc}) during the diffusion process control with the results obtained from the measurements (N_{meas}) of produced waveguides. The quantity of such comparisons are the absolute values of the differences of these values $|N_{calc} - N_{meas}|$ calculated for each row of mode m. They were made for three types of glass substrates and two kinds of admixture ions. The glass substrates were: soda-lime (of Menzel-Glasser company), BK-7 (of Schott company), and Pyrex (Borosilicate 33 of Corning company). The used admixture ions were silver ions Ag^+ (source: silver

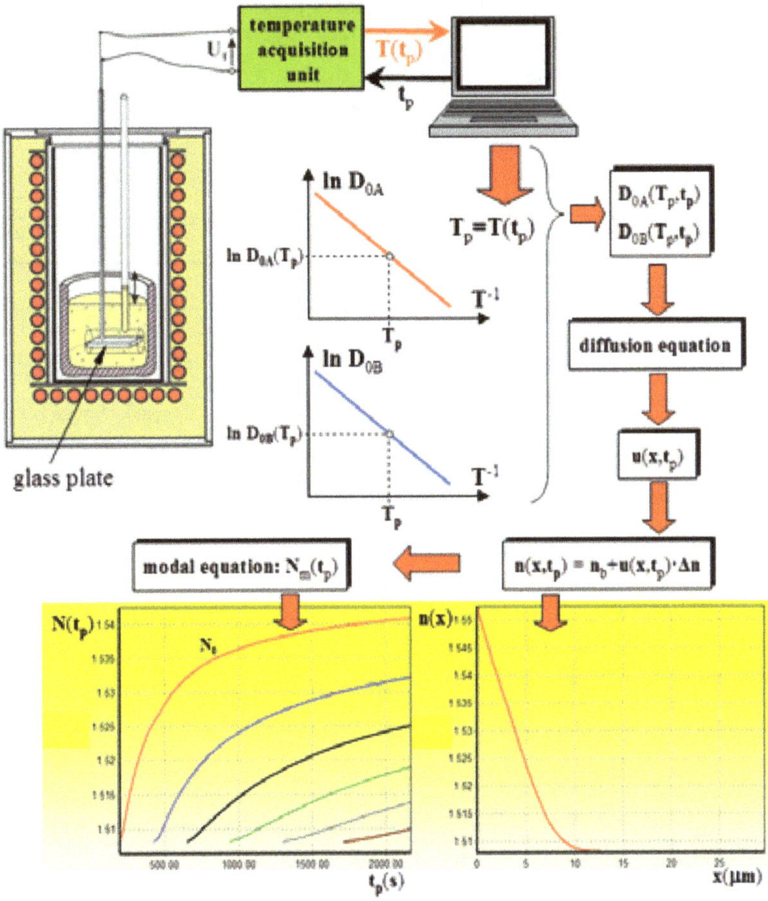

Figure 19. The principle of the diffusion process control in real time based on temperature measurements.

nitrate $AgNO_3$) and potassium ions K^+ (source: potassium nitrate KNO_3). Table 5 summarizes the types of glass and the admixture ions, for which the temperature characteristics of diffusion coefficients were designated. The measurements of effective refractive indices of waveguide modes were performed for a wavelength $\lambda = 677$ nm (TE polarization).

Type of glass	Ion exchange type	Source of ions	Diffusion coefficients				$\lambda = 677$ nm	
			$\Delta Q_A/R$ (K)	$\ln(D_{0A}^*)$	$\Delta Q_B/R$ (K)	$\ln(D_{0B}^*)$	n_b	Δn
Soda-lime	$Ag^+ \leftrightarrow Na^+$	$AgNO_3$	$1.1086 \cdot 10^4$	19.894	$9.2988 \cdot 10^3$	19.017	1.5105	0.0952
BK-7	$K^+ \leftrightarrow Na^+$	KNO_3	$1.5786 \cdot 10^4$	22.740	$1.3376 \cdot 10^4$	19.688	1.5137	0.0079
Pyrex	$Ag^+ \leftrightarrow Na^+$	$AgNO_3$	$1.0771 \cdot 10^4$	18.847	$1.1044 \cdot 10^4$	20.349	1.4678	0.0190

Table 5. The types of glass and admixture ions.

Table 6 summarizes a comparison of the waveguides produced in the soda-lime glass. In this glass the kinetics of $Ag^+ \leftrightarrow Na^+$ ion exchange is the highest among the other glass-admixture systems. A planar 3-modes waveguide is formed during a diffusion process with duration of 4 min. For the $Ag^+ \leftrightarrow Na^+$ ion exchange in this glass the largest absolute differences $|N_{calc} - N_{meas}|$ here are of the order of 10^{-3}.

m	t_{diff} = 3h, T_{ave} = 288°C			t_{diff} = 4h, T_{ave} = 280°C			t_{diff} = 5h, T_{ave} = 260.4°C		
	N_{calc}	N_{meas}	$\|N_{calc}\ N_{meas}\|$	N_{calc}	N_{meas}	$\|N_{calc}\ N_{meas}\|$	N_{calc}	N_{meas}	$\|N_{calc}\ N_{meas}\|$
0	1.5924	1.5939	0.0015	1.5926	1.5938	0.0012	1.5906	1.5905	0.0001
1	1.5823	1.5833	0.0010	1.5826	1.5833	0.0007	1.5794	1.5786	0.0008
2	1.5744	1.5752	0.0008	1.5748	1.5752	0.0004	1.5702	1.5696	0.0006
3	1.5675	1.5682	0.0007	1.5679	1.5683	0.0004	1.5621	1.5614	0.0007
4	1.5611	1.5619	0.0008	1.5616	1.5620	0.0004	1.5544	1.5537	0.0007
5	1.5551	1.5558	0.0007	1.5555	1.5559	0.0004	1.5468	1.5463	0.0005
6	1.5492	1.5499	0.0007	1.5497	1.5501	0.0004	1.5392	1.5388	0.0004
7	1.5433	1.5440	0.0007	1.5438	1.5442	0.0004	1.5313	1.5313	0.0000
8	1.5373	1.5382	0.0009	1.5379	1.5384	0.0005	1.5235	1.5233	0.0002
9	1.5314	1.5323	0.0009	1.5319	1.5326	0.0007	1.5163	1.5162	0.0001
10	1.5254	1.5265	0.0011	1.5259	1.5268	0.0009			
11	1.5196	1.5206	0.0010	1.5201	1.5209	0.0008			
12	1.5143	1.5151	0.0008	1.5147	1.5153	0.0006			
Short time diffusion processes									
m	t_{diff} = 15', T_{ave} = 300°C			t_{diff} = 8', T_{ave} = 300°C			t_{diff} = 4', T_{ave}=302°C		
	N_{calc}	N_{meas}	$\|N_{calc}\ N_{meas}\|$	N_{calc}	N_{meas}	$\|N_{calc}\ N_{meas}\|$	N_{calc}	N_{meas}	$\|N_{calc}\ N_{meas}\|$
0	1.5805	1.5811	0.0006	1.5751	1.5760	0.0009	1.5712	1.5690	0.0022
1	1.5615	1.5621	0.0006	1.5513	1.5520	0.0007	1.5430	1.5389	0.0041
2	1.5454	1.5458	0.0004	1.5295	1.5310	0.0015	1.5154	1.5109	0.0045
3	1.5294	1.5297	0.0003						
4	1.5140	1.5143	0.0003						

Table 6. Diffusion processes in soda-lime glass. Ion exchange $Ag^+ \leftrightarrow Na^+$. Source of admixture ions: $AgNO_3$. λ = 677 nm, polarization TE.

The diffusion processes of $K^+ \leftrightarrow Na^+$ ion exchange are the slowest in the BK-7 glass. Table 7 shows a comparison of the effective refractive indices of the modes for diffusion processes lasting from 48 h to 216 h. The temperatures of these processes were approximately 400°C. In these cases, the differences of effective refractive indices of modes are of the order of 10^{-4}. The same differences occur in the control of the heating processes carried out in BK-7 glass. Table 8 presents the results referring to the heating processes for a waveguide produced in the preliminary diffusion process: t_{diff} = 74.3h, T_{ave} = 401.2°C then subjected to heating. The temperatures of heating processes were at 445°C. The durations of the heating processes were 1 h, 2 h, and 4 h respectively.

m	t_{diff} = 48h, T_{ave} = 401°C			t_{diff} = 74.3h, T_{ave} = 401.2°C			t_{diff} = 216h, T_{ave} = 399°C		
	N_{calc}	N_{meas}	$\lvert N_{calc} N_{meas} \rvert$	N_{calc}	N_{meas}	$\lvert N_{calc} N_{meas} \rvert$	N_{calc}	N_{meas}	$\lvert N_{calc} N_{meas} \rvert$
0	1.5189	1.5193	0.0004	1.5193	1.5194	0.0001	1.5199	1.5203	0.0004
1	1.5171	1.5174	0.0003	1.5177	1.5177	0.0000	1.5187	1.5190	0.0003
2	1.5157	1.5161	0.0004	1.5164	1.5166	0.0002	1.5178	1.5180	0.0002
3	1.5146	1.5150	0.0004	1.5154	1.5156	0.0002	1.5170	1.5173	0.0003
4	1.5139	1.5142	0.0003	1.5146	1.5149	0.0003	1.5163	1.5166	0.0003
5				1.5140	1.5142	0.0002	1.5157	1.5160	0.0003
6							1.5151	1.5155	0.0004
7							1.5146	1.5150	0.0004
8							1.5142	1.5146	0.0004
9							1.5139	1.5203	0.0004

Table 7. Diffusion processes in BK-7 glass. Ion exchange $K^+ \leftrightarrow Na^+$. Source of admixture ions: KNO_3. λ = 677 nm, polarization TE.

Table 9 shows the results referring to the Pyrex glass (Borosilicate 33). Also in this case the differences of calculated and measured effective refractive indices of modes are of the order of 10^{-4}. The kinetics of $Ag^+ \leftrightarrow Na^+$ exchange processes in this glass is moderate. The single-mode waveguides are formed during 1 h and at a temperature of about 290°C.

m	t_{diff} = 74.3h, T_{ave} = 401.2°C + t_{heat} = 1h, T_{ave} = 445.8°C			t_{diff} = 74.3h, T_{ave} = 401.2°C + t_{heat} = 2h, T_{ave} = 445.5°C			t_{diff} = 74.3h, T_{ave} = 401.2°C + t_{heat} = 4h, T_{ave} = 446.3°C		
	N_{calc}	N_{meas}	$\lvert N_{calc} N_{meas} \rvert$	N_{calc}	N_{meas}	$\lvert N_{calc} N_{meas} \rvert$	N_{calc}	N_{meas}	$\lvert N_{calc} N_{meas} \rvert$
0	1.5191	1.5190	0.0001	1.5189	1.5186	0.0003	1.5186	1.5182	0.0004
1	1.5177	1.5176	0.0001	1.5176	1.5173	0.0003	1.5175	1.5171	0.0004
2	1.5165	1.5165	0.0000	1.5165	1.5163	0.0002	1.5165	1.5163	0.0002
3	1.5156	1.5156	0.0000	1.5156	1.5155	0.0001	1.5156	1.5155	0.0001
4	1.5148	1.5149	0.0001	1.5149	1.5147	0.0002	1.5149	1.5149	0.0000
5	1.5142	1.5142	0.0000	1.5142	1.5142	0.0000	1.5143	1.5143	0.0000
6							1.5138	1.5138	0.0000

Table 8. Heating processes in BK-7 glass. Ion exchange $K^+ \leftrightarrow Na^+$. λ = 677 nm, polarization TE.

The results of comparisons of calculated and measured effective refractive indices confirm the effectiveness of the method of control of the diffusion processes. It is based on the best possible determined temperature dependence of the diffusion coefficients of exchanged ions for the given glass-admixture system. The described method allows to guarantee the repeatability of effective refractive indices at the level of $\Delta N \sim 10^{-3}$.

This method can also be used to control the production of two- and three-dimensional structures in gradient glasses.

m	t_{diff} = 21h, T_{ave} = 275.6°C			t_{diff} = 23h, T_{ave} = 281.1°C			t_{diff} = 22h, T_{ave} = 292.1°C		
	N_{calc}	N_{meas}	$\vert N_{calc} N_{meas} \vert$	N_{calc}	N_{meas}	$\vert N_{calc} N_{meas} \vert$	N_{calc}	N_{meas}	$\vert N_{calc} N_{meas} \vert$
0	1.4816	1.4812	0.0004	1.4820	1.4817	0.0003	1.4825	1.4833	0.0008
1	1.4775	1.4771	0.0004	1.4783	1.4779	0.0004	1.4792	1.4795	0.0003
2	1.4742	1.4738	0.0004	1.4753	1.4749	0.0004	1.4766	1.4766	0.0000
3	1.4714	1.4710	0.0004	1.4727	1.4723	0.0004	1.4742	1.4741	0.0001
4	1.4691	1.4689	0.0002	1.4705	1.4700	0.0005	1.4721	1.4719	0.0002
5				1.4687	1.4684	0.0003	1.4703	1.4700	0.0003
6							1.4688	1.4686	0.0002
Short time diffusion processes									
m	t_{diff} = 3h, T_{ave} = 286°C			t_{diff} = 2.9h, T_{ave} = 287°C			t_{diff} = 1h, T_{ave} = 289°C		
	N_{calc}	N_{meas}	$\vert N_{calc} N_{meas} \vert$	N_{calc}	N_{meas}	$\vert N_{calc} N_{meas} \vert$	N_{calc}	N_{meas}	$\vert N_{calc} N_{meas} \vert$
0	1.4780	1.4777	0.0003	1.4781	1.4791	0.0010	1.4750	1.4742	0.0008
1	1.4713	1.4714	0.0001	1.4713	1.4718	0.0005			

Table 9. Diffusion processes in Pyrex glass. Ion exchange $Ag^+ \leftrightarrow Na^+$. Source of admixture ions: $AgNO_3$. λ = 677 nm, polarization TE.

6. Conclusions

The chapter presents selected aspects of the ion exchange in glass processes in terms of changes in the refractive index. The processes of electrodiffusion doping of soda-lime glass with Ag^+ ions have been described. Such processes, which are carried out with a change of polarization of the electric field, enable the production of refraction changes of the glass with refractive index profile of various characters. Using the electrodiffusion processes with a fixed polarization of electric field, the equilibrium concentration of mobile ions in the glass was estimated. These results were confirmed by a method based on glass substrates weighing before and after the diffusion process. The results of experimental studies of stress birefringence in BK-7 glass doped with potassium ions K^+ were also presented. Ion exchange processes $K^+ \leftrightarrow Na^+$ in the glass were carried out during more than 500 h. It also presents the results of changes of these stresses in the long-term processes of glass heating. The final part of the chapter presents the results of a control of the diffusion and heating processes in a real-time implementation of these processes. The described method of controlling these processes allows to obtain a high repeatability of the effects of the ion exchange in glass.

Acknowledgements

This chapter was financed from the funds of the National Science Centre, awarded on the basis of the decision DEC-2011/01/B/ST7/06525.

Author details

Roman Rogoziński*

Address all correspondence to: roman.rogozinski@polsl.pl

Optoelectronics Department, Silesian University of Technology, Gliwice, Poland

References

[1] Cullen T.J., Wilkinson C.D.W., *Radiation losses from singlemode optical Y junctions formed by silver-ion exchange in glass*, Opt. Lett. 10 (4), 134-136 (1984).

[2] Findakly T., Chen B., *Single-mode integrated optical 1 X N star coupler*, Appl. Phys. Lett. 40 (7), 549-550 (1982).

[3] Tervonen A., Honkanen S., Najafi S.I., *Analysis of symmetric directional couplers and asymmetric Mach-Zehnder interferometers as 1.30/1.55 μm dual wavelength demultiplexers/ multiplexers*, Opt. Eng. 32 (9), 2083-2091 (1993).

[4] Buchold B., Glingener C., Culemann D., E.Voges, *Polarization insensitive ion-exchanged arrayed-waveguide grating multiplexers in glass*, Fiber Integr Optics, 17 (4), 279-298 (1998).

[5] Ruschin S., Hurwitz G., Hurwitz T., Kepten A., Arad E., Soreq Y., Eckhouse S., *Glass ion-exchange technology for wavelength management applications*, Proc. SPIE 4944, 150-158 (2003).

[6] Castro J.M., Geraghty D.F., West B.R., Honkanen S., *Fabrication and comprehensive modeling of ion-exchanged Bragg optical add-drop multiplexers*, Appl. Opt. 43 (33), 6166-6173 (2004).

[7] Bucci D., Grelin J., Ghibaudo E., Broquin J.E., *Realization of a 980-nm/1550-nm pump-signal (de)multiplexer made by ion-exchange in glass using a segmented asymmetric Y-Junction*, IEEE Photonics Technol. Lett. 19 (9), 698-700 (2007).

[8] Yip G.L., Finak J., *Directional-coupler power divider by two-step K^+-ion exchange*, Opt. Lett. 9 (9), 423-425 (1984).

[9] Walker R.G., Wilkinson C.D.W., *Integrated optical ring resonators made by silver ion-exchange in glass*, Appl. Opt. 22 (7), 1029-1035 (1983).

[10] Hsiao H.K., Winick K.A., *Planar glasswaveguide ring resonators with gain*, Opt. Express 15 (26), 17783-17797 (2007).

[11] Hinkov V., Sohler W., *Reduced depth polarizer for integrated optics*, Appl. Phys. 14, 229-230 (1977).

[12] Benech P., Persegol D., Saint Andre F., *A glass ion exchanged Mach-Zehnder interferometer to stabilize the frequency of a laser diode*, J. Phys. D 23 (5), 617-619 (1990).

[13] Tervonen A., Pöyhönen P., Honkanen S., Tahkokorpi M., *A guided-wave Mach-Zehnder interferometer structure for wavelength multiplexing*, IEEE Photonics Technol. Lett. 3 (6), 516-518 (1991).

[14] Das S., Geraghty D., Honkanen S., Peyghambarian N., *MMI splitters by ion-exchange in glass, Proc. SPIE* 3936, 239-247 (2000).

[15] Błahut M., Karasiński P., Rogoziński R., *Multimode interference structures made by ion-exchange technique in glass*, Proc. SPIE 5028, 85-89 (2003).

[16] West B., Honkanen S., *MMI devices with weak guiding designed in three dimensions using a genetic algorithm*, Opt. Express 12 (12), 2716-2722 (2004).

[17] Najafi S.I., Wang W.J., Currie J.F., Leonelli R., Brebner J.L., *Fabrication and characterization of neodymium-doped glass waveguides*, IEEE Photonics Technol. Lett. 1 (5), 109-110 (1989).

[18] Aoki H., Ishikawa E., Asahara Y., *Nd3+-doped glass waveguide amplifier at 1.054-μm*, Electron. Lett. 27 (25), 2351-2353 (1991).

[19] Feuchter T., Mwarania E.K., Wang J., Reekie L., Wilkinson J.S., *Erbium-doped ion-exchanged waveguide lasers in BK-7 glass*, IEEE Photon. Technol. Lett. 4 (6), 542-544 (1992).

[20] Miliou N., Cao X.F., Srivastava R., Ramaswamy R.V., *15 dBamplification at 1.06 μm in ion-exchanged silicate glass waveguides*, IEEE Photon. Technol. Lett. 5 (4), 416-418 (1993).

[21] Patel F.D., Di Carolis S., Lum P., Venkatesh S., Miller J.N., *A compact high-performance optical waveguide amplifier*, IEEE Photon. Technol. Lett. 16 (12), 2607-2609 (2004).

[22] Pissadakis S., Ikiades A., Hua P., Sheridan A.K., Wilkinson J.S., *Photosensitivity of ion-exchanged Er-doped phosphate glass using 248 nm excimer laser radiation*, Opt. Express 12 (14), 3131-3136 (2004).

[23] Yliniemi S., Honkanen S., Ianoul A., Laronche A., Albert J., *Photosensitivity and volume gratings in phosphate glasses for rareearth-doped ion-exchanged opticalwaveguide lasers*, J. Opt. Soc. Am. B 23 (12), 2470-2478 (2006).

[24] Ross L., *Ion-exchanged glass waveguide sensors*, SPIE Critical Reviews of Optical Science and Technology CR53, 180-199 (1994).

[25] Parriaux O.M., Roth P., Voirin G., *Multimode glass integrated optics*, SPIE Critical Reviews of Optical Science and Technology CR53, 295-320 (1994).

[26] Yimit A., Rossberg A.G., Amemiya T., Itoh K., *Thin film composite optical waveguides for sensor applications: a review*, Talanta 5 (5), 1102-1109 (2005).

[27] Lambeck P.V., *Integrated optical sensors for the chemical domain*, Meas. Sci. Technol. 17 (8), R93-R116 (2006).

[28] Mazurczyk R., Vieillard J., Bouchard A., Hannes B., Krawczyk S., *A novel concept of the integrated fluorescence detection system and its application in a lab-on-a-chip microdevice*, Sens. Actuators B 118 (1-2), 11-19 (2006).

[29] Saarinen J., Honkanen S., Najafi S.I., Huttunen J., *Double-ion exchange process in glass for the fabrication of computer-generated waveguide holograms*, Appl. Opt. 33 (16), 3353-3359 (1994).

[30] Salmio R.P., Saarinen J., Turunen J., Tervonen A., *Graded-index diffractive elements by thermal ion exchange in glass*, App. Phys. Lett. 66 (8), 917-919 (1995).

[31] Salmio R.P., Saarikoski H., Saarinen J., Westerholm J., Turunen J., *Three-dimensionally modulated graded-index diffractive elements by thermal ion exchange in glass*, Opt. Lett. 22 (9), 591-593 (1997).

[32] Righini G.C., Molesini G., *Design of optical-waveguide homogeneous refracting lenses*, Appl. Opt. 27 (20), 4193-4199 (1988).

[33] Rogozin'ski R. (2012). *Ion exchange in glass - The changes of glass refraction*, Ion Exchange Technologies, Ayben Kilislioglu (Ed.), ISBN InTech, Available from: http://www.intechopen.com/books/ion-exchange-technologies/ion-exchange-in-glass-the-changes-of-glass-refraction.

[34] Terai R., Hayami R., *Ionic diffusion in glasses*, J. Non. Cryst. Solids 18, 217-264 (1975).

[35] Houde-Walter S.N., Moore D.T., *Gradient-index profile control by field-assisted ion exchange in glass*, Appl. Opt., 24 (24), 4326-4333 (1985).

[36] Rogoziński R., *Electrodiffusion processes with the conversion of polarization direction of electric field in the formation of planar waveguide structures using ion exchange technique in glass*, Optica Applicata, 28 (4), 331-343 (1998).

[37] Rogoziński R., *Refractive index profiles of planar waveguides produced in electrodiffusion processes*, Optica Applicata, 34 (4), 489-505 (2004).

[38] Menzel Gläser product information sheet, "Erie Electroverre SA" http://www.menzel.de/fileadmin/Templates/Menzel/pdf/en/EVR_en.pdf.

[39] Rogoziński R., Karasiński P., *Optical waveguides produced in ion exchange process from the solutions of AgNO3-NaNO3 for planar chemical amplitude sensors*, Opto-Electronics Rev., 13 (3), 229-238 (2005).

[40] Schott-optical-glass-pocket-catalog-january-2014-row

[41] Brandenburg A., *Stress in ion-exchanged glass waveguides*, J. Lightwave Tech., 4 (10), 1580-1593 (1986).

[42] Bach H., Neuroth N., *The properties of optical glass*, Berlin: Springer-Verlag Berlin Heidelberg (1995).

[43] Agan S., Ay F., Kocabas A., Aydinli A., *Stress effect in prism coupling measurements of thin polymer films*, Appl. Phys. A, 80, 341-345 (2003).

[44] Rogoziński R., *Investigation of birefringence in planar waveguides produced by ion exchange $K^+ \leftrightarrow Na^+$ in glass BK-7*, Proc. of SPIE, TAL, 5576, 213-218 (2003).

[45] White J.M., Heidrich P.F., *Optical waveguide refractive index profiles from measurement of mode indices: A simple analysis*, Appl. Opt., 15 (1), 151-155 (1976).

[46] Rogoziński R., *Producibility of the ion-exchange method in manufacturing gradient refractive* index in glass, Bull. Acad. Pol. Sci. Biol. - Technical Sciences, 62 (4), 655-665 (2014).

2

Effect of KNO_3 Molten Bath Na Enrichment on the Mechanical Performances of Ion-exchanged Soda-Lime-Silicate Glass

Hamid Hassani and Vincenzo M. Sglavo

Abstract

Systematic doping of pure molten KNO_3 bath by sodium nitride was used to analyze the effect of salt poisoning on the mechanical performances of chemically tempered soda-lime-silicate float glass. The ion-exchange process appears partially influenced by the presence of limited amount of Na in the potassium nitride bath. The interdiffusion coefficient and the penetration depth of K are invariant for $NaNO_3$ content up to 5 wt%; conversely, the surface concentration of potassium scales with the purity of the bath, and this accounts for different compressive residual stress. Nevertheless, the overall reinforcement associated to the final strength does not show clear dependence on the used bath mainly because of the large scatter in the surface flaws sizes, as typically observed in glass.

Keywords: Ion exchange, Chemical tempering, Sodium poisoning, Soda-lime-silicate glass

1. Introduction

The very first records of ion-exchange can be traced back to the Middle Ages, when glass artists used a mixture of silver salts, clay, and natural oil to obtain yellowish color in silicate glasses. The salts mixture was deposited on the glass surface and heated at about 600°C in reducing atmosphere, inducing the diffusion of silver ions into the glass and the formation of silver nanoparticles [1, 2]. Since then, the ion-exchange process was applied without much scientific knowledge. It was in the 20th century that scientist started to understand the surface of chemically tempered glasses and figured out an exchange between monovalent cations in glasses with silver and/or potassium cations in the molten salts [2, 3]. These investigations increased the industrial application of the ion-exchange process, especially with the aim of enhancing the optical and mechanical properties of glass [4].

Glass strength can be largely increased by the ion-exchange process, otherwise called chemical tempering. The exchange of small ions such as Li^+ or Na^+ in an alkali-containing glass, with larger ions such as K^+ from a molten KNO_3 bath at temperatures below the strain point of the glass, is responsible for the creation of bi-axial residual compressive stress in the surface layers of the material. Because glass products generally break due to excessive tension applied at a surface flaw, the introduction of surface compression strengthens the glass component.

During the ion-exchange treatment, the glass matrix is considered as a solid negatively charged structure where some mobile ions (Na^+ in soda-lime-silicate glass) can be replaced by larger monovalent ions (K^+ from molten KNO_3) responsible for the generation of a compressive stress. The replacement takes place through an interdiffusion process, according to the Nernst–Planck equations. The flux of the ion species scales with the interdiffusion coefficient [2]:

$$\bar{D} = \frac{D_{Na} D_K}{D_{Na} N_{Na} + D_K N_K} \tag{1}$$

where N_i is the fractional concentration of alkali ion i (Na or K) and D_i the self-diffusion coefficient. One important parameter in the interdiffusion phenomenon is clearly the concentration of ions in the molten salt. Some studies have shown that the presence of poisoning elements (already present in the salt or coming out from the glass as in the case of Na) can hinder the diffusion of K^+ into the glass and slow down the exchange process, thus reducing the strengthening efficiency. An important issue in industrial practice is also the replacement/renewal of the molten salt which is time and money consuming [5]. Some researchers [3, 6], on the other hand, believe that the ion-exchange process is not affected in the presence of specific amount of poisoning elements.

In the present work, a systematic analysis of the effect of small amount of sodium as poisoning element in the molten bath on the performances of the strengthened soda-lime-silicate float glass was carried out.

2. Experimental procedure

Commercial soda-lime-silicate float glass sheets with nominal thickness of 4 mm were used. Table 1 summarizes the composition of the glass.

SiO_2	Al_2O_3	Na_2O	K_2O	MgO	CaO
71	1	13	1	4	10

Table 1. Composition (wt%) of the float glass

The sheets, obtained from one single original plate, were cut into square samples of 50 mm × 50 mm. The edges of the specimens were rectified and polished with SiC abrasive paper. The samples were then rinsed and cleaned gently in water, avoiding any damage on the glass surface.

A semi-automatic chemical strengthening lab scale furnace was used for the ion-exchange treatment. In each run, 20 samples were placed in a stainless steel basket to be treated at 450 °C for 4 h, with 30 min preheating and 30 min post-cooling above the bath. At least 60 samples were treated in each bath with the same thermal and time conditions.

Pure KNO_3 and $NaNO_3$ salts from Sigma Aldrich were used. Pure potassium nitrate was systematically poisoned by adding a specific amount of $NaNO_3$ as reported in Table 2.

Added $NaNO_3$ (wt%)	0	0.1	0.2	0.5	1	2	5
Bathlable	A	B	C	D	E	F	G

Table 2. Amount of $NaNO_3$ added to pure KNO_3 in the considered salt baths

After each treatment, the samples were rinsed with water and carefully dried.

The surface residual stress and the case depth were optically measured by surface stress-meter (FSM-60LE, Luceo Co., Ltd., Japan). Bi-axial flexural test [7] was used to measure the strength. This was carried out with a ring-on-ring configuration with the upper loading ring and the lower support ring having a diameter of 8 mm and 40 mm, respectively. The actuator speed was $1\frac{mm}{min}$. The strength was determined from the maximum load (F) as:

$$\sigma_F = K\frac{F}{h^2} \tag{2}$$

where h is the thickness of sample and

$$K = \frac{3(1+\nu)}{2\pi}\left(\ln\frac{D_S}{D_L} + \frac{(1-\nu)(D_S^2 - D_L^2)}{0.72D^2(1+\nu)}\right) \tag{3}$$

where D_S and D_L are the radius of the upper and lower supporting ring, respectively, D the sample size (50 mm), and ν the glass Poisson's ratio. A certain number of as-cut samples were also tested in this way for comparison.

Some fragments were collected from the broken samples and used for determining the potassium penetration profile. The fragments were attached on an aluminum disk by conductive adhesive tape and then coated by sputtering with Au-Pd alloy. Clean and flat portions of the fracture surface were analyzed in a Scanning Electron Microscope (SEM) (JSM5500, Jeol,

Japan) and the potassium Kα signal was recorded on a path of ~30 μm long by using the Energy Dispersion X-ray Spectroscopy (EDXS) (EDS2000, IXRF System, USA) probe. The chemical composition of the external surface of the glasses after the ion-exchange process was analyzed in the same way in a region of about 0.5 mm^2.

3. Results and discussion

The residual stress (σ_r) measured on the surface of glasses treated in the different molten salts is reported in Figure 1. The residual stress clearly decreases as $NaNO_3$ content increases, especially above 0.5%.

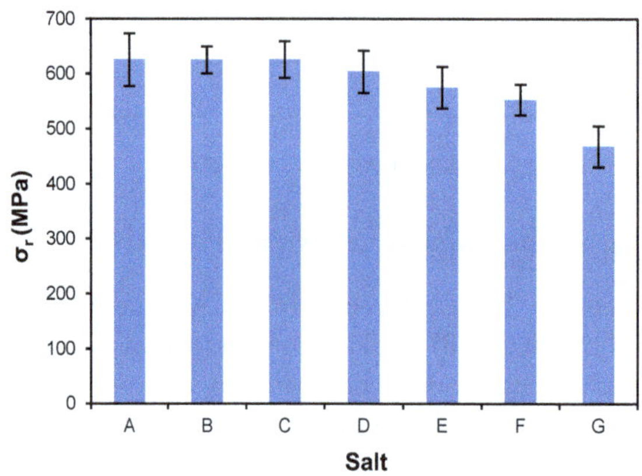

Figure 1. Compressive residual stress on the surface of glass ion exchanged in the different salt baths.

Conversely, the case depth reported in Figure 2 shows a quite different trend and actually seems almost invariant as a function of the sodium content of the salt bath.

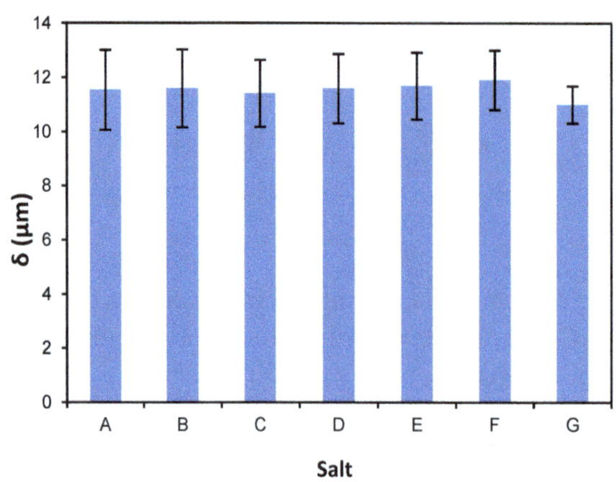

Figure 2. Case depth as a function of the exchanging bath.

The strength of ion-exchanged samples is shown in Figure 3. In spite of the typical large scatter and the relatively short duration of the ion-exchange process (compared to typical industrial duration, usually in excess of 8 h), the strengthening effect is clear in any used salt bath, and the failure stress increases by a factor of about 2.2–2.4 with respect to the as-cut glass. Nevertheless, the large strength scatter does not allow to point out specific trends with respect to the Na content of the exchanging baths.

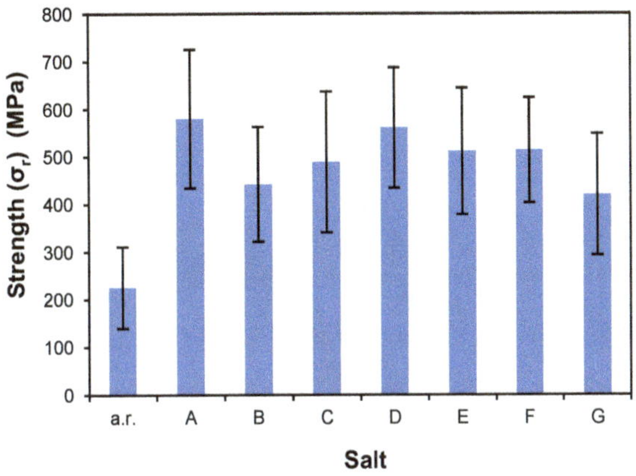

Figure 3. Average flexural strength as a function of the exchanging bath (the standard deviation is also shown).

A more effective representation of the strength data is possible by using Weibull plots where the measured strength data are reported as a function of failure probability. Here, failure probability was calculated as:

$$P = \frac{i - 0.3}{N + 0.4} \tag{4}$$

where i is the rank in the ascending ordered strength distribution and N the total number of samples considered for each condition. The obtained Weibull plots are shown in Figure 4. One can easily observe again the evident strengthening effect in any salt bath with respect to as-cut samples; conversely, no specific trend can be seen with respect to the purity of the used bath.

The relationship between failure probability and tensile stress is typically expressed as:

$$P = 1 - \exp\left[-KS\left(\frac{\sigma}{\sigma_0}\right)^m\right] \tag{5}$$

where S is the surface area of the sample under tensile stress, K the loading factor (whose units are per unit area), m the Weibull modulus, and σ_0 the normalizing stress, representing the scatter of the distribution. Taking twice the natural logarithm of both sides, a linear equation can be obtained, which is functional for the Weibull modulus calculation [8]:

$$\ln\left(\ln\left(\frac{1}{1-P}\right)\right) = m\ln\sigma + \ln\frac{KS}{\sigma_0^m} \tag{6}$$

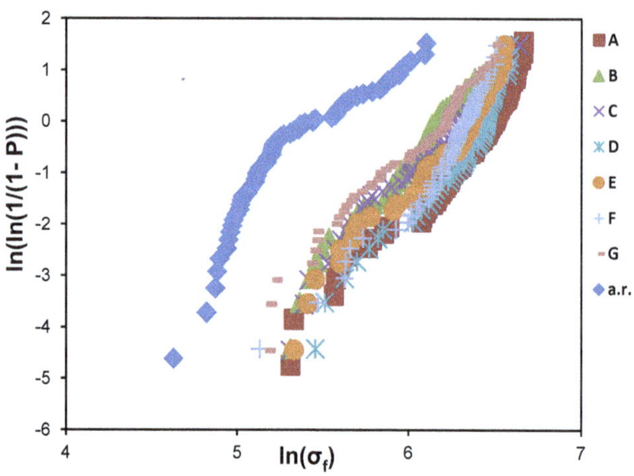

Figure 4. Weibull distributions for treated and untreated(a.r.) glasses.

Table 3 reports the Weibull modulus for the various distributions shown in Figure 4. One can observe that the moduli are also unaffected by changing the purity of the bath.

Salt	A	B	C	D	E	F	G	a.r.
Weibull modulus	3.9	4.1	3.6	4.4	3.9	4.4	3.5	3.2

Table 3. Weibull modulus of the strengthened glasses in different baths along with the as-received glass

The surface concentration of K^+ and Na^+ measured by EDXS is shown in Figure 5. There is a drastic increase in the potassium concentration after the ion-exchange process but the potassium concentration follows substantially the same trend of the surface residual stress (Figure 1), thus indicating a strict correlation between the amount of exchanged ions and the developed "stuffing"/reinforcing effect.

The potassium concentration profiles recorded by the EDXS line analysis always resembled a typical Nernst–Planck diffusion trend, which can be expressed as:

$$C_K^{'}(x,t) = \frac{C_K(x,t)}{C_{K_0}} = erfc\left(\frac{x}{2\sqrt{\bar{D}t}}\right) \tag{7}$$

where x is the distance from the surface, t the ion exchange time, C_{K_0} the potassium concentration on the surface, and $\bar{D}$ the interdiffusion coefficient. The potassium concentration experimental data were fitted by Equation (7) to determine the interdiffusion coefficient and the penetration depth, identified as the distance from the surface where the potassium concentration is lower than 2% with respect to the surface one. Figures 6 and 7 show the obtained results.

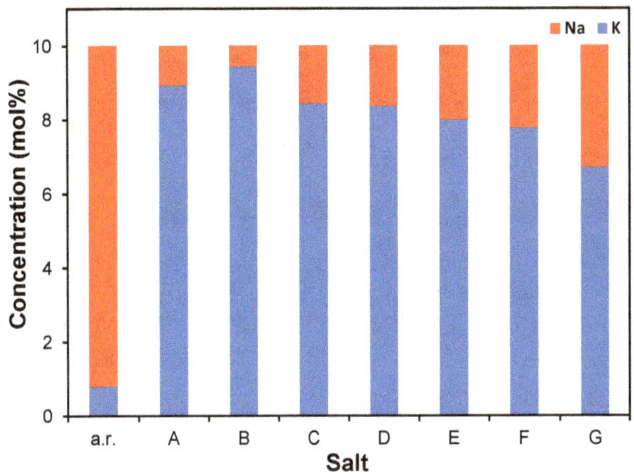

Figure 5. Potassium and sodium concentration on the surface of treated and bare glass.

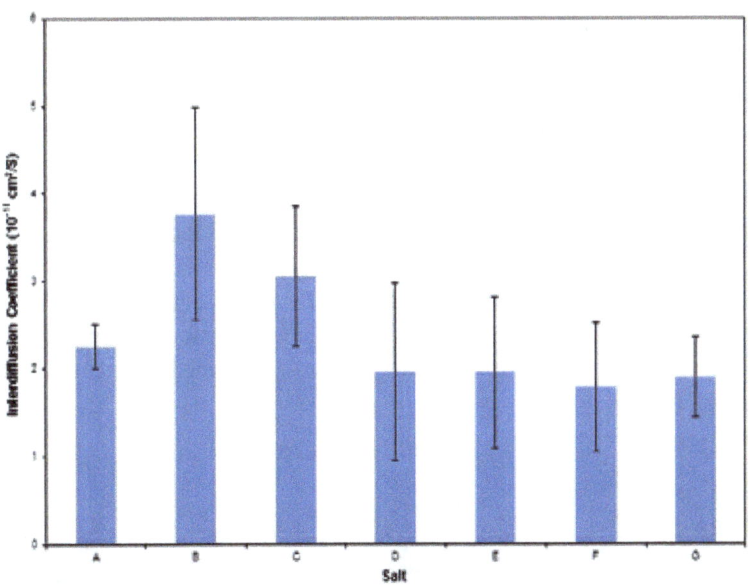

Figure 6. Interdiffusion coefficient for potassium in the different baths.

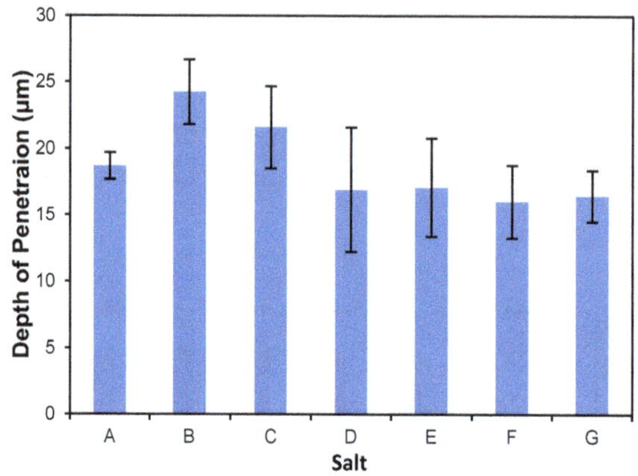

Figure 7. Depth of penetration for potassium during the ion-exchange process in the different baths.

The interdiffusion coefficients are in very good agreement with data reported in previous works [9, 10]. It is also confirmed that the $\bar{D}$ is not really affected by the presence of limited amounts of Na (up to 5%) in the KNO_3 bath [11]. Conversely, the surface concentration and, consequently, the concentration of potassium in the sub-surface layers are lower when the glass is treated in the Na-containing bath. On this basis, the K surface concentration in Equation (7) scales with the concentration in the used salt. Accordingly, the residual stress on the surface (shown in Figure 1) is higher when very pure KNO_3 bath is used, while the case depth does not change to an appreciable extent. Nevertheless, the effect of Na-containing salts on the final strength is substantially negligible, as shown in Figures 3 and 4. Clearly, this is mainly associated to the experimental scatter of the strength measurement, related to the typical dispersion on the surface defect sizes. In addition, due to the limited exchanging time used in the present work, some of the flaws are not "fully" reinforced; as a matter of fact, starting from the strength of the as-cut glass (ranging from ≈ 100 MPa to 400 MPa) and assuming, for simplicity, semicircular surface cracks, one can calculate that flaw sizes vary from ≈5 to ≈80 µm. Therefore, according to the case depth (Figure 2) and K penetration (Figure 7) results, it is evident that only a portion of the surface defects are completely "immersed" in the residual compressive stress field.Deeper defects, in a simplified model that considers flaws as invariant and perfectly closed during the ion-exchange process, are conversely subjected to a residual stress that changes from highly compressive on the surface to slightly tensile at a certain depth (below ≈12 µm). The effect of the initial flaw sizes, i.e. of the surface quality of the bare glass, appears to be more important in the ion-exchange process than the presence of limited amount of Na in the salt bath.

4. Conclusions

The presence of a small amount (up to 5 wt%) of $NaNO_3$ in potassium nitride bath partially influences the strengthening process of soda-lime-silicate glass when the treatment is carried out over a duration of 4 h. The interdiffusion coefficient and the penetration of K in the glass are substantially invariant while the surface concentration of potassium scales with the purity of the bath, and this is responsible for different compressive residual stress, which is higher when pure KNO_3 is used. Nevertheless, the overall reinforcement associated to the final strength does not show clear dependence on the used bath mainly because of the large scatter in the surface flaws sizes, as typically observed in glass. It is clear that the quality of the surface of the bare glass in terms of flaws has a much higher importance than the salt purity in the final mechanical resistance.

Author details

Hamid Hassani* and Vincenzo M. Sglavo

*Address all correspondence to: hamid.hassani@studenti.unitn.it

Department of Industrial Engineering, University of Trento, Trento, Italy

References

[1] Puche-Roig A, Martín VP, Murcia-Mascarós S, Puchades RI. Float glass colouring by ion exchange. Journal of Cultural Heritage, 2008; 9 129-133.

[2] Mazzoldi P, Carturan S, Quaranta A, Sada C, Sglavo VM. Ion exchange process: History, evolution and applications. RIVISTA DEL NUOVO CIMENTO 2013; 36(9) 397-460.

[3] Sglavo VM. Influence of KNO_3 Bath Composition on Ion Exchange Process of Commercial Soda Lime Silicate Float Glass. Ion Exchange Technologies,Intech, 2012.

[4] Gy R. Ion exchange for glass strengthening. Materials Science and Engineering: B, 2008; 149(2) 159-165.

[5] Fu AI, Mauro JC. Mutual diffusivity, network dilation, and salt bath poisoning effects in ion-exchanged glass. Journal of Non-Crystalline Solids, 2013; 363 199-204.

[6] Varshneya AK, Spinelli IM. High-strength, large-case-depth chemically strengthened lithium aluminosilicate glass. American Ceramic Society Bulletin, 2009; 88(5) 27-32.

[7] ASTM and Standard, Standard Test Method for Monotonic Equibiaxial Flexural Strength of Advanced Ceramics at Ambient Temperature, 2005, West Conshohocken, PA, USA: ASTM Book of Standards: USA.

[8] Sglavo VM. Chemical strengthening of soda lime silicate float glass: Effect of small differences in the KNO_3 bath. International Journal of Applied Glass Science, 2014. DOI: 10.1111/ijag.12101.

[9] Shen J, Green DJ, Pantano CG. Control of concentration profiles in two step ion exchanged glasses. Physics and Chemistry of Glasses, 2003; 44(4) 284-292.

[10] Jiang L, Guo X, Li X, Li L, Zhang G, Yan Y. Different K^+–Na^+ inter-diffusion kinetics between the air side and tin side of an ion-exchanged float aluminosilicate glass. Applied Surface Science, 2013; 265 889-894.

[11] Varshneya AK, Milberg ME. Ion exchange in sodium borosilicate glasses. Journal of the American Ceramic Society, 1974; 57(4) 165-169.

3

Ion Exchange Method for Removal and Separation of Noble Metal Ions

Zbigniew Hubicki, Monika Wawrzkiewicz, Grzegorz Wójcik, Dorota Kołodyńska and Anna Wołowicz

Abstract

Ion exchange has been widely applied in technology of chemical separation of noble metal ions. This is associated with dissemination of methods using various ion exchange resins which are indispensable in many fields of chemical industry. Due to small amounts of noble elements in nature and constant impoverishment of their natural raw materials, of particular importance are physicochemical methods of their recovery from the second sources e.g. worn out converters of exhausted gases, chemical catalysts, dental alloys, anodic sludges from cooper and nickiel electrorefining as well as waste waters and running off waters from refineries containing trace amount of noble metals. It should be stated that these waste materials are usually pyro- and hydrometallurgically processed. Recovery of noble metals, from such raw materials requires individual approach to each material and application of selective methods for their removal. Moreover, separation of noble metals, particularly platinum metals and gold from geological samples, industrial products, synthetic mixtures along with other elements is a problem of significant importance nowadays. In the paper the research on the applicability of different types of ion exchangers for the separation of noble metals will be presented. The effect of the different parameters on their separation will be also discussed. The examples of the removal of noble metals chlorocomplexes will also be presented in detail.

Keywords: noble metals, ion exchangers, sorbents, separation

1. Introduction

General characteristics of ion exchangers

Ion exchangers are high molecular substances, most frequently solid, organic or inorganic, insoluble in water and many other solvents and capable of exchange of their own active ions into those coming from the surrounding electrolyte. From the chemical point of view, they are polyacids, polybases or both polyacids and polybases (polyampholyte, amphoteric ion exchangers). Those which exchange cations are called *cation exchangers* and those which exchange anions are called *anion exchangers*. Generally, those exchanging ions are called *ion exchangers*. Some ion exchangers prepared by modification of various types of substances, particularly natural ones, besides capability of exchanging ions exhibit distinct sorption properties [1–5]. The cation exchangers occurring most frequently possess functional groups such as -SO_3H, -COOH and -OH, whereas the anion exchangers possess the primary, secondary and tertiary amine ones and quaternary ammonium ones, quaternary phosphate ones and tertiary sulfone ones. Selective (chelating) ion exchangers and those strongly basic and weakly basic of the polymerization type of the functional trimethylammonium (type 1) and dimethylhydroxyethylammonium (type 2) groups are most widely applied in exchange chromatography. Their affinity mostly depends on the structure, size and change of anion exchanger. Type of functional groups in ion exchangers is decisive about the character of exchange reaction and its applicability.

Besides the general division of ion exchange materials due to the material (organic and inorganic) the skeleton is built, there are many others whose classification is based on the methods of preparation, type of functional groups and skeleton structure. There is still another basis of ion exchangers division conditioned by historical development of this area, i.e. according to their origin – natural, semi-synthetic and synthetic [1–5].

2. Application of ion exchangers of various types in recovery of platinum metal ions from secondary sources

Small amounts of noble elements in nature and constant impoverishment of their natural resources result in significant importance of physicochemical methods used for platinum metal ions recovery from secondary resources e.g. worn car exhaust gas convertor, chemical catalysts, stomatology alloys as well as waste waters and waters of refinery origin. Noble metals recovery from such raw materials requires individual approach and application of selective methods of their removal. Moreover, worse quality of these raw materials makes removal of pure noble metals more difficult. Determination of noble metals, particularly platinum ones in the above-mentioned materials, geological samples and synthetic mixtures together with other elements is of significant importance nowadays. Liquid–solid phase extraction (SPE) has some advantages compared to liquid–liquid extraction. Among others, it is faster, cheaper, uses small amounts of reagents and above all its automatization is easier. Moreover, its simple performance and high enrichment coefficients decide about its common laboratory application. SPE uses solid sorbents which should be characterized by not only high capacity towards metal ions under determination but also large selectivity and suitable sorption and desorption kinetics.

Synthetic ion exchange resins are widely applied in platinum metal ions enrichments. Among them of particular interest are cation exchangers, chelating ion exchangers and anion exchangers of different basicity of functional groups. Ion exchange selectivity depends on the kind and number of functional groups of ion exchanger as well as cross-linking and composition of external electrolyte solution composition from which concentration proceeds. Of these types of ion exchangers, the most effective are monofunctional ones, which ensure the same strength of bonding ions with the ion exchanger surface due to the presence of one type of groups which does not make equilibrium establishment difficult. Complexity of platinum carrying samples for enrichment makes it necessary to separate a component under determination due to large interference of components present in the analyte. In the chloride systems, ion exchange enables platinum metal ions separation not only from their mixtures but also from other metals. In the hydrogen chloride acid solutions, most platinum metal ions are present in the form of anion chloride complexes; therefore they are retained by strongly basic anion exchangers. Anion exchangers enable selective removal of platinum metal ions from solutions of other metals; however, there appear some problems with their elution due to strong sorption of platinum metal ions. It can often occur that recovery is not quantitative or rendered difficult. Such situation arises because of stable ion pairs formation between anionic complexes and the quaternary ammonium groups of the anion exchanger. Reduction of noble metal ions can proceed in the ion exchanger phase which also affects incomplete recovery. Some difficulties with determination can result from different behaviour of new and 'old' solutions which is connected with the hydrolysis process in the solution. This reflects mainly to rhodium or iridium chloride complexes because they have the greatest tendency towards hydrolysis. Using cation exchangers, rare earth elements, transient metals as well as alkali metals and alkaline earth family form weakly anionic or stable cationic complexes therefore they are retained on the cation exchange deposit whereas platinum metals go through the column not being retained by the cation exchangers making separation of the above-mentioned metal ions possible.

3. Application of cation exchangers in concentration and removal process of platinum metals

There are numerous reports in the literature about cation exchangers application in concentration and separation of trace amounts of platinum metal ions. Ion exchange is widely applied also for control of bound and free platinum contents in the serum added in the cis-platinum form. After ultrafiltration, ethylenediamine was added to form complexes which are sorbed on cation exchange disks. Platinum ions were desorbed from the disks by means of 5 M HCl and determined using the AAS method. The detection limit was 35 $\mu g/dm^3$ [6].

Besides the above-described procedures, there are separation methods using cation exchangers. As commonly known platinum metals tend to form anion complexes in the chloride systems, therefore partition coefficient values should not be high [7–9]. Much higher values of partition coefficients of platinum metal ions can be obtained by the addition of thiourea which results in cation complexes formation [9].

The polystyrenesulfone cation exchanger Dowex 50Wx8 in the hydrogen form was used for the determination of metals from the platinum group in ores and concentrates. Before ion exchange stage, noble metals in metallic copper were collected. The obtained alloy was digested in *aqua regia*. The pH of the solution was made 1 using hydrochloric acid. Ions of metals such as Cu(II), Ni(II) and Fe(III) were sorbed on the cation exchangers, whereas noble metals were not retained. The deposit was washed in HCl solution of pH = 1. The separated noble metals were determined gravimetrically and spectrophotometrically [10–14]. A similar technique was applied for the determination of noble metals and for the collection of Cu-Ni-Fe and ferronickel alloys [15–24]. Also the cation exchanger Dowex 50Wx8 was used for noble metal ions removal. Before the ion exchange stage, collection was made using fused tin. The obtained alloy was digested in the HCl-H_2O_2 solution. Tin(IV) was removed by evaporation from the HCl-HBr solution, and the obtained solution was evaporated dry several times in the presence of 12 M HCl [25]. The removal of noble metals from ores, the cation exchange followed by anion exchange method was applied. Cu(II), Ni(II) and Fe(III) ions are adsorbed at pH = 1.5 on the cation exchanger Dowex 50Wx8, whereas the noble metal ions pass on the column filled with the strongly basic anion exchanger in the chloride form Amberlite IRA-400. Before sorption on the anion exchange column, the solution is evaporated dry in the presence of sodium chloride, then it is dissolved in 12 M HCl and diluted to weakly acidic reaction. Under such conditions, rhodium(III) ions pass through the column, and the ions of other noble metals are retained on the anion exchanger bed [26]. A similar method was applied for the determination of iridium in the flotation concentrate [27]. The application of cation exchange for the determination of platinum metals on the meteorites is quite interesting. In this method, two-stage adsorption on the cation exchanger was used. Non-noble metal ions were adsorbed at pH 1.5 on the first column. Then they were desorbed by means of 3 M HCl. The obtained eluant was evaporated dry and the residue was dissolved in diluted HCl and the pH value was brought to 1.5. The other stage of non-noble metal ions adsorption was conducted in the same way as the first one. Application of the second stage allows avoiding errors connected with co-adsorption of noble metal ions [28, 29]. For determination of platinum and palladium in the copper and nickel stone, there were used two ion exchangers: the cation exchanger Dowex 50 in the hydrogen form and the anion exchanger Amberlite IRA-400 in the chloride form. Before the ion exchange process, the sample was melted with SnO_2 to collect noble metals. Tin(IV) was evaporated from the mixture of HCl and HBr acids. After bringing the solution to pH 1.5, Cu(II), Ni(II) and Fe(III) cation were adsorbed on the cation exchanger bed. The obtained eluant containing Pt(IV) and Pd(II) was evaporated from NaCl and dissolved in HCl, next it was passed through the anion exchanger bed. At first, there were eluted Pd(II) ions by means of 12 M HCl and then Pt(IV) ions using 2.4 M $HClO_4$. Both elements were determined using the spectrophotometric method [30]. The residue in the copper-nickel stone was determined using the cation exchanger Bio-Rad AG50W-x8 and the chromatographic column Porasil C impregnated by means of TBP (tri-n-butyl phosphate). The sample was melted with Na_2O_2, then digested in HCl and the acid concentration was brought to 0.1 M. Non-noble metal ions were separated on the cation exchanger and platinum metals were sorbed on the chromatographic column. Pt(IV) and Pd(II) ions were eluted by means of TBP in toluene but Rh(III) and Ir(IV) ions using water. Noble metals were determined gravimetrically and by means of AAS [31, 32]. Platinum alloys were determined in a two-stage separation process of noble metal ions. The sample was digested in *aqua regia*, next it was evaporated dry and the residue was diluted with HCl up to concentration about 0.1 M. The weakly acidic cation exchanger Amberlite IRC-50 of carboxylic groups in the sodium form was used in the first column. Palladium(II) ions were sorbed (probably precipitated in the hydrated oxide form [33]) on this ion exchanger, whereas Rh(III) and Pt(IV) passed to the eluant which was next put through the column with the strongly basic anion exchanger Dowex 2 in the chloride form. The adsorbed Rh(III) and Pt(IV) ions were eluted with 2 M HCl and 7 M HCl, respectively. The ion exchange technique was also applied for the determination of trace amounts of noble metals in common metals of high purity such as Fe, Ni, Cu, Mn and Al. The sample can be digested or melted with alkalis depending on its kind. The two-column cation–anion exchanger system was used for separation of noble metal ions. On the cation exchanger Dowex 50x8, there were adsorbed ions of metals such as Fe(III), Ni(II), Mn(II), Cu(II) and Al(III) from the 90% v/v ethanol + 10% v/v 1 M HCl solution. Then by evaporation, the medium changed from chloride to nitrate(V) one. In the other column on the anion exchanger Dowex 1x 8 in the nitrate form, there were sorbed noble metal ions from the aqueous solution of pH 6. Noble metals were analyzed in the resin phase by means of the radioisotope technique [34]. The similar method was used for the determination of noble metals in atmospheric dusts melting them with Na_2O_2 [35]. The cation exchanger Dowex 50x8 was used for the separation of copper(II) from the 0.03 M hydrochloric acid solution from the noble metal ions such as Pt(IV), Pd(II), Au(III) and Rh(III) [36, 37]. Ion exchange combined with extraction was applied for the determination of noble metals present in uranium alloys which can be uncoupled. Uranium can be separated from rhodium by extraction with 30% solution of TBP in CCl_4. Then after complete removal of chlorides, fluorides and nitrates by evaporation with chloric acid, rhodium(III) cations were sorbed by cation exchanger Dowex 50Wx8 from 0.3–0.9 % $HClO_4$ solution. Rhodium(III) was eluted form the cation exchanger by washing the column with 6 M hydrochloric acid solution and then determined spectrophotometrically using the method with $SnCl_2$ [38].

Separation of noble metals can be conducted on the cation exchangers from the thiourea systems. Separation of microquantities of various pairs of noble metal ions was made using the polystyrene sulfone cation exchanger Bio-Rad AG 50Wx4 in the hydrogen form. Pd(II) and Au(I), Pd(II) and Pt(II) ions as well as the Rh(III), Au(I), Pt(II) and Ag(I) mixture were separated using the acetate-thiourea solutions in the hydrochloric or hydrobromic acid medium [39]. Platinum(II) and palladium(II) ions were separated from aluminium ions using also the cation exchanger Bio Rad AG50Wx4 in the hydrogen form. Aluminium ions as well as Fe, Zn, Pb, U, Ni, Co and Sr ones do not form cationic complexes under experimental conditions, therefore only noble metals are sorbed from 0.1 M thiourea solution in 1.5 M hydrochloric acid solution. 2% Br_2 and 1.5 M HCl solution was used for elution of Pt(II) and Pd(II) ions. 0.87 M HBr–0.01 M thiourea solution in 90% acetone proved to be the best eluant towards Pt(II) ions. The presence of Cu(II) and Hg(II) ions is not recommended because of possible co-adsorption with noble metals [40]. It is relatively difficult to separate rhodium(III) ions from platinum(IV) ones in the chloride medium, therefore in some cases it is necessary to change the medium into the nitrate one. Before the separation of platinum(IV) and rhodium(III) ions on the polystyrene-sulfone cation exchanger Varion KS in the hydrogen form, the chloride complexes were in contact with sodium hydroxide at pH 13 for four hours. Then the obtained solution was acidified with 4 M HNO_3 to pH 2. Under such conditions, rhodium(III) ions occur in the cation form and platinum(IV) ions in the anion form. Platinum(IV) ions are not retained by the cation exchanger. Rhodium(III) ions can be eluted with 1 M hydrochloric acid from the cation exchanger [41].

4. Application of chelating ion exchangers for concentration and removal of platinum metal ions

Chelating ion exchangers also called complexing ion exchangers are formed by building organic reagents containing organic groups into the ion exchange resin skeleton. Owing to that they possess active chemical groups capable of selective/specific interactions with metal ions in the solution forming chelating complexes when a metal ion can bind with two or a larger number of donor atoms of their functional groups. These ion exchangers are characterized by high selectivity and their sorption capacities depend, among others, on the kind of functional groups, their reciprocal position and spatial configuration (steric effects) and also on physicochemical properties of the polymer matrix [42, 43].

On the huge number of chelating ion exchangers, on a large laboratory and industrial scale, there are produced ion exchangers of functional dithizone, thiourea, isothiourea, aminophosphonic, phosphonic, thiol, amidooxime, aminoacetate, dithiocarbamate, iminodiacetate, thiosemicarbamate groups as well as chelating ion exchangers containing triisobutylphosphine sulfides [44-64].

5. Chelating ion exchangers of functional dithizone groups

Chelating ion exchangers of functional dithizone (diphenyl carbamate) groups are widely applied in concentration, separation and recovery of noble metal ions [44–54].

Grote and Kettrup [45–48], by conversion of the ion exchanger of functional dehydrodithizone groups, prepared a chelating resin containing dithizone groups. It was used on both sorption and separation of 27 noble and non-noble ions from acids (HCl, HNO_3).They showed very high values of partition coefficients of noble metal ions of the order 10^4–10^6 (Pd(II)–7.7×10^5; Pt(IV)–3.4×10^5; Au(III)–2.1×10^5 (0.01 M HCl)) in the whole range of hydrochloric acid (0.01–6 M) as well as Ag(I) (1.1×10^3) and Hg(II) (5.8×10^4) ions in diluted nitric(V) acid solutions compared to the values of partition coefficients for non-noble metal ions. Also high values of ion exchange capacities of platinum metals and gold (Au(III) 0.74; Pd(II) 0.68; Pt(II) 0.39; Pt(IV) 0.31; Os(IV) 0.12; Ir(IV) 0.14; Ir(III) 0.02; Ru(III) 0.14; Rh(III) 0.16 mmol/g of the dithizone resin) indicate large selectivity of the resin with dithizone functional groups towards noble metal ions as well as possibility of their application in separation from other non-noble metals. Satisfactory results were obtained using the chelating ion exchanger with dithizone functional group in concentration and recovery of Au(III), Pt(IV) and Pd(II) ions originating from the extraction of sulfide ores, stones and enriched ores. Elution of the above-mentioned ions was conducted by means of 2 M chloric(VII) acid and 5% thiourea solution [44]. There was also made a thorough analysis of desorption of single noble metal ions and their mixtures from the resin with the dithizone functional groups using the following eluents: HCl, $HClO_4$, NH_4NO_3, NaSCN, $(NH_2)_2CS$. Palladium(II) and platinum(IV) ions retained on this resin can be qualitatively desorbed using the thiourea solution [46].

Similar investigations using polyvinylpyridine resin of functional dithizone groups in Pd(II) and Pt(IV) ions concentration in the presence of Au(III), Ni(II) i Hg(II) ions were carried out by Shah and Devi [49]. The values of maximal ion exchange capacities towards palladium and platinum ions were 100 and 250 mg/g of resin, respectively. Separation of the above-mentioned ions from nickel, gold and mercury ions (Pd(II)-Ni(II); Pt(IV)-Au(III); Pt(IV)-Ni(II); Pd(II)-Pt(IV)-Ni(II); Pt(IV)- Au(III)-Hg(II)) was conducted using various eluants 0.1 M HCl + 1% $(NH_2)_2CS$ (elution of Pd(II)), 0.1 M HCl + 5 % $(NH_2)_2CS$ (elution of Pt(IV)), 0.2 M CH_3COOH (elution of Ni(II)), 5 M HCl + 1 M HNO_3 (elution of Au(III)) and 0.5 M HNO_3 + 2 % NH_4NO_3 (elution of Hg(II)).

Modification of the commercially available polyacrylate matrix Diaion HP-2MG with dithizone resulted in the preparation of the selective sorbent towards Pd(II) and Pt(IV) ions. Chwastowska et al. [51] used the above-mentioned sorbent for removal of Pd(II) and Pt(IV) ions from the environmental samples, among others, from road dusts, soil and grass collected from fast traffic routes. After proper preparation, among others, drying (673 K, 1 h) and digestion in *aqua regia*, the geological samples were analyzed using the GF AAS technique. The detection limit (LOD) for the determined metal ions was 1 ng/g for Pt(IV) and 0.2 ng/g for Pd(II). Metal ions desorption was run in two ways: using thiourea solution (possibility of determination of both elements in the eluant by the GFAAS technique) and concentrated HNO_3 solution (possibility of directed determination of only Pd(II) ions in the eluant). The obtained ion exchange capacity towards both metal ions was about 0.16 mmol/g of resin.

Chelating resin formed by immobilization of sulfonated dithizone on the anion exchanger Amberlite IRA-400 was applied in concentration of heavy metal ions, i.e. Pd(II), Ni(II), Co(II), Cu(II) and Pb(II) in the water samples collected in Japan. The first four elements were determined by the GFAAS technique but Pb(II) by HGAAS. The affinity series of the studied ions towards the present resin is as follows: Pd(II) > Cu(II) > Co(II) > Pb(II) > Ni(II) [52].

6. Ion exchangers of functional thiourea and phosphonic groups and their derivatives

Of significant importance are the results of studies on the application of commercial ion exchangers Tulsion of the functional isothiourea and phosphonic groups for removal of Pd(II) ions from nitric(V) acid solutions (0.1–4 M). High values of partition coefficients (particularly in diluted HNO_3 (0.1 M) solutions) decreased with the increasing concentration of nitric(V) acid for the ion exchangers Tulsion CH-95, Tulsion CH-96 and Tulsion CH-97. The values of partition coefficients (particularly in diluted HNO_3 (0.1 M) solutions) which were >10 000 cm^3/g, 1650 cm^3/g, 5210 cm^3/g for Tulsion CH-95, Tulsion CH-96 and Tulsion CH-97, respectively, decreased with the increasing nitric acid(V) concentration. The selectivity series of the Tulsion ion exchangers towards Pd(II) ions is as follows: *Tulsion CH-95 > Tulsion CH-97 > Tulsion CH-96.*

The ion exchangers of the functional thiourea and isothiourea groups, commercially known as Srafion NMRR, Lewatit TP-214 and Purolite S-920, are widely used for the concentration of platinum metal ions [53-61]. Srafion NMRR (Ionac SR-3) exhibits great affinity for Au(III) and Ag(I) as well as Pd(II), Pt(IV), Rh(III), Ru(III), Ir(III) and Os(VI) ions [53-61]. It is applied in the separation of noble metal ions [57], similar to the ion exchangers Monivex which develop high sorption capacity values particularly in the largely acidic solutions (2–6 M HCl) [67]. Lewatit TP-214 was used for the removal Pd(II) ions from two component Pd(II)–Zn(II); Pd(II)–Cu(II)) and multicomponent (Pd(II)–Cu(II)–Zn(II)) chloride solutions [61] and for the removal of Pd(II) ions from chloride solutions (0.1–8.0 – 0,001 M Pd(II)) and chloride–nitrate ones (0.1÷0.9 M HCl – 0.9÷0.1 M HNO_3– 0,0011 M Pd(II))) [55]. The total ion exchange capacity of Lewatit TP-214 towards Pd(II) ions was from 0.97 mmol/g–3 M HCl to 1.16 mmol/g–0.1 M HCl [55].

Zuo and Muhammed prepared a large number of ion exchangers of the functional thiourea groups by modification of macroporous polystyrene matrices of Bonopore, Amberlite XAD-2, Amberlite XAD-4 as well as the weakly basic anion exchanger Amberlite IR-45 [62]. Newly prepared ion exchangers were used in sorption of Au(III), Ag(I), Cu(II) and Fe(III) ions from one-component chlorides solutions (2 M HCl) and their mixtures (Au(III), Ag(I), Fe(III), Cu(II)), as well as in sorption of platinum metal ions e.g. Pd(II), Pt(IV), Rh(III), etc. The studied ion exchangers are characterized by high selectivity towards noble metal ion, particularly gold and silver in the presence of Cu(II) and Fe(III).

Separation of Pd(II), Cu(II) and Zn(II) ions using the melamine-formaldehyde-thiourea (MFT) resin was presented in [63]. There was proved high selectivity of this resin towards Pd(II) ions, and the obtained values of sorption capacity were 15.29 mg Pd(II)/g (static method) and 1580 μg Pd(II)/g resin (dynamic method). Desorption of ions with studied resin was conducted using 0.5 M thiourea, 0.5 M HCl solutions and the acidified thiourea (0.5 M $(NH_2)_2CS$–0.5 M HCl).

7. Ion exchangers of functional thiol groups

The chelating ion exchangers of functional thiol and methylene thiol groups with the commercial names: Chelite S, Duolite GT-73, Imac GT-73, Duolite GT-74, Purolite S-924, Spheron Thiol 1000 and Tulsion CH-97 are widely applied in sorption and separation of noble metal ions [55, 56, 64-69]. Introducing 8-hydroxchinoline (HOxn) and sodium salicylate groups to the methacrylate polymer matrix generates two new chelating ion exchangers Spheron Oxine 1000 and Spheron Salicyl 1000, respectively, which similar to Spheron Thiol 1000 are applied in the separation of Pd(II) and Cu(II) ions. From the determined ion exchange capacities, the affinity series of Spheron ion exchangers towards Pd(II) ions is as follows: *Spheron Thiol 1000 (0.949 mmol/g) > Spheron Oxine 1000 (0.536 mmol/g) > Spheron Salicyl (0.094 mmol/g)* [68]. Elution of Pd(II) and Cu(II) ions form the ion exchanger Spheron Oxine 1000 was conducted using 0.5 M thiourea solution (pH = 2) and 2 M HCl [70].

8. Ion exchangers of functional imidazole groups

Many authors prove high reactivity of imidazole groups and their derivatives towards metal ions, particularly noble metals ions. For example, VBC-DVB (VBC – vinylbenzene chloride, DVB – divinylbenzene) – recovery of Pd(II) and Pt(IV) ions from chloride solutions (1–2 M HCl) and their separation from Cu(II), Zn(II) and Ni(II); sorption capacity, 1.6–1.7; 1.4–1.5 and 1 mmol of metal/g for the 0.1 M HCl, 1 M HCl and 2 M HCl solutions, respectively; desorption of Pd(II) and Pt(IV) ions, 0.1–0.3 M thiourea solution (desorption effectiveness >98% for Pd(II); 85% for Pt(IV)) [71, 72]; epoxide resin of the functional imidazole groups concentration of trace (20 ng/dm^3) amounts of Au(III), Pd(IV) and Ru(III) ions. Noble metal ions can be enriched in a quantitative way (recovery 94.5–100%, pH = 4), desorption: acidified thiourea solutions (16 cm^3 6 M HCl, 0.2 g thiourea), recovery 96–99.5% [73], polystyrene-divinylbenzene resin of functional benzimidazole group – separation of heavy metal ions that is Pd(II), Ag(I) and Hg(II) from medical alloys, geological materials, different kinds of waste waters and sludges; sorption capacity: 1 mmol/g for Ag(I) (pH = 4–6), 0.62 mmol/g for Pd(II) (pH = 5–6) and 0.83 mmol/g for Hg(II) (pH = 4–6); desorption: 5% thiourea solution in 0.1 M $HClO_4$ (recovery: Ag(I) and Hg (II) – 100%, Pd(II) – 60%) and 12 M hydrochloric acid solution (recovery: Ag(I) – 45 %, Pd(II) – 100 %, Hg(II) – 65 %) [74]; polystyrene-divinylbenzene resin of Im-NO_3 (Im-imidazole) groups – removal of Pd(II) (0.00062 M) form nitric(V) acid solutions (1–5 M HNO_3); sorption capacity depends, among others, on the molar ration of initial reagents of chloromethylated resin (I) and 1-methylimidazole (II) at the reagents ratio I:II = 1:1.5 this capacity is the highest (4.06 mmol/g), the total capacity determined form the Langmuir 88 mg/g (3 M HNO_3) [75]; resin of functional aminoethylene imidazole (IEA) – concentration of trace amounts of Au(III), Pt(IV), Ir(IV), Pd(II) ions and their separation from Cu(II), Fe(III), Zn(II), Ni(II), Mn(II), Cr(III), Ca(II) and Mg(II) ions; sorption capacity in 2 M HCl solution: 4.0 mmol/g – Au(III), 1.57 mmol/g – Pt(IV), 2.26 mmol/g – Pd(II) and 1.85 mmol/g – Ir(IV) [76].

9. Ion exchangers of functional thiosemicarbazide, piperazine and tetrazine groups

Table 1 presents the characteristic of chosen chelating ion exchangers of thiosemicarbazide, piperazine and tetrazine functional groups widely applied in the concentration and removal of platinum group ions [77–80].

Ion exchanger	Application	Additional Information	Ref.
Resin of thiosemicarbazide functional groups	Separation of Pd(II), Rh(III), Pt(IV), Ru(III), Ir(III), Au(III), Os(VI), as well as Cu(II), Bi(III), Hg(II), Fe(III) and Al(III)	Sorption capacities (mmol/g): Pd(II) – 0.78 (pH = 0); Pt(IV) – 0.71 (pH = 0); Ru(III) – 0.685 (1.5 M HCl); Rh(III) – 0.615 (2 M HCl); Ag(I) – 0 (pH = 0) and increase with increasing solution pH, Ir(III) – 0 (pH = 0); Au(III) and Os(VI) are reduced during the contact with the resin; Cu(II), Bi(III), Hg(II), Fe(III) and Al(III) are not retained by the resin; Desorption: 4 M and 9 M HCl as well as 5 % N,N-diphenylthiourea in ethanol	[77]
Resin of amine and 1,2,4,5-tetraazine functional groups	Preconcentration and separation of noble metal ions in the amount of 0.001 M in the presence of Zn(II), Co(II), Cu(II), Fe(III), Ni(II) and Cd(II) from acidic solutions of the proportion of noble metal to base metal: 1:1, 1:10, 1:100	Selectivity series: Pd(II) > Au(III) >> Ir(IV) > Os(IV) > Pt(IV) > Ru(III) > Rh(III); Desorption: acidified 5% $(NH_2)_2CS$	[79]
Resin of 1-(2-aminethylene)-piperazine functional groups	Pre-concentration and recovery of Au(III), Pd(II), Ru(III), Os(VI), Pt(IV) and Ir(IV) in the presence of Zn(II), Fe(III), Cu(II) and Ni(II) from chloride solutions	Sorption capacities for Au(III), Pd(II), Ru(III), Os(VI), Pt(IV) and Ir(IV) equalled to 5.38, 3.67, 3.46, 3.10, 2.46 and 2.24 mmol/g, respectively	[80]

Table 1. Examples of the chelating ion exchangers of thiosemicarbazide, piperazine and tetrazine functional groups and their characteristic.

10. Ion exchangers of functional (amino)pyridine groups

Resins of pyridine [81] and α-aminopyridine groups on the polyphenylethylene support [82] were successfully used for the separation of technetium and platinum metals from diluted chloride solutions and in concentration of Pt(IV), Pd(II) and Ir(III) (*on-line* FAAS). Technetium ions were eluted using concentrated solutions of hydrochloric acid but platinum metals (Ru(III), Rh(III) and Pd(II)) by means of 0.1 M NH_2CSNH_2 in 1 M HCl [81] or 0.5 M HCl – 0.5 M $HClO_4$ – 0.5 M $Mg(ClO_4)_2$ [82]. Most interferences coming from noble-metal ions were eliminated using 2 M HCl and those from iron (> 200 mg Fe/dm^3) adding 0.2 M EDTA [82].

11. Ion exchangers of functional amine and guanidine groups

Resins of functional amine and guanidine functional groups were successfully used in sorption of Pd(II), Pt(IV) and Au(III) ions from chlorides solutions (initial concentration of each metal was 24 mM). The highest selectivity towards the above-mentioned ions and the possibility of practical application of their recovery are exhibited by the ion exchangers A2 (Au – 99.7; Pt – 108; Pd – 54.3 mg/g), B2 (Au – 94.1; Pt – 104.8; Pd – 48.7 mg/g) and AG2 (Au – 85,1; Pt – 64; Pd – 49.2 mg/g) and BG2 (Au – 74.3; Pt – 76.2; Pd – 48.5 mg/g). The sorption mechanism of Pd(II), Pt(IV) and Au(III) chloro complexes is the combination of coordination interactions of electron nitrogen pair and electrostatic interactions of these ions with the protonated group $-NH_2$ [83].

In the case of the ion exchangers of the same functional groups: amine (D1, D2, D3, D4) and guanidine (D1G, D2G, D3G, D4G) but on the basis of copolymer VBC/AN/DVB (copolymer D), the studies of Pd(II), Pt(IV), Au(III), Cu(II), Ni(II) and Fe(III) ion removal from single and multi-component chloride solutions (0.1–3.0 M HCl) showed that the highest selectivity towards metal ions is exhibited by the resin D4 (sorption capacity 190 mg Au(III)/g, 245 mg Pt(IV)/g and 280 mg Pd(II)/g of the resin), and the selectivity series towards Pd(II) ions is as follows: D4 (117 mg/g) >> D1 > D4G > D1G > D2 = D3G > D2G > D3 (25 mg/g). Recovery of noble metals from multi-component solutions with ten greater excess of Cu(II), Ni(II) and Fe(III) ions was > 95 % [84].

12. Ion exchangers of functional hydrazone and hydrazine groups

Modification of the macroporous copolymer ACR/ST/DVB (where ACR – acrolein, ST – styrene, DVB – divinylobenzene) using isonicotinic acid hydrazide and hydrazine hydrate results in preparation of new chelating ion exchangers: P-NHZ and P-THZN, respectively [85–87]. These resins exhibited high selectivity towards noble metal ions which allows their separation from common metal ions. The maximal values of sorption capacities obtained in 0.1 M HCl are 3.77 mmol Au(III)/g, 1 mmol Pt(IV)/g, 0.79 mmol Pd(II)/g, 0.72 mmol Ir(III)/g, 0.51 mmol Ru(III)/g, 0.39 mmol Rh(III)/g for the resin P-NHZ and 4.2 mmol Au(III)/g, 1.36 mmol Pd(II)/g, 0.5 mmol Pt(IV)/g, 0.4 mmol Ir(III)/g, 0.31 mmol Rh(III)/g, 0.29 mmol Ru(III)/g for the resin P-THZN, whereas the sorption capacities for common metal ions (Cu(II), Fe(III), Co(II), Ni(II), Cr(III), Zn(II), Mn(II) and Al(III)) are small, in the range 0–0.13 mmol/g for the resin P-NHZ and 0.03–0.32 mmol/g for the resin P-THZD. Desorption of the above-mentioned ions from the P-NHZ and P-THZN was achieved using a 5% thiourea solution in 0.1 M HCl. Quantitative desorption was possible only in the case of Au(III) and Pd(II) from the resin P-NHZ. Another advantage of the ion exchanger P-NHZ over P-THZN is fast kinetics of sorption and easy elution of adsorbed ions [87].

Ge et al. [86] proved high selectivity of the P-NHZ resin towards noble metal ions and possibility of its exploitation for separation and concentration of trace amounts of Pd(II) and Pt(IV) contained in road dust samples. Sorption of the above-mentioned ions on the P-NHZ resin can be conducted also from the HNO_3, HF and H_3BO_3 solutions of the concentration 0.08-1.2 M.

13. Ion exchangers of functional formazane groups

Chelating ion exchangers of functional formazane groups [88–91] based on the styrene divinylbenzene skeleton [58], amino- and chloromethylated polystyrene [91], cellulose [58, 91], saccharose [65] and methacrylate [58] were applied for sorption and separation of platinum metals, gold, mercury, silver and copper. The sorption capacity towards Pd(II), Ag(I) and Hg(II) ions was in the range 0.05–0.06 mmol/g, 0.4–0.9 mmol/g and 0.6–0.8 mmol/g, respectively, for formazane ion exchanger based on cellulose, aminomethylated polystyrene; however, their sorption capacity towards Cu(II) ions was close to zero.

14. Ion exchangers of functional cyclane groups

For separation and recovery of Pd(II), Pt(IV) and Au(III), Jermakowicz-Bartkowiak [92] used the resin A7C prepared by the reaction of cyclane (1,4,8,11-tetraazacyclotetradecane) with the copolymer of vinylbenzyl and divinylbenzene chloride (VBC/DVB) obtaining the following sorption capacities: 47.4 (S0), 47.4 (S1) mg Au(III)/g, 48.1 (S0), 47.5 (S1) mg Pt(IV)/g, 26.0 (S0), 25.6 (S1) mg Pd(II)/g resin, where S1 – the multifunctional system, S0 – the single component system (static method) and 0.92 mmol Au(III)/g, 0.66 mmol Pt(IV)/g, 0.9 Pd(II) mmol/g dry resin (dynamic method). Desorption of Au(III), Pt(IV) and Pd(II) ions from the resin was made using a 5 % thiourea solution in 0.1 M hydrochloric acid with the yield 98, 97 and 98%.

15. Ion exchanger Metalfix Chelamine

The studies of the use of the ion exchanger Metalfix Chelamine of tetraethylenepentamine ligands (1,4,7,10,13 – pentaazatridecane) in the process of recovery, concentration and separation of platinum metal ions, gold and their accompanying elements from the environmental samples, anodic sludges from the electrorefining of copper ores, spent car exhaust gas catalysts, etc. [93–98].

16. Polyorgs ion exchangers

Polyorgs type chelating ion exchangers [58, 99–103] are used for separation, concentration and removal of palladium(II) ions and other noble metals from deposits, rocks, ores, minerals and industrial waste waters. The Polyorgs sorbents were prepared by introducing e.g. imidazole, pirazole, mercaptobenzothiazole, amidooxime groups to the macroporous copolymers (polystyrene, polyvinyl, polyacrylonitrile) and other matrices. Sorbents of this type are characterized by high chemicals stability in strong acid and alkaline solutions as well as high thermal resistance and can be applied in the whole pH range. The sorbents of Polyorgs type (11-n, 15-n, 17-n and 33-n) were also used for filling fibres e.g. polyacrylonitrile (PAN), cellulose, polyvinyl (PVA) ones. In the literature, such sorbents are called filled fibrous sorbents (FFS) [99, 102, 103]. Main advantages of FFS are good kinetics of sorption and ease in their separation from solution which make them more attractive than sorbents in the form of powder or grains.

17. Ion exchangers of functional iminodiacetate groups

Recovery of palladium ions from chloride and chloride-nitrate(V) solutions using the ion exchanger Amberlite IRC-718 of functional iminodiacetate groups and polystyrene skeleton was conducted by Hubicki et al. [104]. The largest total ion exchange capacities of Amberlite IRC-718 were obtained in the 0.1M HCl (1.099 mmol/g) and 0.9M HCl–0.1M HNO_3 (0.693 mmol/g) solution. The additional of 1M $AlCl_3$, 1M $CuCl_2$ or 1M $NiCl_2$ results in significant drop of ion exchange capacity.

Studies of Amberlite IRC-718 selectivity towards Pd(II), Pt(IV) and Au(III), Ti(IV), Ag(I), Al(III), Co(II) and Fe(III) were carried out also by Park and co-workers [105]. The optimal sorption conditions for Pd(II), Pt(IV) and Au(III) ions are pH = 0.1–4 and a flow rate ≤ 2 cm^3/min. Quantitative desorption of noble metal ions was conducted using 20 cm^3 of 0.25 M thiourea and a flow rate 1 cm^3/min. The ion exchange capacity of the studied ion exchanger was 0.34 mmol Pt(IV)/g; 0.69 mmol Au(III)/g and 0.55 mmol Pd(II)/g of the resin. From a practical point of view, the chelating ion exchanger Amberlite IRC-718 can find application in removal, concentration and separation of noble metal ions from the solutions originating from hydrometallurgical processing of car exhaust gas catalysts, anodic sludges, wastewaters as well as in removal and determination of gold(III) ions in printed computer plates PCP [105].

18. Copolymers modified with triisobutylphosphine sulfide

Of special importance are the studies by Sanchez et al. [106–109] of building triisobutylphosphine sulfide (TIBPS, Cyanex 471X) into the styrene-divinylbezene (4% DVB) copolymer in order to obtain a resin concentrating noble metal ions. Polymers are characterized by high selectivity towards Au(III) and Pd(II) ions (sorption from chloride solutions) as well as a lack of selectivity towards Ir(IV), Rh(III), Pt(IV) and Fe(III), Cu(II) or Ni(II) [106]. Based on the obtained values of sorption capacities (mmol/g) towards Pd(II) ions, the affinity series of the studied resins is as follows: Pd(II): Polymer 2 (0.07 mmol/g) < Polymer 1 (0.11 mmol/g) < Polymer 5 (0.17 mmol/g) < Polymer 3 (0.51 mmol/g) < Polymer 4 (0.72 mmol/g) [106].

With the temperature increase in the range 298–333K, there is observed the increase in Au(III) ions sorption compared to that of Pd(II) ions. The sodium nitrate(III) and thiourea solutions were used as eluents. Sodium nitrate(III) (2 M, pH 4.7) desorbs 75% of Pd(II) practically not washing out Au(III) from the ion exchanger, whereas thiourea (0.5 M, [H^+] 0.1 M) enables almost quantitative recovery of gold(III) containing trace amounts of palladium (II) ions.

19. Ion exchangers of cyclic ligands

Kałędowski and Trochimczuk [110, 111] synthesized polymers containing calixpyroles preparing among others the resin B4 (covalently bonded thiophene kalix[4]pyrol[2] with the vinylbenzene chloride and divinylbenzene (VBC-DV, 0.5% DVB) of the expanded gel structure) selective towards noble metal ions. The maximal sorption capacity values determined from the Langmuir adsorption isotherms are 0.26 mmol/g; 0.55 mmol/g; 0.6 mmol/g; 0.61 mmol/g and about 1.7 mmol/g for Ag(I), Pt(II), Pt(IV), Pd(II) and Au(III), respectively. The resin B4 can be applied for the selective removal of Au(III) from the solutions originating from processing of ores, galvanic sludges containing other noble metals. The metals adsorbed on the resin are quantitatively washed out using 2 M solution of HCl or 2 M HNO_3-Pt(IV), 0.5 M thiourea + 0.1 M HCl-Ag(I) and 5% solution of KCN-Au(III). In the case of palladium(II), the recovery was not quantitative (>57%, 0.5M thiourea in 0.1M HCl).

Garcia et al. [112] used the polystyrene-based resin of 15-membered triolefinic azamacrocycle rings in Pd(II) and Pt(IV) ions sorption from aqueous and aqueous dioxane solutions. The addition of dioxane (10% v/v) results in increasing polymer swelling and effectiveness of palladium(II) ions sorption from 48% to 73%. The total sorption capacities for Pd(II) and Pt(IV) ions were 0.36 mmol/g and 0.28 mmol/g of the resin, respectively. The presence of Cu(II) and Ni(II) ions affects insignificantly the selectivity of the resin towards noble metals. 0.5 M thiourea in 0.1 M HCl was used as an eluent of Pd(II) and Pt(IV) ions with 100% and 79% recovery, respectively.

20. PAN-ATAL, PS-BMT ion exchangers

Chen and Zhao [113] prepared the chelating resin PAN-ATAL by immobilization of cross-linked polyacrylnitrile (PAN) (7% DVB) with 2-amino-2-thiazole (ATAL) selective towards noble metals, i.e Pd(II), Ru(II), Ir(IV) and Rh(III) ions, for which the sorption capacity values were 230.7 mg Pd(II)/g, 147.1 mg Ir(IV)/g, 137.6 mg Ru(IV)/g and 72.1 mg Rh(III)/g of the resin. The affinity of chelating macroporous resin PS-BMT (where PS – the cross-linked chloromethylated polystyrene (10% DVB), BMT – 2,5-dimercapto-1,3,4-thiadiazole) for Pd(II), Au(III) and Pt(IV) ions from chloride solutions was studied by Qu et al. [114]. The sorption capacity for Au(III) ions 5.8 mmol/g is much higher than for Pd(II) ions 0.19 mmol/g and Pt(IV) ions 0.033 mmol/g which is associated with gold(III) chlorocomplexes coordination by two donor N and S atoms of the resin and reduction of gold(III) to a metallic form. However, in the case of $PdCl_4^{2-}$ and $PtCl_6^{2-}$ complexes, only one donor atom – sulfur atom takes part in bonding.

21. Chelating fibres

It is worth presenting also the studies of using chelating fibres for the removal of platinum elements, for example, those Gong [115] and Li et al. [116] on application of fibres of functional imidazole groups in noble metal ions sorption. Bilba et al. [117] used chelating polyacrylamidoxime fibres in concentration and recovery of Pd(II) ions from chloride solutions. The attention should be also paid to the investigation by Gong and Wang [118, 119] as well as Chang et al. [120] on concentration of trace amounts of Au(III), Pd(II), Pt(IV) and Ir(IV) by means of chelating polyacrylacylaminothiourea fibres. Poly (acrylamidrazonehydrazide) [121] and poly (acryl-p toluenesulfonamideamidine-p-toluenesulfonylamide) [122] fibres were applied in quantitative concentration and separation of Au(III) and Pd(II) as well as Ru(III), Rh(III), Au(III) and Pd(IV) ions in the column system.

22. Amphoteric ion exchangers and anion exchangers

Of a large group of ion exchangers, anion exchangers of different basicity (strong, average and weak basic) of functional groups are applied in ion exchange chromatography of noble metal ions. Strongly basic anion exchangers possessing well-dissociated functional groups capable of anion exchange of even weak acids, e.g. quaternary ammonium groups, are widely applied in the whole pH range. This group includes types 1 and 2 strongly basic anion exchangers of functional trimethylammonium groups (type 1) and dimethylhydroxyethylammonium groups (type 2). Weakly basic anion exchangers possess poorly dissociated functional groups i.e. primary-, secondary- and tertiary amine groups. There is also a group of amphoteric ion exchangers which, depending on solution pH, are able to exchange anions or cations. They are polyacids and polybases so-called polyampholites, e.g. of COO^- and $-N^+(CH_3)_3$ groups (snake in cage polymers).

Application of amphoteric ion exchangers for removal of trace amounts of Pd(II), Pt(IV) and Au(III) ions, among others, geological materials was studied by Chajduk-Maleszewska and Dybczyński [123], Dybczyński et al. [124] and Samczyński et al. [125] and Hubicki et al. [104]. Duolite ES 346 containing the functional amidoxime groups was successfully applied for recovery and separation of noble metal ions. There was proved high selectivity of amphoteric ion exchange Duolite ES 346 in Pd(II) ions sorption from the chloride (0.1–6M HCl–0.0011M Pd(II) and chloride-nitrate(V) (0.1–0.9 M HCl–0.9–0.1M HNO_3–0.0011M Pd(II)) systems.The total sorption capacity towards Pd(II) ions is 1.099 mmol/g (0.1 M HCl) and 1.545 mmol/g (0.1M HCl–0.9M HNO_3). The processes of sorption and separation of trace amounts of Pd(II), Au(III) and Pt(IV) from ammonium and aqueous-non-aqueous solutions on Duolite ES 346 were also conducted [123, 124]. Also high selectivity of this ion exchanger towards noble metal ions was proved. Ions desorption was achieved using the solutions: 2 M HCl (elution of Pt and other metal ions, 8 M NH_4OH–0.01 M NH_4Cl-CH_3OH (1:5), desorption temperature 323 K (elution of Au(III) and 0.3 M $CS(NH_2)_2$ in 2 M HNO_3 (elution of Pd(II)).

The amphoteric vinylpiridine ion exchangers VP-14K, ANKF-5 and the anion exchanger AN-251M were used for recovery of Pd(II) ions from spent car exhaust gas convertors subjected to extraction with the NaCl (2–2.3 M) solution acidified with hydrochloric acid (0.5–2 M) at 353 K, the extraction time was 4h. These ion exchangers were characterized by high affinity towards palladium(II) ions and their recovery was 98–99%. The sorption capacity of the anion exchanger AN-251M towards Pd(II) ions and the aminophosphonic ion exchanger ANKF-5 was comparable (2.4-2.5 mmol/g) and much larger than that of the ion exchanger VP-14K (1.4 mmol/g) so the ion exchangers AN-251M and ANKF-5 can be recommended for this type of application.

The strongly basic gel anion exchanger Dowex 1x10 (Cl^- form, grain size 100-200 mesh) was successfully applied for removal of Pd(II) and Pt(IV) ions from the dust collected in Germany from street and fast traffic roads (Saarbrücken, motorway A-1, A-61, road B-262). Quantitative desorption of sorbed metal ions took place using the 0.1 M thiourea solution in 0.1 M HCl at the increased temperature 333 K enabling reduction of the eluent volume by half. The matrix ions, i.e. Cd, Cu and Fe, were not retained on the anion exchanger but Ni, Pb and Zn sorbed at 8–15 %. Elimination of interferences during noble metals determination was achieved by using the reagents masking the matrix ions even before the sorption process, e.g. xylene orange ($C_{31}H_{32}N_2Na_4O_{13}S$) [126]. Application of ion exchange technique for the determination of platinum(II) ions in biological tissues gives interesting results. The tissues with the cis-dichloro-diamineplatinum(II) were irradiated with neutron in the reactor. The sample irradiation was mineralized by means of HNO_3-H_2SO_4-H_2O_2 mixture. Then platinum ions were sorbed on the anion exchanger Dowex 1x8 in the chloride form with 6 M hydrochloric acid solution. Platinum was determined using the radiometric method [127]. Platinum and rhodium contained in ores were determined after separation on the anion exchanger Dowex 1x8 in the chloride form. The ore was digested in *aqua regia*. Next the solution was passed through two columns. In the first one, platinum ions were sorbed from 9 M HCl solution. In the other one, rhodium was sorbed in the form of a complex with zinc(II) chloride from 0.5 M HCl solution. ^{104}Rh was determined directly in the ion exchange phase using the radiometric method [128].

For removal and determination of platinum from geological materials, a technique using the anion exchanger Rexyn 201 was proposed. Sorption was performed from 0.5 M of hydrochloric acid solution containing Ir(IV), Pt(IV), Pd(II) and Au(III) ions. Elution was carried out by means of 0.1 M solution of thiourea in 0.1 M HCl. Ir(III) was eluted using 6 M HCl. Platinum metals and gold were determined radiometrically [129]. The same methods were applied for the determination of platinum in carbons [130]. Somewhat modified technique was used for the determination of platinum metals in meteorites. Modification consisted in the change of the anion exchanger Rexyn 201 on Deacidite FF in the chloride form [131].

Bio-Rad AG1x8 (100–200 mesh) was characterized by high selectivity for Pd(II), Pt(IV) and Au(III) ions (the partition coefficient values were 10^6, 10^4, 10^3 for Au(III), Pt(IV) and Pd(II) ions, respectively), and therefore it could be applied for removal of noble metal ions from the environmental and geological samples, among others, from rocks, ores as well as dust and road dust [132].

Similar studies of application of anion exchangers Amberlite IRA-900 (macroporous, polystyrene, strongly basic anion exchanger of type 1, 16–50 mesh) [133–135] and Amberlite IRA-410 (gel, polystyrene, strongly basic of type 2, 16–50 mesh) [135] for recovery and removal of trace amounts of Pd(II), Pt(II), Ru(III), Rh(III), Au(III) and Ir(IV) ions from chloride and radioactive nitrate waste waters were carried out by the Els et al. [133, 134] and El-Said et al.

[135]. Selectivity of the anion exchanger Amberlite IRA–900 for Pd(II) ions depends on the concentration of Cl^- ions in the solution. Quantitative sorption of palladium(II) ions from the chloride solutions is obtained at [Cl^- < 0.25 M. Sorption capacity of the anion exchanger Amberlite IRA-900 for Pd(II) ions in 0.2 M HCl solution was 0.0017 mmol/dm^3 ([Pd^{2+}] 350 ppm) [133]. Selectivity of the anion exchanger Amberlite IRA-900 towards noble metal ions changes in the series: *Au(III) > Pt(II) > Pd(II) > Ru(III) > Ir(IV) > Rh(III)* [134].

Satisfactory results were obtained using the above-mentioned anion exchanger for separation of Pd(II) and Ni(II), Sr(II), Rh(III), Eu(III), Ce(III), Ru(III), U(VI), Fe(III), Cr(III), Al(III), Ca(II) and Cs(I) from the radioactive nitrate waste waters [135].

The anion exchangers of quaternary ammonium groups: Purolite A-850 and Amberlite IRA-958 of polyacrylate skeleton, Lewatit MP 500A of polystyrene-divinylbenzene skeleton as well as Varion AP of functional pyridine groups and polystyrene-divinylbenzene skeleton exhibit high selectivity for Pd(II) ions from chloride and chloride-nitrate(V) solutions. Sorption capacities towards Pd(II) ions are 0.0282 g/cm^3 (in 0.1 M HCl) and 0.0005 g/cm^3 (in 6 M HCl) for Amberlite IRA-958 as well as 0.0408 g/cm^3 (in 0.1 M HCl) and 0.005 g/cm^3 (in 6 M HCl) for Purolite A-850. The addition of Zn(II) and Al(III) to the solution largely decreases selectivity of most anion exchangers for Pd(II) ions. The exception is Varion AP, whose selectivity changes insignificantly despite the presence of Al(III) ions [136, 137].

Due to modification of the macroporous polystyrene-divinylbenzene resin Amberlite XAD-1, the ion exchanger containing functional tertiary amine groups was obtained. This ion exchanger was used for separation of noble metal ions. Separation of Rh(III), Pd(II) and Pt(IV) ions mixture was achieved using suitable eluents:

Rh(III) – 1M HCl or 1M NaCl in 0.1 M HCl

Pd(II) – 0.05 M $NaClO_4$ in 1 M HCl or 1 M NaCl + 0.025 M $NaClO_4$

Pt(IV) – 0.1 M $NaClO_4$ in 1 M HCl or 1 M NaCl + 0.15 M $NaClO_4$.

The method is quick (about 30–40 minutes) and allows separation of 1.18–11.8 μg amounts of noble metal ions [138].

Among weakly basic anion exchangers of special interest is the macroporous polystyrene-divinylbenzene anion exchanger of functional dimethylamine groups Amberlite-93 used for recovery of Pd(II), Pt(II) and Rh(III) ions from spent car exhaust gas convertors. Rhodium(III) was desorbed from the anion exchanger as the first using 6 M hydrochloric acid solution, then palladium(II) was desorbed using 1% ammonia solution at room temperature. Platinum(II) was washed out with the ammonia solution of the concentration 5% (at increased temperature). Separation of palladium from platinum from the eluant solution can be achieved reducing to the metallic form or precipitating $(NH_4)_2PdCl_4$ and $(NH_4)_2PtCl_6$ using hydrochloric acid. The presented method of selective removal of platinum metals using Amberlite IRA-93 can be regarded as an effective technique for separation of these ions on a laboratory and commercial scale [139].

The weakly basic Amberlite IRA 67 is applied for selective removal of microquantities of platinum(IV) ions from the acid solution containing $CuCl_2$, $FeCl_3$, $NiCl_2$, $AlCl_3$ and $ZnCl_2$. In chloride solutions, the above components can partly form anions, which reduces the sorption capacity of weakly basic anion exchangers. The effect of the above-mentioned macrocomponents on decrease of sorption capacity towards platinum(IV) ions can be presented in the series: $CuCl_2 \approx FeCl_3 \approx NiCl_2 < AlCl_3 < ZnCl_2$ [140, 141]. A similar series can be determined for the anion exchanger Duolite S 37, which contains secondary and tertiary functional groups added to the phenol-formaldehyde skeleton [142].

23. Low-cost sorbents

In the literature, there are many examples of the alternative sorbents for noble metals removal produced from renewable and low-cost resources [143]. Among them, those based on bacteria, fungi and algae as well as agriculture and seafood wastes (coffee, green tea, tea, yuzu, aloe, wheat and barley straw, maize crop, coconut shell, rise husk, etc.) have been investigated. One of the low-cost sorbents is chitosan (CS) [144-155]. It is a kind of abundant natural polysaccharide. Chitosan is produced by the alkaline deacetylation of chitin, the most abundant biopolymer in nature after cellulose. It is extracted from shrimp and crab shells. Due to large availability of functional groups such as amino and hydroxyl ones, CS has been proved to be very efficient for the recovery of several toxic metal ions such as Cu(II), Cd(II) and Pb(II) and strategic metal ions such as Pt(IV) and Pd(II) [144]. It should be mentioned that sorption properties of CS are due to its composition and presence of active and functional groups. CS is characterized by its high percentage of nitrogen present in the form of amine groups, which are responsible for metal ion binding through chelation mechanisms. Due to the fact that it is protonated in acidic solutions, it is also capable of sorbing metal ions through anion exchange mechanisms. It should be mentioned that chitosan protonation in acidic solution causes the polymer to dissolve (except in sulphuric(VI) acid solutions). For the sorption of some metal ions (for example noble metals), sulphuric(VI) acid cannot be used for pH control due to reduction of sorption efficiency [145].

In the case of chitosan derivatives obtained by glutaraldehyde cross-linking (GA), poly(ethyleneimine) grafting through glutaraldehyde linkage (PEI) or thiourea grafting (T), noble metal ions can be sorbed. The reasons for grafting new functional groups are (i) to increase the density of sorption sites, (ii) to change the pH range for metal ions sorption and (iii) to change the sorption sites and/or the uptake mechanism in order to increase sorption selectivity for the noble metals [146]. Such derivatives were used for palladium and platinum removal [147]. It was found that the maximum adsorption capacity occurred at pH 2.0 for both Pt(IV) and Pd(II) species. The material selectively adsorbs Pt(IV) and Pd(II) from binary mixtures with Cu(II), Pb(II), Cd(II), Zn(II), Ca(II) and Mg(II). The isotherm adsorption equilibrium was well described by the Langmuir isotherms with the maximum adsorption capacity of 129.9 mg/g for Pt(IV) and 112.4 mg/g for Pd(II), which was relatively high compared with the glycine chitosan derivative (122 mg/g for Pt(IV) and 120 mg/g for Pd(II)). The results show that 0.5 M EDTA-0.5 M H_2SO_4 solution can effectively desorb Pt(IV) and Pd(II) (>97%) metal ions from the adsorbent material. The high percentage of desorption obtained when the 0.5 M EDTA–

0.5 M H_2SO_4 solution was used and can be explained by both stable complexes and the electrostatic interactions between the Pt(IV) and Pd(II) species.

In the paper [143], rubeanic acid was grafted on chitosan through the reaction with glutaraldehyde as the linker to obtain the sorbent for Au(III) with the thiol functional groups. It was found that the maximum sorption capacity was high and equal to 600 Au(III) mg/g. The speciation of gold in the chloride and hydroxide chloride systems appears to be a predominant parameter influencing the removal process. The optimum pH range was between 2 and 3 for glutaraldehyde cross-linked chitosan. However, the sorption capacity strongly decreases with the increasing pH. It was found that in the case of the grafting of sulfur compounds on chitosan derivatives, the partial change in the sorption mechanism occurs. In this case, metal ion chelation with sulfur compounds is weakly sensitive to the pH change [148]. Sorption capacity in the HCl system reaches 2 mmol/g (180 mg/g) and is slightly lower than for the chitosan derivatives obtained by grafting of pyridyl groups (6 mmol/g). Increasing chloride concentration involves a significant decrease in sorption capacity.

The removal of Au(III), Pt(IV) and Pd(II) onto the glycine modified cross-linked chitosan resin was investigated in the paper by Ramesh et al. [146]. The results show that the optimum pH appeared to be 1.0–4.0, and the maximum percentage removal was obtained at pH 2.0 for Au(III), Pt(IV) and Pd(II). The pH_{ZPC} was found to be 5.1. At $pH < pH_{PZC}$, the surface of modified chitosan resin is positively charged, whereas at a $pH > pH_{PZC}$, the surface of modified chitosan resin is negatively charged. Due to the positive surface charge of sorbent at pH lower than pH_{PZC}, it attracts the chlorocomplexes of platinum, palladium and gold, resulting in the greater amounts of adsorption at low pH. The authors proposed the following mechanism of the sorption process:

$$R\text{-}NH_2 + HCl \rightleftharpoons RNH_3^+Cl^-$$

$$R\text{-}NH_3^+ + HCl \rightleftharpoons RNH_3^+Cl^- + H^+$$

$$2R\text{-}NH_3^+Cl^- + \left[PtCl_6\right]^{2-} \rightleftharpoons (RNH_3^+)_2\left[PtCl_6\right]^{2-} + 2Cl^-$$

$$2R\text{-}NH_3^+Cl^- + \left[PdCl_4\right]^{2-} \rightleftharpoons (RNH_3^+)_2\left[PdCl_4\right]^{2-} + 2Cl^-$$

$$R\text{-}NH_3^+Cl^- + \left[AuCl_4\right]^- \rightleftharpoons RNH_3^+\left[AuCl_4\right]^- + Cl^-$$

The results also demonstrated that the amount of adsorption was decreased with the increasing chloride ion concentration. This is because of strong interaction between the chloride ions and precious metal ions to form chlorocomplexes. The 0.7 M thiourea-2 M HCl solution was the most effective for the desorption of Au(III), Pt(IV) and Pd(II).

In the case of chitosan derivatives, noble metal ions are sorbed according to several kinetic models based on pure sorption, pure reduction and dual sorption-reduction mechanisms [149, 150]. Moreover, the optimum acid pH for noble metal ions sorption depends on the metal. For platinum and palladium, it was equal to 2. For CS cross-linked by glutaraldehyde (CS-GA) for hydrochloric acid solutions of palladium at pH 2, sorption reached the same level as achieved at pH 1 (capacity strongly decreased) [151]. However, in the sulphuric(VI) acid solutions, the sorption capacity remains almost unchanged. It is well known that pH has a critical effect on the speciation of the metal in solution because the distribution of metal species depends on pH. Other parameters which affect the sorption efficiency are connected with the nature of the sorbent (ionic charge), chemistry of the metal ion: ionic charge, ability to be hydrolysed as well as metal concentration and the composition of the solution and the form of polynuclear species [151]. Sorption kinetics is controlled by particle size, cross-linking ratio and palladium concentration. In hydrochloric acid solutions, equilibrium is achieved at 24 h contact. For chitosan-cellulose fibres, it was found that incorporation of cellulose fibres improves the binding efficiency of chitosan towards Ag(I). The sorption capacity was close to 220 mg/g. This is much higher than for the pure chitosan (140 mg/g) [152]. The cellulose fibres contribute to dispersion of the chitosan chains that are more accessible and available for silver. It is also possible to modify the chitosan structure by introducing cross-linking structure, blending chitosan with synthetic polymers such as poly(vinyl alcohol) (PVA) – a non-toxic, water-soluble synthetic polymer with good physical and chemical properties and film-forming ability. It is also possible to apply the sol-gel process to develop organic–inorganic hybrid materials [153–155]. For this aim, clays and silicas are frequently used. Clays are composed of silicate layers which form three-dimensional structures after hydrated in water. They have negative charge and can interact with chitosan. Also silicas are characterized by several advantages which are among others surface stability in the acidic medium and highly developed surface, acceptable kinetics, thermal stability, resistance to microbial attack and low cost should be mentioned [156]. Chemically modified silicas (CMSs) with the functional groups covalently bound to the surface such as polyamines, particularly, linear polyhexamethylene guanidine (PHMG) with convenient amine group configurations and nitroso-R salt (NRS) were used in palladium sorption [157, 158]. Complex of palladium(II) with the ratio Pd:NRS = 1:2 formed the SiO_2–PHMG–NRS. The other examples are presented in Table 2.

Functional group	Metal ion	Ref.
Thiosemicarbazide	Pd(II)	[159]
Dithizone	Ag(I)	[160]
2-Amino-1-cyclopentene-1-dithiocarboxylic acid	Ag(I) and Pd(II)	[161]
N-(3-triethoxysilylpropyl)-4,5-dihydroimidazoleuene	Pd(II), Pt(II)	[162]
Amidinothiourea	Ag(I), Au(III) and Pd(II)	[162]
3-(1-thioureido)propyl	Ag(I), Au(III) and Pd(II)	[163]

Table 2. Organofunctionalized silicas in pre-concentration of noble metal ions.

24. Conclusions

Ion exchange has been widely applied in the technology of chemical separation of noble metal ions. This is associated with the dissemination of methods using various ion exchange resins which are indispensable in many fields of chemical industry. Due to small amounts of noble metals in nature and constant impoverishment of their natural sources, of particular importance are physicochemical methods of their recovery from the secondary sources as well as waste waters.

Recovery of noble metals, from such raw materials, requires individual approach to each material and application of selective methods for their removal. Moreover, separation of noble metals, particularly platinum metals and gold from geological samples, industrial products and synthetic mixtures along with other elements, is a problem of significant importance nowadays.

Author details

Zbigniew Hubicki*, Monika Wawrzkiewicz, Grzegorz Wójcik, Dorota Kołodyńska and Anna Wołowicz

*Address all correspondence to: zbigniew.hubicki@poczta.umcs.lublin.pl

Department of Inorganic Chemistry, Faculty of Chemistry, Maria Curie-Sklodowska University, Lublin, Poland

References

[1] Bogoczek R., Surowiec J. Obecny stan rozwoju wymieniaczy jonowych. Przemysł Chemiczny 1978;57 51-56.

[2] Abrams I.M., Millar J.R. A history of the origin and development of macroporous ion-exchanger resins. Reactive and Functional Polymers 1997;35 7-22.

[3] Bogoczek R., Surowiec J. Procesy jonitowe w technologii chemicznej. Przemysł Chemiczny 1978;57 217-221.

[4] Minczewski J., Chwastowska J., Dybczyński R. Separation and preconcentration methods in inorganic trace analysis. Chichester: John Wiley & Sons; 1982.

[5] Leszczyńska M. Badania procesu sorpcji jonów palladu(II) z chlorkowych i chlorkowo – azotanowych układów modelowych na jonitach i sorbentach różnego typu. PhD thesis UMCS Lublin; 2005.

[6] Bannister S.J., Chang Y., Sternson L.A., Repta A.J. Atomic absorption spectrophotometry of free circulating platinum species in plasma derived from cis-dichlorodiammineplatinum(II). Clinical Chemistry 1978;24 877-880.

[7] Marczenko Z., Kasiura K., Krasiejko M. Cation-exchange separation and colorimetric determination of some elements in trace analysis of platinum-rhodium (10%) alloys. Microchimica Acta 1969;57 625-633.

[8] MacNevin W.M., McKay E.S. Separation of rhodium from platinum, palladium, and iridium by ion exchange. Analytical Chemistry 1957;29 1220-1223.

[9] Weinert C.H.-S.W., Strelow F.W.E. Cation-exchange behaviour of the platinum group and some other rare elements in hydrobromic acid-thiourea-acetone media. Talanta 1983;30 755-760.

[10] Marks A.G., Beamish F.E. Separation of rhodium and iridium from base metals by ion exchange. Analytical Chemistry 1958;30 1464-1466.

[11] Banbury L.M., Beamish F.E. Fire assay collection of platinum and palladium by copper. Fresenius Journal of Analytical Chemistry 1965;211 178-187.

[12] Banbury L.M., Beamish F.E. Fire assay collection of rhodium by copper. Fresenius Journal of Analytical Chemistry 1966; 218 263-272.

[13] Agrawal K.C., Beamish F.E. Fire assay collection of iridium by copper. Fresenius Journal of Analytical Chemistry 1965; 211 265-274.

[14] Tertipis G.G., Beamish F.E. Separation of rhodium from iridium by copper powder. Analytical Chemistry 1960;32 486-489.

[15] Van Loon J.C., Beamish F.E. Inclusion of osmium in assay method for six platinum metals by iron-copper-nickel collection. Analytical Chemistry 1965;37 113-116.

[16] Van Loon J.C., Beamish F.E. A fire assay for osmium in sulfide concentrates. Analytical Chememistry 1964;36 872-875.

[17] Gupta J.G.S., Beamish F.E. New fire assay methods for the analysis of iridosmines. Analytical Chemistry 1962;34 1761-1764.

[18] Tertipis G.G., Beamish F.E. New fire assay for iridium. Analytical Chemistry 1962;34 108-110.

[19] Tertipis G.G., Beamish F.E. The recovery of rhodium, iridium, palladium and platinum from ores and concentrates by wet assay and a comparison with recovery by fire assays. Talanta 1963;10 1139-1151.

[20] Zachariasen H., Beamish F.E. Separation of ruthenium from base metals by cation exchange. Analytical Chemistry 1962; 34 964-966.

[21] Sant B.R., Beamish F.E. New fire assay method for rhodium. Analytical Chemistry. 1961;33 304-305.

[22] Plummer M.E.V., Beamish F.E. Determination of platinum and palladium in ores and concentrates new fire assay method. Analytical Chemistry 1959;31 1141-1143.

[23] Plummer M.E.V. Fire assay for platinum and palladium in ores and concentrates. Analytical Chemistry 1959;31 254-258.

[24] Coburn H.G., Beamish F.E., Lewis C.L. Determination of platinum and palladium in ferronickel assay buttons. Analytical Chemistry 1956;28 1297-1300.

[25] Faye G.H., Moloughney P.E. The tin-collection scheme for the determination of platinum-group metals, gold and silver. Talanta 1972;19 269-284.

[26] Faye G.H., Inman W.R. New fire assay method for rhodium in ores and concentrates. Analytical Chemistry 1962;34 972-974.

[27] Faye G.H., Inman W.R., Moloughney P.E. New fire assay method for iridium. Analytical Chemistry 1964;36 366-368.

[28] Sen Gupta J.G. Determination of microgram amounts of the six platinum-group metals in iron and stony meteorites. Analytica Chimica Acta 1968;42 481-488.

[29] Sen Gupta J.G. Abundances of the six platinum metals in some iron and stony meteorites: Relationships to the theories of evolution of parent bodies of meteorites and the origin of these elements. Chemical Geology 1968;3 293-305.

[30] Faye G.H., Inman W.R. The determination of platinum and palladium in copper-nickel matte and ore concentrates: A fire assay method using tin as a collector. Analytical Chemistry 1961;33 278-283.

[31] Pohlandt C., Steele T.W. Chromatographic separation and determination of noble metals in matte-leach residues. Talanta 1974;21 919-925.

[32] Coombes R.J., Chow A. A comparison of methods for the determination of platinum in ores. Talanta 1979;26 991-998.

[33] Blasius E., Rexin D. Trennung der Platinelemente unter Anwendung von Ionenaustauschern Fresenius Journal of Analytical Chemistry 1961;179 105-116.

[34] Samadi A.A., Grynszpan R., Fedoroff M. Behaviour of trace impurities during chemical dissolution of a metal: Platinum elements in aluminium and nickel. Talanta 1976;23 829-833.

[35] Schutyser P., Govaerts A., Dams R., Hoste J. Neutron activation analysis of platinum metals in airborne particulate matter. Journal of Radioanalytical and Nuclear Chemistry 1977;37 651-660.

[36] Govaerts A., Gijbels R., Hoste J. The determination of Pd, Pt, Au, Ag and Ir in copper by neutron activation analysis. Analytica Chimica Acta 1975;79 139-148.

[37] François J.P., Gijbels R., Hoste J. A radiochemical study of the fire assay of rhodium with copper and the subsequent rhodium-copper separation. Talanta 1974;21 780-783.

[38] Karttunen J.O., Evans H.B. Cation exchange separation and spectrophotometric determination of microgram amounts of rhodium in uranium-base fissium alloys. Analytical Chemistry 1960;32 917-922.

[39] Weinert C.H.-S.W., Strelow F.W.E. Cation-exchange behaviour of the platinum group and some other rare elements in hydrobromic acid-thiourea-acetone media. Talanta 1983;30 755-760.

[40] Siegfried C.H., Weinert W., Strelow F.W.E., Böhmer R.G. Cation-exchange in thiourea-hydrochloric acid solutions. Talanta 1983;30 413-418.

[41] Brajter K., Kleyny K., Vorbrodt Z. Studies on optimization of conditions for separating rhodium and platinum by cation-exchange. Talanta 1980;27 433-435.

[42] Kołodyńska D. Żywice chelatujące w procesie usuwania jonow metali ciężkich w obecności czynnika kompleksującego z wód i ścieków. Przemysł Chemiczny 2009;88 182-189.

[43] Hubicki Z., Leszczyńska M., Łodyga B., Łodyga A. Recovery of palladium (II) from chloride and chloride-nitrate solutions using ion exchange resins with S-donor atoms. Desalination 2007;207 80-86.

[44] Kantipuly C, Katragadda S, Chow A, Gesser H.D. Chelating polymers and related supports for separation and preconcentration of trace metals. Talanta 1990;37 491-517.

[45] Grote M., Kettrup A. Ion-exchange resins containing s-bonded dithizone and dehydrodithizone as functional groups: Part 3. Determination of gold, platinum and palladium in geological samples by means of a dehydrodithizone resin and plasma emission spectrometry. Analytica Chimica Acta 1987;201 (C) 95-107.

[46] Grote M., Kettrup A. Ion-exchange resins containing s-bonded dithizone and dehydrodithizone as functional groups: Part 2. Desorption properties and development of separation procedures for gold and platinum group metals. Analytica Chimica Acta 1985;175 (C) 239-255.

[47] Grote M., Kettrup A. Separation of noble metals by ion exchangers containing S-bonded dithizone and dehydrodithizone as functional groups. In: Ion Exchange Technology. London: Ellis Horwood Limited;1984 618-625.

[48] Grote M., Kettrup A. Ion-exchange resins containing S-bonded dithizone and dehydrodithizone as functional groups: Part 1. Preparation of the resins and investigation of the sorption of noble metals and base metals. Analytica Chimica Acta 1985;172 (C) 223-239.

[49] Shah R., Devi S. Preconcentration and separation of palladium(II) and platinum(IV) on a dithizone anchored poly(vinylpyridine)-based chelating resin. Analytica Chimica Acta 1997;341 217-224.

[50] Patterson J.W., Passino R. Metals Speciation, Separation and Recovery. Chelsea: Lewis Publishers Inc.; 1990.

[51] Chwastowska J., Skwara W., Sterlińska E., Pszonicki L. Determination of platinum and palladium in environmental samples by graphite furnace atomic absorption spectrometry after separation on dithizone sorbent. Talanta 2004;64 224-229.

[52] Chikuma M., Aoki H., Tanaka H. Determination of metal ions in environmental waters by flameless atomic absorption spectrometry combined with preconcentration using a sulfonated dithizone-loaded resin. Analytical Sciences 1991;7 1131-1134.

[53] Hubicki Z., Wawrzkiewicz, M., Wołowicz A. Application of ion exchange methods in recovery of Pd(II) ions - a review. Chemia Analityczna 2008;53 759-784.

[54] Hubicki Z., Leszczyńska M. Zastosowanie jonitów modyfikowanych w procesie odzysku jonów Pd(II). Przemysł Chemiczny 2005;84 750-755.

[55] Hubicki Z., Leszczyńska M., Łodyga B., Łodyga A. Recovery of palladium(II) from chloride and chloride-nitrate solutions using ion-exchange resins with S-donor atoms. Desalination 2007;207 80-86.

[56] Blokhin A.A., Abovskii N.D., Murashkin Yu.V. Ion-exchange recovery of palladium(II) from multicomponent chloride solutions. Russian Journal of Applied Chemistry 2007;80 1058-1062.

[57] Szymanowski J. Rozwój technologii wydzielania i rozdziału metali szlachetnych. Rudy Metale 1989;34 54-58.

[58] Mayasoedova G.V., Antokol'skaya I.I., Savvin S.B. New chelating sorbents for noble metals. Talanta 1985;32 1105-1112.

[59] Jermakowicz-Bartkowiak D. Preparation, characterisation and sorptive properties towards noble metals of the resins from poly(vinylbenzyl chloride) copolymers. Reactive and Functional Polymers 2005;62 115-128.

[60] Warshawski A., Fieberg M.M.B., Mihalik P., Murphy T.G., Ras Y.B. The separation of Platinum group metals (PGM) in chloride media by isothiouronium resins. Separation and Purification Methods 1980;9 209-265.

[61] Labošowá L., Štofková M., Kračunovská J. The study of possibilities of selective recovery of palladium(II) from chlorides solutions by ion exchange resin Lewatit TP-214. Acta Metallurgica Slovaca 2006;12 235-241.

[62] Zuo G., Muhammed M. Thiourea-based coordinating polymers: synthesis and binding to noble metals. Reactive Polymers 1995;24 165-181.

[63] Brinci E., Gülfen M., Aydin A.O. Separation and recovery of palladium(II) from base metal ions by melamine–formaldehyde–thiourea (MFT) chelating resin. Hydrometallurgy 2009;95 15-21.

[64] Saha B., Iglesias M., Cumming I.W., Streat M. Sorption of trace heavy metals by thiol containing chelating resins. Solvent Extraction and Ion Exchange 2000;18 133-167.

[65] Iglesias M., Anticó E., Salvado V. Recovery of palladium(II) and gold(III) from diluted liquors using the resin Duolite GT-73. Analytical Chimica Acta 1991;381 61-67.

[66] Pohl P., Prusisz B. On the applicability of Duolite GT-73 to column preconcentration of gold and palladium prior to determination by inductively coupled plasma-atomic emission spectrometry. Microchimica Acta 2005;150 159-165.

[67] Slovák Z., Smarž M., Dočekal B., Slováková S. Analytical behaviour of hydrophilic glycolmethacrylate gels with bound thiol groups. Analytica Chimica Acta 1979;111 243-249.

[68] Anticó E., Masana A., Salvadó V., Hidalgo M., Valiente M. Adsorption of palladium by glycolmethacrylate chelating resins. Analytica Chimica Acta 1994;296 325-332.

[69] Venkatesan K.A., Selvan B.R., Antony M.P., Srinivasan T.G., Rao P.R.V. Extraction of palladium from nitric acid medium by commercial resins with phosphinic acid, methylene thiol and isothiouranium moieties attached to polystyrene-DVB. Journal of Radioanalytical and Nuclear Chemistry 2005;266 431-440.

[70] Anticó E., Masana A., Salvadó V., Hidalgo M., Valiente M. Separation of Pd(II) and Cu(II) in chloride solutions on a glycol methacrylate gel derivatized with 8-hydroxyquinoline. Journal of Chromatography A 1995;706 159-166.

[71] Trochimczuk A.W. Chelating resins with N-substituted diamides of malonic acid as ligands. European Polymer Journal 1998;34 1657-1662.

[72] Parodi A., Vincent T., Pilśniak M., Trochimczuk A., Guibal E. Palladium and platinum binding on imidazole-resin. Hydrometallurgy 2008;92 1-10.

[73] Chang X., Su Z., Yang D., Gong B., Pu Q., Li S. Synthesis and efficiency of a spherical macroporous epoxy-imidazole complexing resin for preconcentrating trace noble metal ions. Analytica Chimica Acta 1997;354 143-149.

[74] Das D., Das A.K., Sinha Ch. A new resin containing benzimidazolylazo group and its use in the separation of heavy metals. Talanta 1999;48 1013-1022.

[75] Venkatesan K.A., Selvan B.R., Antony M.P., Srinivasan T.G., Vasudeva Rao P.R. Extraction of palladium (II) from nitric acid medium by imidazolium nitrate immobilized resin. Hydrometallurgy 2007;86 221-229.

[76] Su Z.X., Pu Q.S., Luo X.Y., Chang X.J., Zhan G.Y., Ren F.Z. Application of a macroporous resin containing imidazoline groups to preconcentration and separation of gold, platinum and palladium prior to ICP AES determination. Talanta 1995;42 1127-1133.

[77] Siddhanta S., Das H.R. Separation and concentration of some platinum metal ions with a new chelating resin containing thiosemicarbazide as functional group. Talanta 1985;32 457-460.

[78] Chen Y.Y., Liang C., Chao Y. Synthesis and characterization of polyacrylonitrile-thiosemicarbazide resin and its sorption behavior for Rh(III) Ru(IV) Pd(II) and Ir(IV) ions. Reactive and Functional Polymers 1998;36 51-58.

[79] Topp K.D., Grote M. Synthesis and characterization of a 1,2,4,5-tetrazine-modified ion-exchange resin. Reactive and Functional Polymers 1996;31 117-136.

[80] Yi-Yong Ch., Xing-Zhong Y. Synthesis and properties of 1-(2-aminoethyl)piperazine resin used in the sorption of the platinum group and gold ions. Reactive Polymers 1994;23 165-172.

[81] Suzuki T., Fujii Y., Yan W., Mimura H., Koyama S., Ozawa M. Adsorption behaviour of VII group elements on tertiary pyridine resin in hydrochloric acid solution. Journal of Radioanalytical and Nuclear Chemistry 2009;282 641-644.

[82] Di P., Davey D.E. On-line preconcentration and separation of palladium, platinum and iridium using α-amino pyridine resin with flame atomic absorption spectrometry. Talanta 1995;42 685-692.

[83] Jermakowicz-Bartkowiak D. Preparation, characterisation and sorptive properties towards noble metals of the resins from poly(vinylbenzyl chloride) copolymers. Reactive and Functional Polymers 2005;62 115-128.

[84] Jermakowicz-Bartkowiak D., Kolarz B.N., Serwin A. Sorption of precious metals from acid solutions by functionalised vinylbenzyl chloride–acrylonitryle–divinylbenzene copolymers bearing amino and guanidine ligands. Reactive and Functional Polymers 2005;65 135-142.

[85] Zhang B.W., Zhang Y., Grote M., Kettrup A. Studies on macroporous cross-linked polyacrolein-styrene resin I. Synthesis of polyacrylic aldehyde-hydrazone and polyacrolein-phenylhydrazone resins and their chelating properties for gold and platinum group metals. Reactive Polymers 1994;22 115-125.

[86] Ge X., Wendler I., Schramel P., Kettrup A. Application of polyacrolein–isonicotinic acid hydrazone resin for the separation and concentration of Pd and Pt in road dust. Reactive and Functional Polymers 2004;61 1-10.

[87] Ge X.P., Zhang B.W., Grote M. Polyacrolein-isonicotinic acid hydrazone and polyacrylic acid-thiohydrazide resins-synthesis and sorption properties for precious and base metals. Microchimica Acta 1998;129 303-310.

[88] Grote M., Wigge P., Kettrup A. Formazane als funktionelle gruppen chelatbildender ionenaustauscher. Frezenius Zeitschrift für Analytische Chemie 1982;310 369-377.

[89] Grote M., Schwalk A., Hüppe U., Kettrup A. Formazans as functional groups of chelating ion-exchangers. V. Noble metal coordination and ligand-conversion of chelat-

ing formazan functionalized supports. Frezenius Zeitschrift für Analytische Chemie 1983;316 247-252.

[90] Grote M., Schumacher U. Bipolar ion-exchange resins based on functional acidic tetrazolium groups - their synthesis, structure and properties. Reactive and Functional Polymers 1997;35 179-196.

[91] Grote M., Kettrup A. Formazane als funktionelle gruppen chelatbildender ionenaustauscher. Frezenius Zeitschrift für Analytische Chemie 1979;295 366-370.

[92] Jermakowicz-Bartkowiak D. A preliminary evaluation on the use of the cyclam functionalized resin for the noble metals sorption. Reactive and Functional Polymers 2007;67 1505-1514.

[93] Iglesias M., Anticó E., Salvadó V. Characterisation of Metalfix-Chelamine and its application in precious metal adsorption. Solvent Extraction and Ion Exchange 2000; 18 965-979.

[94] Pohl P., Prusisz B., Zyrnicki W. Application of Metalfix Chelamine prior to the determination of noble metals by the inductively coupled plasma atomic emission spectrometry. Talanta 2005;67 155-161.

[95] Iglesias M., Anticó E., Salvadó V. On-line determination of trace levels of palladium by flame atomic absorption spectrometry. Talanta 2003;59 651-657.

[96] Muzikar M., Fontàs C., Hidalgo M., Havel J., Salvadó V. A preconcentration system using polyamine Metalfix-Chelamine resin for the on-line determination of palladium(II) and platinum(IV) by inductively coupled plasma optical emission spectrometry. Talanta 2006;70 1081-1086.

[97] Coedo A.G., Dorado M.T., Padilla I., Alguacil F. Preconcentration and matrix separation of precious metals in geological and related materials using metalfix-chelamine resin prior to inductively coupled plasma mass spectrometry. Analytica Chimica Acta 1997; 340 31-40.

[98] Lopes C. M.P.V., Almeida A.A, Lúcia M.M.F.S., Lima J.L.F.C. Determination of Rh, Pd and Pt in urine samples using a pre-concentration sequential injection analysis system coupled to a quadrupole-inductively coupled plasma-mass spectrometer. Analytica Chimica Acta 2007;600 226-232.

[99] Myasoedova G.V. POLYORGS as complexing sorbents for preconcentration of trace metals. Fresenius Journal of Analytical Chemistry 1991;341 586-591.

[100] Mokиodoeva O.B., Myasoedova G.V., Kubrakova I.V. Preconcentration of noble metals with the Polyorgs 4 complexing sorbent under the action of microwave irradiation. Journal of Analytical Chemistry 2007;62 406-410.

[101] Mokиodoeva O.B., Myasoedova G.V., Kubrakova I.V. Sorption preconcentration in combined methods for the determination of noble metals. Journal of Analytical Chemistry 2007;62 607-622.

[102] Myasoedova G.V., Zaharchenko E.A., Mokиodoeva O.B., Kubrakova I.V., Nikashina V.A. Sorption preconcentration of Platinum-Group Metals with filled fibrous Polyorgs sorbents. Journal of Analytical Chemistry 2004;59 536-540.

[103] Myasoedova G.V., Mokиodoeva O.B., Kubrakova I.V. Trends in sorption preconcentration combined with metal determination. Analytical Sciences 2007;23 1031-1039.

[104] Hubicki Z., Leszczyńska M., Łodyga B., Łodyga A. Palladium(II) removal from chloride and chloride-nitrate solutions by chelating ion-exchangers containing N-donor atoms. Minerals Engineering 2006;19 1341-1347.

[105] Park Ch., Chung J.S., Cha K.W. Separation and preconcentration method for palladium, platinum and gold from some heavy metals using Amberlite IRC 718 chelating resin. Bulletin of the Korean Chemical Society 2000;21 121-124.

[106] Sànchez J.M., Hidalgo M., Valiente M., Salvado V. New macroporous polymers for the selective adsorption of gold(III) and palladium(II). I. The synthesis, characterization, and effect of spacers on metal adsorption. Journal of Polymer Science Part A: Polymer Chemistry 2000;38 269-278.

[107] Sànchez J.M., Hidalgo M., Salvado V. A comparison of the separation behavior of some new coordinating resins and commercial quaternary ammonium resins with reference to their separation of gold(III) and palladium(II) in hydrochloric acid media. Solvent Extraction and Ion Exchange 2004;22 285-303.

[108] Sànchez J.M., Hidalgo M., Salvado V. The selective adsorption of gold (III) and palladium (II) on new phosphine sulphide-type chelating polymers bearing different spacer arms equilibrium and kinetic characterization. Reactive and Functional Polymers 2001;46 283-291.

[109] Sànchez J.M., Hidalgo M., Salvado V. Synthesised phosphine sulphide-type macroporous polymers for the preconcentration and separation of gold(III) and palladium(II) in a column system. Reactive and Functional Polymers 2001;49 215-224.

[110] Kałędkowski A., Trochimczuk A.W. Chelating resin containing hybrid calixpyrroles: new sorbent for noble metal cations. Reactive and Functional Polymers 2006;66 957-966.

[111] Kałędkowski A, Synteza i właściwości polimerów zawierających kalikspirole. PhD thesis PW Wrocław; 2006.

[112] Garcia L., Torrent A., Anticó E., Fontàs C., Roglans A. Selective Pd(II) and Pt(IV) sorption using novel polymers containing azamacrocycle functional groups. Reactive and Functional Polymers 2008;68 1088-1096.

[113] Chen Y., Zhao Y. Synthesis and characterization of polyacrylonitrile-2-amino-2-thiazoline resin and its sorption behaviors for noble metal ions. Reactive and Functional Polymers 2003;55 89-98.

[114] Qu R., Sun Ch., Ji Ch., Wang Ch., Xu Q., Lu S., Li Ch., Chen G. Synthesis and characterization of polystyrene supported 2,5-dimercapto-1,3,4-thiodiazole and its sorption behaviour for Pd(II), Pt(IV) and Au(III). Journal of Applied Polymer Science 2006;101 631-637.

[115] Gong B. Synthesis of polyacrylaminoimidazole chelating fiber and properties of concentration and separation of trace Au, Hg and Pd from samples. Talanta 2002;57 89-95.

[116] Li Y., Liu R., Zhang B. Application of an imidazole group containing chelating fiber for the determination of trace noble metals in superhigh-temperature alloys. Fresenius Journal of Analytical Chemistry 2000;366 821-824.

[117] Bilba N., Bilba D., Moroi G. Synthesis of polyacrylamidoxime chelating fiber and its efficiency in the retention of palladium ions. Journal of Applied Polymer Science 2004;92 3730-3735.

[118] Gong B., Wang Y. ICP AES determination of traces of noble metal ions pre-concentrated and separated on a new polyacrylaminothiourea chelating fibre. Analytical and Bioanalytical Chemistry 2002;372 597-600.

[119] Gong B., Wang Y., Sun Y.P., Zhao J.G. Synthesis of polyacrylaminothiourea chelating fibre and properties of concentration and separation of trace noble metal ions from samples. Chinese Journal of Chemistry 2002;20 63-67.

[120] Chang X., Su Z., Zhan G., Luo X., Gao W. Synthesis and efficiency of a polyacrylacylithiourea chelating fibre for the preconcentration and separation of trace amounts of gold, palladium and ruthenium from solution samples. Analyst 1994;119 1445-1449.

[121] Chang X., Zhan G., Luo X., Su Z. Synthesis and efficiency of poly(acrylamidrazone-hydrazide) chelating fibre for preconcentrating and separating trace gold and palladium from solution. Microchimica Acta 1994;112 245-251.

[122] Zheng H., Chang X., Lian N., Mao J., Wang S., Dong Y. Synthesis and application of a novel poly(acryl-*p*-toluenesulfonamideamidine-*p*-toluenesulfonylamide) chelating fiber for preconcentration and separation of trace noble metal ions from solution samples. Microchimica Acta 2005;149 259-266.

[123] Chajduk-Maleszewska E., Dybczyński R. Selective separation and preconcentration of trace amounts of Pd on Duolite ES 346 resin and its use for the determination of Pd by NAA. Chemia Analityczna 2004;49 281-297.

[124] Dybczyński R., Hubicki Z., Kulisa K. Ion exchange behaviour of 23 elements and amphoteric properties of chelating resin Duolite ES 346 containing amidoxime groups. Solvent Extraction and Ion Exchange 1988;6 699-724.

[125] Samczyński Z., Danko B., Dybczyński R. Application of Chelex 100 ion exchange resin for separation and determination of palladium, platinum and gold in geological and industrial materials by neutron activation analysis. Chemia Analityczna 2000;45 843-856.

[126] Kovacheva P., Djingova R. Ion-exchange method for separation and concentration of platinum and palladium for analysis of environmental samples by inductively coupled plasma atomic emission spectrometry. Analytica Chimica Acta 2002;464 7-13.

[127] Wilkinson B. W., Toth-Allan J. Neutron activation analysis using daughter activity. Nuclear Technology 1972;13 103-110.

[128] Steinnes E. Determination of rhodium in platinum bearing ores by neutron activation analysis. Radiochemical and Radioanalytical Letters 1975;21 233-239.

[129] Crocket J.H., Keays R.R., Hsieh S. Determination of some precious metals by neutron activation analysis. Journal of Radioanalytical Chemistry 1968;1 487-507.

[130] Crocket J.H., Keays R.R., Hsieh S. Precious metal abundances in some carbonaceous and enstatite chondrites. Geochimica Et Cosmochimica Acta 1967;31 1615-1623.

[131] Crocket J.H. Some aspects of the geochemistry of Ru, Os, Ir and Pt in iron meteorites. Geochimica Et Cosmochimica Acta 1972;36 517-535.

[132] Enzweiler J., Potts P.J. The separation of platinum, palladium and gold from silicate rocks by the anion exchange separation of chloro complexes after a sodium peroxide fusion: an investigation of low recoveries. Talanta 1995;42 1411-1418.

[133] Els E.R., Lorenz L., Aldrich C. The recovery of palladium with the use of ion exchange resins. Minerals Enginering 1997;10 1177-1181.

[134] Els E.R., Lorenz L., Aldrich C. The adsorption of precious metals and base metals on a quaternary ammonium group ion exchange resin. Minerals Engineering 2000;13 401-414.

[135] El-Said N., Mekhail A., Khalifa S.M., Aly H.F. Separation of strontium from simulated waste in K_2SO_4/nitrate medium containing Sr^{2+}, Eu^{3+}, Ce^{3+}, Pd^{2+}, Rh^{3+}, Ru^{3+}, $UO^{2+}{}_2$, Fe^{3+}, Cr^{3+}, Ni^{2+}, Al^{3+}, Ca^{2+}, and Cs^+ by strongly basic anion exchange. Journal of Radioanalytical and Nuclear Chemistry 1996;208 243-255.

[136] Hubicki Z., Leszczyńska M. Studies of sorption of Pd(II) microquantities on strongly basic polyacrylate anion exchangers. Desalination 2005;175 289-295.

[137] Hubicki Z., Leszczyńska M. Sorption of palladium (II) chloride complexes on weakly, intermediate and strongly basic anion exchangers. Desalination 2005;175 227-236.

[138] Pohlandt C.,Fritz J.S. Separation of metal ions on a new amide resin. Journal of Chromatography 1979;176 189 -197.

[139] Gaita R., Al-Bazi S.J. An ion-exchange method for selective separation of palladium, platinum and rhodium from solutions obtained by leaching automotive catalytic converters. Talanta 1995;42 249-255.

[140] Hubicki Z., Wójcik G. Studies of the selective removal of micro-quantities of platinum(IV) ions from macro-quantities of model solutions of aluminium, copper, iron, nickel and zinc chloride on anion-exchangers of various types. Adsorption Science and Technology 2006;24, 559-569.

[141] Hubicki Z., Wójcik G. Studies of the selective removal of microquantities of platinum(IV) ions from model chloride solutions onto ion exchangers containing functional tertiary amine and polyamine groups. Adsorption Science and Technology 2004;22 627-637.

[142] Hubicki Z., Wójcik G. Studies of removal of platinum (IV) ion microquantities from the model solutions of aluminium, copper, iron, nickel and zinc chloride macroquantities on the anion exchanger Duolite S 37. Journal of Hazardous Materials 2006;136 770-775.

[143] Arrascuea M.L., Garciaa H.M., Horna O., Guibal E. Gold sorption on chitosan derivatives. Hydrometallurgy 2003; 71 191-200.

[144] Kołodyńska D. Chitosan as an effective low-cost sorbent of heavy metal complexes with polyaspartic acid. Chemical Engineering Journal 2011; 173 520-529.

[145] Guibal E., Vincent T., Larkin A., Tobin J.M. Chitosan sorbents for platinum sorption from dilute solutions. Industrial and Engineering Chemistry Research 1999; 38 4011-4022.

[146] Ramesh A., Hasegawa H., Sugimoto W., Maki T., Ueda K. Adsorption of gold(III), platinum(IV) and palladium(II) onto glycine modified crosslinked chitosan resin, Bioresource Technology. 2008; 99 3801-3809.

[147] Zhou L., Liu J., Liu Z. Adsorption of platinum(IV) and palladium(II) from aqueous solution by thiourea-modified chitosan microspheres. Journal of Hazardous Materials 2009; 172 439-446.

[148] Chassary P., Vincent T., Guibal E. Metal anion sorption on chitosan and derivative materials: A strategy for polymer modification and optimum use. Reactive and Functional Polymers 2004; 60 137-149.

[149] Guibal E., Roussy J., Le Cloirec P. Photochemical reaction of uranium with glucosamine, acetylglucosamine and related polymers: chitin and chitosan. Water 1996; 22 19-25.

[150] Chassary P., Vincent T., Marcano J.S., Macaskie L.E., Guibal E. Palladium and platinum recovery from bicomponent mixtures using chitosan derivatives. Hydrometallurgy 2005; 76 131-147.

[151] Ruiz M., Sastre A., Zikan M.C., Guibal E. Palladium sorption on glutaraldehyde-crosslinked chitosan in fixed-bed systems. Journal of Applied Polymer Science 2001; 81 153-165.

[152] Guibal E., Cambe S., Bayle S., Taulemesse J.M., Vincent T. Silver/chitosan/cellulose fibers foam composites: from synthesis to antibacterial properties. Journal of Colloid and Interface Science 2013; 393 411-420.

[153] Da Costa Neto B.P., Da Mata A.L.M.L., Lopes M.V., Rossi-Bergmannd B., Ré M.I. Preparation and evaluation of chitosan–hydrophobic silica composite microspheres: Role of hydrophobic silica in modifying their properties. Powder Technology 2014; 255 109-119.

[154] Parida U.K., Nayak A.K., Binhani B.K., Nayak P.L. Synthesis and characterization of chitosan-polyvinyl alcohol blended with cloisite 30B for controlled release of the anticancer drug curcumin. Journal of Biomaterials and Nanobiotechnology. 2011; 2 414-425.

[155] Wang X., Du Y., Luo J., Lin B., Kennedy J.F., Chitosan/organic rectorite nanocomposite films: structure, characteristic and drug delivery behavior, Carbohydrous Polymers; 2007 69 41-49.

[156] Repo E., Warchoł J., Bhatnagar A., Sillanpää M. Heavy metals adsorption by novel EDTA-modified chitosan–silica hybrid materials. Journal of Colloid Interface Science 2011; 358 261-267.

[157] Losev V.N., Didukh S.L., Trofimchuk A.K., Leshchenko V.N. Palladium(II) and cobalt(II) sorption by silica gel sequentially modified by polyhexamethylene guanidine and a nitroso-R salt. Mendeleev Commun. 2009; 19 167-169.

[158] Jal P.K., Patel S., Mishra B.K. Chemical modification of silica surface by immobilization of functional groups for extractive concentration of metal ions, Talanta 2004; 62 1005-1028.

[159] Mahmoud M.E. Comparison of metal uptake properties of silica gel-bound ion exchangers and some amine derivatives, Analytical Letters 1996; 29 1791-1804.

[160] Zaporozhets O., Petruniock N., Sukhan V. Determination of Ag(I), Hg(II) and Pb(II) by using silica gel loaded with dithizone and zinc dithizonate. Talanta 1999; 50 865-876.

[161] Seshadri T., Kettrupt A. Synthesis and characterization of silica gel ion-exchanger bearing 2-amino-1-cyclopentene-1-dithio-carboxylic acid (ACDA) as chelating compound, Fresenius' Zeitschrift für analytische Chemie 1982; 310, 1-5.

[162] Zhang S., Pu Q., Liu P., Sun Q., Su Z., Synthesis of amidinothioureido-silica gel and its application to flame atomic absorption spectrometric determination of silver, gold and palladium, Analytica Chemica Acta, 2002, 452, 223-230.

[163] Liu P., Pu Q., Su Z., Synthesis of silica gel immobilized thiourea and its application to the on-line preconcentration and separation of silver, gold and palladium, Analyst, 2000, 125, 147-150.

4

Anion Exchange Resins as Effective Sorbents for Removal of Acid, Reactive and Direct Dyes from Textile Wastewaters

Monika Wawrzkiewicz and Zbigniew Hubicki

Abstract

Coloured wastewaters are a consequence of batch processes in both dye-manufacturing and dye-consuming industries. Dyes are widely used in a number of industries, such as textile and leather dyeing, food, cosmetics, paper printing, gasoline, with the textile industry as the largest consumer. Dyeing as a fundamental operation during textile fibre processing causes the production of more or less coloured wastewaters, depending on the degree of fixation of dyes on substrates, which varies with the nature of substances, desired intensity of coloration, and application method. Dye bearing effluents are considered to be a very complex and inconsistent mixture of many pollutants ranging from dyes, dressing substances, alkalis, oils, detergents, salts of organic and inorganic acids to heavy metals.Thus after dyeing wastewaters are characterized not only by intensive and difficult for removal colour but also by high pH, suspended and dissolved solids, chemical and biochemical oxygen demands. Ion exchange is a very versatile and effective tool for treatment of aqueous hazardous wastes including dyes. The role of ion exchange in dye effluents treatment is to reduce the magnitude of hazardous load by converting them into a form in which they can be reused, leaving behind less toxic substances in their places or to facilitate ultimate disposal by reducing the hydraulic flow of the stream bearing toxic substances. Another significant feature of the ion exchange process is that it has the ability to separate as well as to concentrate pollutants. Taking into account high capacity and selectivity of ion exchange resins for different dyes, they seem to be proper materials for dyes sorption from textile effluents. The aim of the paper is to study the removal of the acid, reactive and direct textile dyes such as C.I. Acid Orange 7, C.I. Reactive Black 5 and C.I. Direct Blue 71 on the commercially available anion exchangers (Lewatit MonoPlus MP 62, Lewatit MonoPlus MP 64, Lewatit MonoPlus MP 500, Lewatit MonoPlus M 500, Amberlite IRA 67, Amberlite IRA 478RF, Amberlite IRA 458 and Amberlite IRA 958) differing not only in basicity of the functional groups but also in composition and structure of the matrix. Comparison of the sorption parameters obtained by the batch method taking into account influence of phase contact time, dyes initial concentration and solution pH were discussed in detail. Desorption conditions depending on the dyes sorption mechanism were also presented. Influence of the auxiliaries typically present in textile effluents such as inorganic electrolytes and different surfactants on the amounts of dyes retained by the anion exchangers was presented. The adsorption behaviour of the polyacrylic Amberlite IRA 958 demonstrates that it can be a promising adsorbent for the textile wastewater treatment. The results obtained with raw textile wastewaters purification confirmed this statement.

Keywords: anion exchangers, dyes sorption, removal, textile effluents

1. Introduction

Wastewaters originating from the textile, cellulose, paper, chemical, tanning, food and cosmetic industries containing dyes are a hazardous source of natural environment contamination. They are troublesome in purification process due to complex structure of dye molecules. Even small amounts of dye are undesirable as they colour water making it look unaesthetic and disturb life processes in water. Most dyes do not undergo biodegradation, deteriorate light penetration into water and inhibit photosynthesis, increase chemical and biological demand for oxygen. Some dyes are toxic and sometimes even carcinogenic and mutagenic towards living organisms and they should be carefully removed [1]. Purification of wastewaters containing dyes becomes more and more important and is aimed at avoiding potential threat for the environment and legal consequences. Decolourization of these wastewaters before their reaching water outlets is a must. Conventional methods of wastewaters purification do not remove colour completely. Therefore it is advisable to work out a more effective methods of wastewaters purification which would reduce the amount of discharged impurities but also contribute to recovery of water and raw materials used in technological processes. The difficulties in elaboration of both simple and economical methods of dyes removal are due to frequent changes in technology of their production as well as the use of various dyes in technological processes. In the case of impurities whose creation cannot be prevented, there should be applied highly effective technologies for their rendering harmless which can be combined into multi-stage purification systems. Such possibilities are provided by adsorption processes using, among others, ion exchangers allowing not only to separate substances dissolved based on selective interactions but also to concentrate amounts of impurities and formed closed circulation of water in the technological process.

2. Dyes – Definition and classification

Dyes are commonly defined as organic compounds of natural or synthetic origin which absorb visible light selectively in the range 400–700 nm, being capable of dying various materials (fabrics, paper, leather, wood, plastic, food, cosmetics) [1]. Dye molecules possess chromophoric groups, e.g. –N=N–, $–NO_2$, –NO, –CH=CH–, owing to which these compounds selectively absorb electromagnetic radiation in the visible range and auxochromic one (e.g. $–NH_2$, –OH, –OR) responsible for dyes affinity for dyed materials [1, 2]. Dye molecules can also contain other substituents which impart specific properties affecting their solubility or stability improvement.

There are two basic ways of dyes classification. Based on the chemical structure of dye molecule, there can be distinguished carbocyclic and heterocyclic dyes. Considering the occurrence of chromophore, carbocyclic dyes include, among others, azo, nitro, anthraquinone ones and heterocyclic dyes include xanthene, acridine, indigoid ones. Using the chemical classification of dyes is convenient while discussing methods of their synthesis, chemical structure, dependence between the structure and different properties. Technical classification of dyes is based on the way of dyeing taking into account their solubility and chemical properties (Table 1) [2]. It is useful in the case of different applications of dyes. Not all dyes are universal. The way of their application depends on different methods of dye bonding with the material. Thus dyes are divided into reactive, direct, acid, basic, ice, vat, mordant, suspended, sulfurous, oxidative, pigments and lakes. From a practical point of view, there are two groups of dyes. The first one are dyes soluble in water with a created coloured cation (basic dyes) or anion (acid, acid-chromic, reactive, direct, and metal complex dyes). The second group includes dyes insoluble in water (suspended and those formed on the fibre, e.g. ice, oxidative, mordant) as well as pigments. Among them, there are dyes whose salts dissolve in water (sulfur and vat ones) and those which dissolve in organic solvents, e.g. fatty dyes.

Class of dyes	Chemical character of dyes	Exemplary dyes	Kind of coloured material
Acid	Organic salts of sulfonic and carboxylic acids soluble in water	Anthraquinone, azo, triphenylmethane, nitro, xanthene	Protein fibres (mainly wool), polyamide fibres, paper, leather, cosmetics, food
Basic	Salts of organic bases soluble in water	Arylmethane, xanthene, azo, acridine,anthraquinone	Wool, natural silk, polyester and polyacrylnitrile fibres
Direct (substantive)	Salts of organic sulfonic acids soluble in water	Phthalocyanines, stilbene, oxazine, azo	Cellulose fibres (cotton, linen), leather
Reactive	Salts of organic bases and acids soluble in water	Azo, anthraquinone, formazine, oxazine	Cellulose and protein fibres
Mordant	Soluble in water, form complex with metals	Nitroso dyes, e.g. alizarin and its derivatives	Cellulose and protein fibres
Vat	Insoluble in water	Anthraquinone, indigo	Cellulose fibres
Sulfur	Insoluble in water	-	Cellulose fibres
Dispersive	Insoluble or sparsely soluble in water	Azo, anthraquinone, nitro, styrole	Synthetic mainly poliester, polyamide, polyacrylnitrile, acetic

Table 1. Technical classification of dyes

There is a special register called *Colour Index* for dyes first published in 1924. The dual classification system *Colour Index* includes: *Colour Index Generic Name (CIGN)*, e.g. Disperse Yellow 1, which is the name describing the kind of dye as regards its technical application and colour as well as chronological notation in the register and for the known and published chemical structure, so-called *Colour Index Constitution Number (CICN)* in this case C.I. 10345 [3]. Their largest use was recorded in 2008 in Asia, particularly in China and India (Figure 1) [4]. The yearly world production of dyes is 700, 000–1000, 000 tones, which corresponds to over 100, 000 commercial products whereby azo dyes constitute 70% [5-11].

Natural dyes such as indigo, woad, or madder lost their position due to dynamic development of textile industry in the XVIIIth century. In the second half of the XIXth century, there started a synthesis of intermediate products and new dyes not having equivalents in nature of various colours, shades, and high quality (resistance to chemical factors, light, friction).

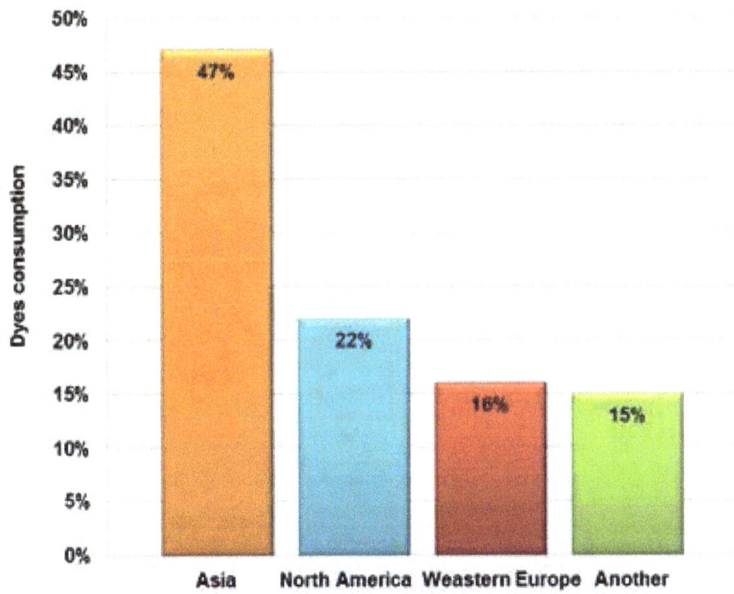

Figure 1. World dyes demand by the regions in 2008

The first synthetic dyes were fuchsine (prepared in 1855 by J. Natanson) and mowein (prepared in 1856 by W. Perkin) [1]. Development of dyes synthesis is promoted by their huge application in many fields of industry. The large tonnage consumer of dyes is textile industry. Moreover, they are applied in electronic industry for the production of liquid crystal and electrochromium visual indicators as well as in technology of optical recording, photographic (for creating coloured pictures), reprography (ink, non-carbon paper), food dyeing as well as for dyeing biological material, paper, leather, wood and cosmetics and also as indicators (e.g. redox, pH).

3. Sources of dyes in wastewaters and wastewater characterization

The basic sources of dyes in industrial wastewaters are complex technological processes in organic dye production plants, textile and paper plants as well as furrier and tanning ones. Waste products containing dyes from plastic production, food processing, petroleum, polygraphic, cosmetic, photographic and electronic industries are in somewhat lower amounts. Wastewaters from intermediate products and organic dyes production usually contain various chemical compounds which occur not only in the form of aqueous solutions but also in liquids, emulsions suspensions and pitch sparsely soluble in water. Besides residues of raw and intermediate materials such as benzene, aniline, phenol, amine, nitro compounds, alcohols, esters, salts, inorganic acids (mainly HNO_3, H_2SO_4), they contain those of ready-made products or dyes. The weight amounts of unreacted substrates, by-products or auxiliary compounds largely exceed the amounts of the main product. They are characterized by intense colouring despite insignificant dye concentration as well as smell and taste due to the presence of nitro compounds. Some aniline dyes are distinctly visible even at the concentration about 40 mg/L [12]. Table 2 presents the characteristics of wastewaters from organic dyes production [12].

Indicator	**Kind of produced dye**			
	Pigment yellow 10G	**Acid orange**	**Lake claret BLC**	**Azo Dyes**
Reaction pH	4.3	8.8	4.7	2.6
COD (mg/L)	-	83,735	2269	19,410
Sulfates (mg SO_4^{2-}/L)	620	1887	16,527	5921
Chlorides (mg Cl^-/L)	975	153,500	17,750	20,500
Dry residue (mg/L)	11,910	393,600	49,400	377,800
Suspensions (mg/L)	3858	-	-	-

Table 2. Characteristics of wastewaters from organic dyes production plant

Amount and composition of wastewaters originated from textile plants depend on many factors, among others, on the kind of fibre or fabric, way of dyeing (kind of used dyes) as well as apparatus. Their largest amounts have arisen during chemical treatment of textile products. Textile plants in the three-shift (24 h/day and night) system equipped with 40 dyeing machines generate enormous amounts of effluents 2400–5200 m^3/day and night assuming the average capacity of dyeing machine 200–6500 L and taking into account 7–10 fillings (baths) of the machine during one dyeing process [13].

Processes of chemical treatment of textile products are a separate branch of producing of above wares likewise mechanical treatment [14]. Their main aim is to give textile products suitable properties facilitating their further treatment and desirable usable features like shape stability, resistance to outer factors' action (washing, friction, dirt, sweat) as well as surface appearance (smooth, creased, shiny, dull, colorful). The above mentioned properties of textile products are obtained during their treatment using various chemical substances: alkalis, acids, salts, surface active substances, oxidizers, reducers, dyes, thickeners, water, solvents and many others. The amounts of substances added to dyeing baths depending on colouring intensity are given in Table 3 [13, 15].

Depth of shade	Dyeing bath composition (g/kg textiles)			
	Dyes	Organic auxiliaries	Inorganic auxiliaries	Electrolytes NaCl, Na_2SO_4
Light shade	0.5–4	0–30	50–250	90–400
Medium shade	5–30		30–150	600–700
Deep shade	30–80	0–35		800–1500

Table 3. Load of impurities carried in dyeing bath

These compounds are a main load in textile wastewaters; and besides dyes, they remain in wastewaters in the concentration close to the initial one (part is deposited on the textile product) as suitable dying conditions are created. The average water consumption in all processes of chemical treatment is from about 150 to over 300 liters per one kilogram of fibre causing formation of the same amount of strongly contaminated wastewaters [14]. Beside dyes and auxiliaries, textile industry wastewaters contain specific contaminations such as fat, wax, dextrin, starch, dressing or casein. Organization for Economic Cooperation Development estimated that 7–20% of acid dyes, 5–20% of direct dyes and 20–50% of reactive dyes were lost in the effluents in Europe [16-19]. A large percentage of pollution generated by the textile industry can be attributed to salts, sizing agents, preparation agents, detergents and organic acids [16], see Figure 2.

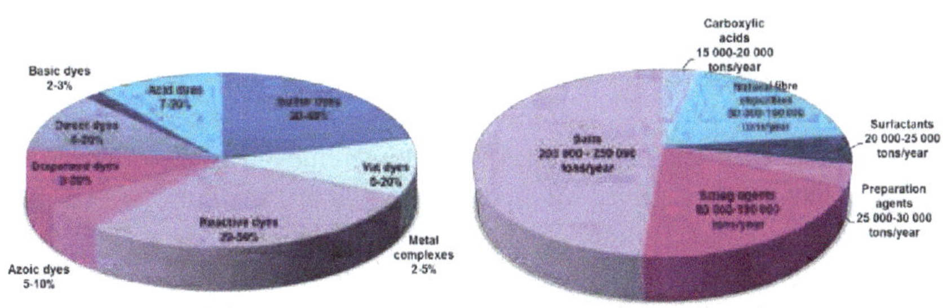

Figure 2. Percentage of non-fixed dyes and auxiliaries that may be discharged in the textile effluent (based on the data presented in the paper [16])

For example, reactive dyeing of 1 kg of cotton requires about 150 L of water, 0.6–0.8 kg of NaCl and about 40 g of reactive dye. One can easily imagine the total amount of generated pollution.

According to Epolito et al. [19] under typical dyeing conditions, up to 50% of the initial dye concentration remains in the spent dye bath in its hydrolyzed form. Thus after dyeing wastewaters are characterized not only by intensive and difficult colour removal but also by high pH, suspended and dissolved solids, chemical and biochemical oxygen demands. Typical characteristics of raw textile wastewaters are presented in Table 4 [12–14, 20, 21].

Indicator	Range of indicator fluctuations in the textile industry wastewater impurities			
	Cotton	Wool	Silk	Flax
Grade of dye	Reactive, vat	Acid	Acid, reactive, direct	Direct
Colour threshold*	1:10–1:250	1:10–1:300	1:8–1:400	1:2–1:150
Reaction pH	6.1–11.2	4.1–8.8	3.7–9.5	2.2–11
BOD (mg O_2/L)	50–620	60–455	50–700	30–1800
Permanganate value (mg O_2/L)	80–450	52–495	32–412	40–3600
Sulfates (mg SO_4^{2-}/L)	50–690	32–423	70–520	35–150
Chlorides (mg Cl^-/L)	80–650	44–530	40–1400	20–360
Sulfides (mg H_2S/L)	2.2–30	1–20	1–40	-
Total chromium (mg Cr/L)	-	1–5	-	-
Detergents (mg/L)	1–70	5–124	4–35	-
Fat (mg/L)	5–15	10–150	-	-
Dry residue (mg/L)	300–3500	478–2120	400–4100	400–2450
Total suspension (mg/L)	50–530	69–403	40–330	20–350
Temperature (°C)	30–50	25–50	36–40	25–45

*Colour threshold – given according to the platinum scale (e.g. in mg Pt/L) or in a descriptive way giving dilution extent at which specific colour disappears

Table 4. Composition of textile wastewaters

Wastewaters coming from tannery, besides impurities removed from leather (hair, blood, epidermis, fat tissue) contain significant amounts of chemicals: sulfuric acid and hydrogen chloride, lime, soda, sodium sulfide, chromium (III) compounds, detergents and organic solvents [14]. Water consumption in tanneries and furriery plants is from 30 to 81 m^3 and from 8.5 to 400 m^3, respectively, for 1000 leathers depending on their kind. Concentration of organic impurities from furriery plants is somewhat smaller than from tannery. Dyes contained in these wastewaters come from tanning, dyeing and rising of raw material and their concentration is 1 kg/m^3 [21]. They constitute 17–32.5% of the total amount of wastewater generated by these plants [12]. The composition of wastewaters from the tannery and furriery plants is presented in Table 5 [21].

Wastewaters from paper factories containing dyes particularly aniline and sulfur ones are generated from the paper treatment process consisting in the addition of fillers and dyes to the bleached cellulose material. As a result of multi-stage purification of wastewaters from the cellulose plants, there is obtained a low value of BOD (biochemical oxygen demand) – 4 mg/L but COD (chemical oxygen demand) is maintained on the level 75 mg/L. These wastewaters contain a small amount of suspended matter – 5 mg/L, but they are characterized by intensive colour – 40 mg Pt/L [21].

Indicator	Tannery	Furriery plant
Reaction pH	7.8–9.8	3–10
Chlorides (mg Cl^-/L)	1600–4000	30000
Sulfates (mg SO_4^{2-}/L)	500–2000	1000
Chromium(III) (mg/L)	30–80	3000
Chromium(VI) (mg/L)	300–500	500
Surface active substances (mg/L)	60–200	2000
COD (mg O_2/L)	1000–9000	-
BOD (mg O_2/L)	500–4500	-

Table 5. Concentration of impurities in wastewaters from the tannery and furriery plants

4. Impact of dyes on natural environment

Wastewaters containing dyes are troublesome in purification processes due to a complex structure of dye molecules. Even small amounts of dyes (of a few ppm order) are undesirable; they colour water, making it look unaesthetic and disturb life processes in water. Most dyes do not undergo biodegradation, deteriorate penetration of light into water and inhibit photosynthesis processes, increase chemical and biological demand for oxygen. Some dyes exhibit toxic and even cancerogenic as well as mutagenic action towards living organisms and therefore they should be carefully removed [5, 17]. Most dyes have a harmful effect (directly or indirectly) on fish. Direct activity consist in colouring of water and changing its composition, which significantly deteriorates the living conditions of fish and plankton, but indirect activity consist in poisonous properties of many dyes. Studies of dyes toxicity for fish seem to be particularly interesting not only with respect for estimation of water purity but also for the fact that they are a valuable source of food for people. Based on the ETAD (Ecological and Toxicological Association of Dyes and Organic Pigments Manufacturers) tests made for 3000 commonly used dyes in 27 cases, the registered LC_{50} (Lethal $Dose_{50}$) values were of the order 0.05 mg/L (compared with, e.g. for DDT [dichlorodiphenyltrichloroethane] the LC_{50} value was 0.006 mg/L) [22]. It proved that 98% of studied dyes reveal LC_{50} toxicity over 1 mg/L. High toxicity was found in C.I. Basic Violet 1 with LC_{50} equal to 0.05 mg/L and C.I. Basic Yellow 37 with LC_{50} equal to 0.8 mg/L [22].

5. Methods of coloured wastewaters purification

The dye removal technologies can be divided into three categories: biological, chemical and physical. Possible decolourization methods of textile wastes along with advantages and disadvantages are listed in Figure 3 [23]. Because of high cost and disposal problems, many of these conventional methods for treating dye wastewaters have not been widely applied on a large scale in the textile industry. At present, there is no single process capable of treatment, mainly due to the complex nature of effluents, combination of the above mentioned techniques provides effective treatment of coloured wastewaters [12, 13, 17]. According to Babu et al. [24], more than 100 references in the bibliographical review of textile wastewater treatment prove that combination techniques permit not only the reduction of suspended solids, organic substances and colour but also the recovery of process chemicals. Currently, the main methods of textile dye treatment are by physical and chemical means with research concentrating on cheap and effective sorbents such as sandy soils, hen feathers, bottom ash, rice husk ash, orange peel, sugarcane dust, etc. However, it should be stressed that they are characterized by relatively low sorption capacity towards dyes compared with activated carbons or ion exchangers and what is more they need to be dumped.

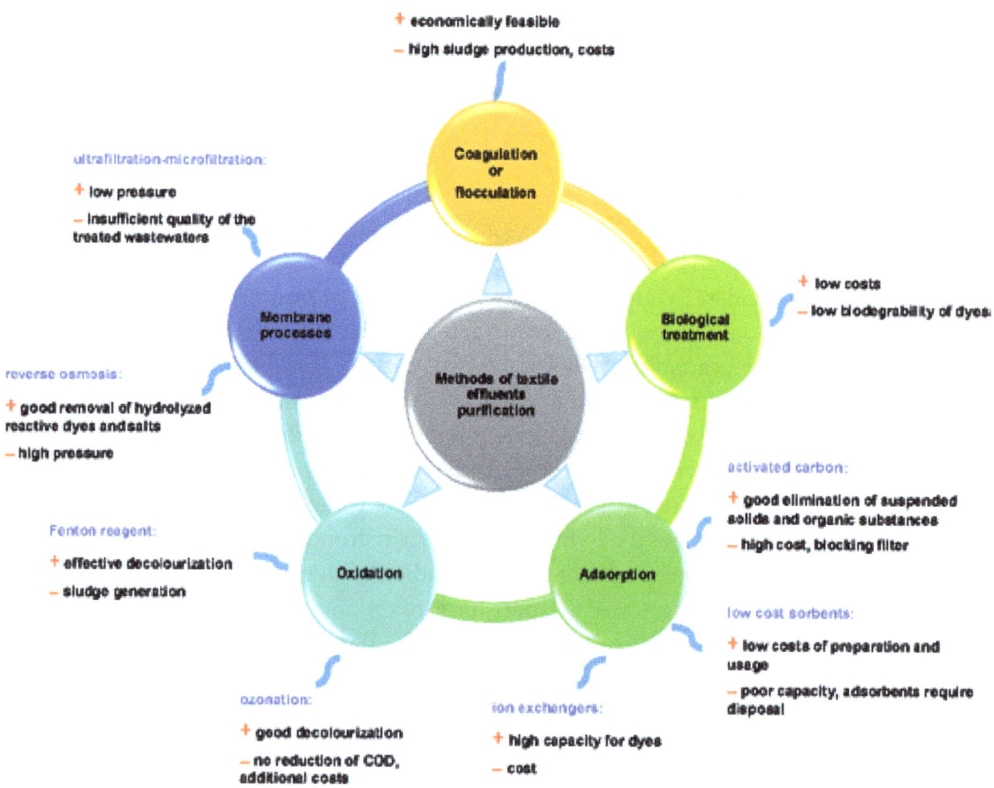

Figure 3. Advantages and disadvantages of the current methods of dye removal from industrial effluents

Ion exchange is a very versatile and effective tool for treatment of aqueous hazardous wastes. The role of ion exchange in dye effluents treatment is to reduce the magnitude of hazardous load by converting them into a form in which they can be reused, leaving behind less toxic substances in their places or to facilitate ultimate disposal by reducing the hydraulic flow of the stream bearing toxic substances. Another significant feature of the ion exchange process is that it has the ability to separate as well as to concentrate pollutants.

Ion exchange resins known as reactive polymers are highly ionic, covalently cross-linked, insoluble polyelectrolytes, usually supplied as beads. Ion exchange resins have been classified based on the charge of the exchangeable counter-ion (cation exchanger or anion exchanger) and the ionic strength of the bound ion (strong exchanger or weak exchanger). Thus, there are four primary types of ion exchange resins: (a) strong cation exchange resins, containing $-SO_3^-H^+$ groups or the corresponding salts, (b) weak cation exchange resins, containing $-COO^-H^+$ groups or the corresponding salts, (c) strong anion exchange resins of quaternary ammonium groups (type I resins contain $-CH_2N(CH_3)_3^+Cl^-$ groups and type II resins contain $-CH_2N(CH_3)_2(CH_2CH_2OH)^+Cl^-$ groups), (d) weak anion exchange resins of primary ($-NH_2$), secondary (=NH) or tertiary-amine (≡N) functional groups in the chloride or hydroxide form.

The resin beads have either a dense internal structure with no discrete pores (gel resins, also called microporous) or a porous, multichannelled structure (macroporous or macroreticular resins). They are commonly prepared from styrene and various levels of the cross-linking agent – divinylbenzene, which controls the particles' porosity. Popular ion exchangers available on the market are those of acrylic, epoxy-amine and phenol-formaldehyde matrices. The common choice is between styrene-divinylbenzene or acrylic-divinylbenzene copolymer. Disregarding structural features (gel or macroporous) for the time being, the acrylic matrix is more elastic than the rigid styrene-based copolymer. However, the elastic resilience of acrylic matrix could be of concern where the columns of resin operate under a high net compression force.

The internal structure of the resin beads, i.e. whether microporous (gel-type) or macroporous, is important in the selection of an ion exchanger. Macroporous resins, with their high effective surface area, facilitate the ion exchange process. They also give access to the exchange sites for larger ions, can be used with almost any solvent, irrespective of whether it is a good one for the uncross-linked polymer, and take up the solvent with little or no change in volume. They make more rigid beads, facilitating the ease of removal from the reaction system. In the case of the microporous resins, since they have no discrete pores, solute ions diffuse through the particle to interact with the exchange sites. Despite diffusion limitations on the reaction rates, these resins offer certain advantages: they are less fragile, requiring less care in handling, react faster in functionalization and application reactions as well as possess higher loading capacities [25].

Taking into account high capacity and selectivity of ion exchange resins for different dyes, they seem to be proper materials for dyes sorption from textile effluents. Applicability of the anion exchange resins in the removal of acid, reactive, direct dyes widely used in the textile industry, from aqueous solutions and wastewaters, was confirmed in some papers [2, 15, 20, 23, 25–28].

6. Experimental

In the paper, the results of the sorption of three textile dyes such as C.I. Acid Orange 7, C.I. Reactive Black 5 and C.I. Direct Blue 71 on the commercially available anion exchangers are summarized and discussed based on the data presented in papers [2, 15, 20, 23, 25–31].

The essential physicochemical properties of these resins (produced by Lanxess, Germany or Dow Chemical Company, USA) are given in Table 6 and Figure 4. The resins were washed with both 1 M HCl and distilled water in order to remove impurities and change the ionic form to the chloride one. The resins were dried at room temperature to the constant mass.

Molecular structures of the above mentioned dyes are presented in Figure 5. The dyes were purchased from Sigma-Aldrich (Germany) and used without further purification. These dyes were selected for the studies because they are extensively used in the textile industry. C.I. Acid Orange 7 is applied for fibers such as silk, wool and nylon using neutral to acid dye baths. Direct Blue 71 is used for cotton, paper, leather, wool, silk and nylon dyeing. Reactive dyes, e.g. RB5, are by far the best choice for dyeing of cotton and other cellulose fibers.

The other chemicals used were produced by Sigma-Aldrich (Germany) and were of analytical grade.

Commercial name	Functional groups	Matrix composition and structure	Total capacity (eq/L)	Operating temp. (°C)	Beads mean size (mm)
Lewatit MonoPlus MP 62	$-N(CH_3)_2$	S-DVB, m	1.7	70	0.47
Amberlyst A 23	$-N(CH_3)_2$	PF, m	>1.8	80	0.47–0.74
Amberlite IRA 67	$-N(CH_3)_2$	A-DVB, g	>1.6	60	0.55–0.75
Lewatit MonoPlus MP 64	$-N(CH_3)_2$ and $-N^+(CH_3)_3$	S-DVB, m	1.3	70	0.59
Amberlite IRA 478RF	$-N(CH_3)_2$ and $-N^+(CH_3)_3$	A-DVB, g	>1.15	35	0.78–0.98
Amberlite IRA 900	$-N^+(CH_3)_3$	S-DVB, m	1.0	60	0.65–0.82
Lewatit MonoPlus MP 500	$-N^+(CH_3)_3$	S-DVB, m	>1.1	45	0.63
Lewatit MonoPlus M 500	$-N^+(CH_3)_3$	S-DVB, g	1.1	70	0.64
Amberlite IRA 458	$-N^+(CH_3)_3$	A-DVB, g	1.25	35	0.6–0.9
Amberlite IRA 958	$-N^+(CH_3)_3$	A-DVB, g	>0.8	80	0.63–0.85
Amberlite IRA 910	$-N^+(CH_3)_2C_2H_4OH$	S-DVB, m	1.1	60	0.53–0.8
Lewatit MonoPlus M 600	$-N^+(CH_3)_2C_2H_4OH$	S-DVB, g	1.3	30	0.62

where: [green] - weakly basic anion exchangers, [blue] - intermediate base anion exchangers, [yellow] - strongly basic anion exchangers, S-DVB – styrene-divinylbenzene, A-DVB – acrylic-divinylbenzene, PF- phenol-formaldehyde, m – macroporous, g - gel

Table 6. Properties of applied anion exchangers

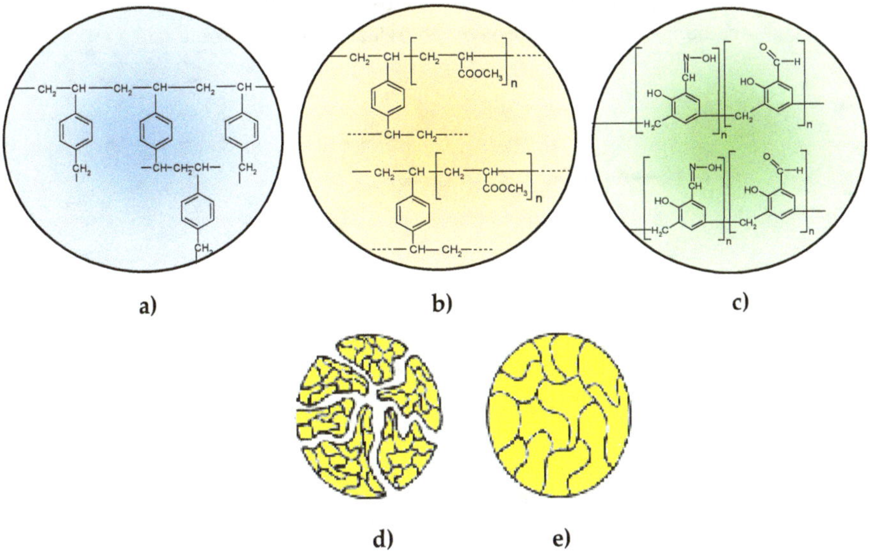

Figure 4. Composition of resin matrices: a) styrene-divinylbenzene skeleton, b) acrylic-divinylbenzene skeleton, c) phenol-formaldehyde skeleton, and their structure: d) macroporous, e) gel

The sorption studies were performed by the batch method. The dye solutions (50 mL) were shaken with the dry anion exchanger (0.5 g) in conical flasks using a thermostated laboratory shaker Elphin (Poland) at 20°C. The experiments were conducted in the two parallel series with the reproducibility 5%. The amount of dye adsorbed after time t, q_t (mg/g), was calculated from Equation 1:

$$q_t = \frac{(C_0 - C_t)}{w} \times V \tag{1}$$

where: C_0 and C_t (mg/L) are the liquid-phase concentrations of dye at the time t=0 and after time t, respectively, V (L) is the volume of solution and w (g) is the mass of dry anion exchanger.

To test the influence of shaking speed on dye removal, preliminary experiments were carried out by varying the shaking speed from 140 to 200 rpm. The best results were obtained for the shaking speed 180 rpm. Therefore, 180 rpm was used in all batch experiments. To evaluate the kinetics of the sorption process, 50 mL solutions of 100 mg/L (or 500 mg/L or 1000 mg/L) dye concentration and 0.5 g of the anion exchanger samples were used. The shaking time was varied from 1 to 12 h, respectively (e.g. up to equilibrium). All the kinetic studies were carried out at the natural pHs (pH 4.98–5.83) of solutions (pH-meter; CX-742 Elmetron, Poland). The dyes concentration after the sorption was measured spectrophotometrically at the maximum absorbance wavelengths. Absorption spectra of raw textile wastewaters of different composition were recorded for the predetermined time interval of decolourization using a spectrophotometer Specord M-42 (Carl Zeiss, Germany).

C.I. Acid Orange 7; C.I. No 15510; MW = 350.32 g/mol
sodium salt of 4-(2-hydroxynaphthylazo)benzenesulfonic acid

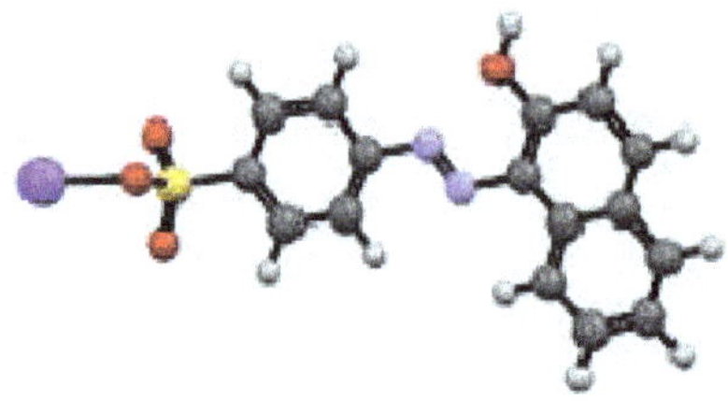

C.I. Reactive Black 5; C.I. No 20505; MW = 991.82 g/mol
tetrasodium salt of 4-amino-5-hydroxy-3,6-bis((4-((2-(sulfooxy) ethyl)sulfonyl)phenyl)azo)-2,7-naphthalenedisulfonic acid

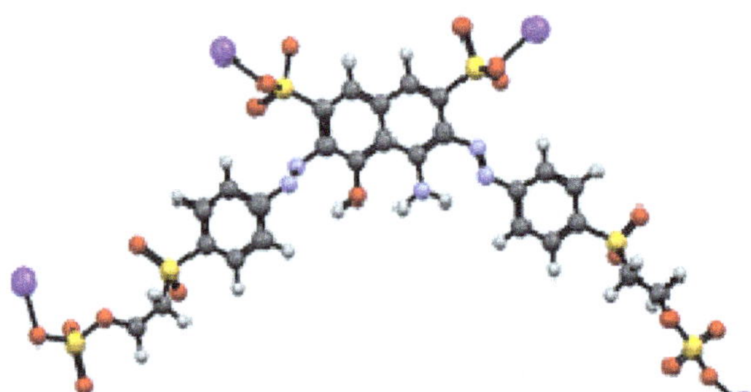

C.I. Direct Blue 71; C.I. No 34140; MW = 1029.88 g/mol
tetrasodium 3-[(E)-{4-[(E)-{4-[2-(6-amino-1-oxo-3-sulfonatonaphthalen-2(1H) ylidene)hydrazino]-6-sulfonatonaphthalen-1-yl}diazenyl]naphthalen-1-yl}diazenyl]naphthalene-1,5-disulfonate

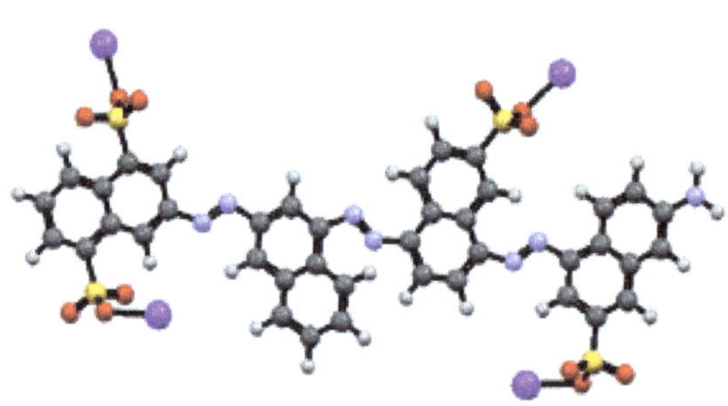

Figure 5. Dyes characteristics

Sorption isotherm studies were carried out analogously using dyes solutions of the increasing initial concentration at 20°C for 24 h. The effects of salts and surfactant additions on dyes' uptake at equilibrium were studied by shaking the anion exchanger (0.5 g) with the 100 mg/L dye solution containing different amounts of salts (Na_2SO_4, Na_2CO_3, NaCl) or surfactants (sodium dodecyl sulfate (SDS) and cetyltrimethylammonium bromide (CTAB)). The dyes concentration after the sorption was measured spectrophotometrically at the maximum absorbance wavelengths depending on the system.

Regeneration tests for the anion exchange resin were conducted with different regeneration agents (1 M HCl, 1 M NaOH, 1 M KSCN, 1 M NaCl, 1 M Na_2SO_4, 1 M Na_2CO_3, 90% methanol, 1 M KSCN or 1 M NaOH or 1 M HCl in 90% methanol). The loaded resin containing 10 mg/g of dye was put into flasks in contact with 50 mL of different eluting agents. The flasks were agitated for 3 h and the dye concentrations in the solution were determined at the maximum absorbance wavelength in order to calculate the desorption percentage (%).

The above methods were described in detail in [2, 15, 20, 23, 25–31].

7. Results and discussion

7.1. Adsorption equilibrium and retention mechanisms

The equilibrium sorption isotherms are one of the most important data to understand the mechanism of the sorption. They describe the ratio between the quantity of sorbate retained by the sorbent and that remaining in the solution at the constant temperature at equilibrium and are important from both theoretical and practical points of view. The parameters obtained from the isotherm models provide important information not only about the sorption mechanisms but also about the surface properties and affinities of the sorbent. The best known adsorption models in the linearized form for the single-solute systems are:

a. the Langmuir isotherm

$$\frac{C_e}{q_e} = \frac{1}{Q_0 b} + \frac{C_e}{Q_0} \tag{2}$$

where Q_0 (mg/g) is the Langmuir monolayer sorption capacity, b (L/mg) is the Langmuir adsorption constant, calculated from the intercepts and slopes of straight lines of plot of C_e/q_e vs C_e.

The Langmuir isotherm is applied to homogeneous adsorption based on the following assumptions: (a) all the adsorption sites are identical; (b) each site retains one molecule of the given compound; (c) all sites are energetical and sterical independent of the adsorbed quantity [32].

b. the Freundlich isotherm

$$\log q_e = \log k_F + \frac{1}{n}\log C_e \tag{3}$$

where: k_F (mg/g) is the Freundlich adsorption capacity, n is the Freundlich constant related to the surface heterogeneity, determined from the slope and intercept of the linear plot of $log\ q_e$ vs $log\ C_e$.

The Freundlich isotherm assumes heterogeneous surface with a non-uniform distribution of heat of adsorption [33].

The applicability of isotherm equations was compared in this paper by judging the determination coefficients (R^2). The calculated constants of these models were largely dependent on the type of the anion exchanger (Tables 7–9). Not only the basicity of the resin but also the matrix composition and structure were the determining factors of sorption. It could be stated generally that the Langmuir isotherm model fits better than the Freundlich one. However, the comparison of the R^2 values indicates that the experimental data of C.I. Direct Blue 71 adsorption onto Amberlite IRA 67, Amberlite IRA 900 and Lewatit MonoPlus MP 62 were more appropriate for the Freundlich isotherm model. It could be in agreement with the aggregation tendency of the C.I. Direct Blue 71 dye, especially at high concentrations, and with the possibility of the dye uptake through many types of interactions. Similar observations were described for C.I. Basic Blue 9 sorption onto Purolite C 145 and Purolite C 107E resins by Suteu et al. [34].

Anion exchanger	Langmuir isotherm model			Freundlich isotherm model			Ref.
	Q_0 (mg/g)	b (L/g)	R^2	k_F	n	R^2	
Lewatit MonoPlus MP 62	79.9	0.03	0.993	43.2	9.2	0.896	[27]
Amberlyst A 23	80.2	0.03	0.996	39.7	8.6	0.872	This study
Amberlite IRA 67	168.2	0.002	0.996	41.2	7.3	0.892	[29]
Lewatit MonoPlus MP 64	586.5	0.68	0.999	168.8	5.6	0.794	[27]
Amberlite IRA 478RF	1279.2	0.166	0.999	213.0	2.95	0.996	[25]
Amberlite IRA 900	1289.7	0.007	0.999	207.6	0.421	0.943	[26]
Lewatit MonoPlus MP 500	979.0	0.17	0.993	157.7	2.5	0.956	[27]
Lewatit MonoPlus M 500	215.9	0.129	0.999	154.7	21.0	0.427	This study
Amberlite IRA 458	1211.3	0.066	0.999	198.6	0.221	0.842	[29]
Amberlite IRA 958	1370.4	0.063	0.999	161.7	0.345	0.872	[30]
Amberlite IRA 910	1012.6	0.015	0.998	159.1	0.521	0.900	[26]
Lewatit MonoPlus M 600	174.1	0.987	0.998	38.3	36.9	0.789	This study

Table 7. Comparison of the isotherms parameters for C.I. Acid Orange 7 sorption on the studied anion exchangers

Anion exchanger	Langmuir isotherm model			Freundlich isotherm model			Ref.
	Q_0 (mg/g)	b (L/g)	R^2	k_F	n	R^2	
Lewatit MonoPlus MP 62	796.1	0.049	0.990	183.5	4.96	0.925	[23]
Amberlyst A 23	282.1	0.204	0.999	125.6	8.77	0.937	[23]
Amberlite IRA 67	66.4	0.503	0.998	26.4	7.10	0.914	[23]
Lewatit MonoPlus MP 64	592.7	0.242	0.999	84.3	3.45	0.717	This study
Amberlite IRA 478RF	150.4	0.057	0.999	93.0	16.1	0.906	[25]
Amberlite IRA 900	1351.8	0.039	0.999	179.4	6.92	0.845	[31]
Lewatit MonoPlus MP 500	1170.5	0.416	0.996	312.1	3.24	0.955	This study
Lewatit MonoPlus M 500	17.4	0.099	0.998	3.83	3.69	0.789	This study
Amberlite IRA 458	1329.5	0.037	0.999	241.3	4.92	0.823	[30]
Amberlite IRA 958	1655.2	0.468	0.999	263.2	3.21	0.815	[30]
Amberlite IRA 910	1219.9	0.021	0.998	172.4	4.62	0.794	[30]
Lewatit MonoPlus M 600	4.6	0.236	0.896	0.15	1.2	0.629	This study

Table 8. Comparison of the isotherms parameters for C.I. Reactive Black 5 sorption on the studied anion exchangers

Anion exchanger	Langmuir isotherm model			Freundlich isotherm model			Ref.
	Q_0 (mg/g)	b (L/g)	R^2	k_F	n	R^2	
Lewatit MonoPlus MP 62	470.6	0.0003	0.874	1.47	0.86	0.931	[20]
Amberlyst A 23	456.2	0.079	0.994	22.2	2.69	0.645	This study
Amberlite IRA 67	90.9	0.005	0.909	1.02	1.57	0.998	[20]
Lewatit MonoPlus MP 64	420.4	0.021	0.997	24.2	2.35	0.510	[20]
Amberlite IRA 478RF	41.8	0.0047	0.974	4.67	3.6	0.969	[25]
Amberlite IRA 900	778.2	0.0001	0.828	0.21	1.18	0.901	[20]
Lewatit MonoPlus MP 500	523.6	0.0023	0.959	2.65	1.39	0.874	[20]
Lewatit MonoPlus M 500	2.02	0.165	0.692	0.14	1.06	0.509	[20]

Anion exchanger	Langmuir isotherm model			Freundlich isotherm model			Ref.
	Q_0 *(mg/g)*	*b (L/g)*	R^2	k_F	*n*	R^2	
Amberlite IRA 458	48.4	0.034	0.982	5.6	2.84	0.650	[20]
Amberlite IRA 958	1630.6	0.0001	0.987	0.65	1.26	0.965	[20]
Amberlite IRA 910	663.7	0.008	0.945	0.19	1.02	0.900	This study
Lewatit MonoPlus M 600	1.89	0.154	0.713	0.11	1.01	0.603	This study

Table 9. Comparison of the isotherms parameters for C.I. Direct Blue 71 sorption on the studied anion exchangers

Based on the values of the monolayer sorption capacities determined from the Langmuir isotherm model, the studied anion exchangers can be arranged in the following sorption effectiveness series as:

- for C.I. Acid Orange 7

Amberlite IRA 958 > Amberlite IRA 900 > Amberlite IRA 478RF > Amberlite IRA 458 > Amberlite IRA 910 > Lewatit MonoPlus MP 500 > Lewatit MonoPlus MP 64 > Lewatit MonoPlus M 500 > Lewatit MonoPlus M 600 > Amberlite IRA 67 > Amberlyst A 23 ≈ Lewatit MonoPlus MP 62

- for C.I. Reactive Black 5

Amberlite IRA 958 > Amberlite IRA 458 > Amberlite IRA 900 > Amberlite IRA 910 > Lewatit MonoPlus MP 500 > Lewatit MonoPlus MP 62 > Lewatit MonoPlus MP 64 > Amberlyst A 23 > Amberlite IRA 478RF > Amberlite IRA 67 > Lewatit MonoPlus M 500 > Lewatit MonoPlus M 600

- for C.I. Direct Blue 71

Amberlite IRA 958 > Amberlite IRA 900 > Amberlite IRA 910 > Lewatit MonoPlus MP 500 > Lewatit MonoPlus MP 62 > Amberlyst A 23 > Lewatit MonoPlus MP 64 > Amberlite IRA 67 > Amberlite IRA 458 > Amberlite IRA 478RF > Lewatit MonoPlus M 500 ≈ Lewatit MonoPlus M 600

as far as their applicability is concerned in removal of these dyes.

Under experimental conditions, sorption of the dyes occurred between the sulphonic groups of dyes (e.g. $R_1(SO_3^-)_4(Na^+)_4$) and the functional groups of the weakly ($\overline{RCH_2NH(CH_3)_2{}^+CI^-}$) or strongly ($\overline{RCH_2N(CH_3)_3{}^+CI^-}$) basic anion exchangers in the chloride form [25]:

$$4\overline{RCH_2NH(CH_3)_2{}^+CI^-} + R_1(SO_3^-)(Na^+) \rightleftarrows \overline{[RCH_2NH(CH_3)_2^+]_4}(SO_3^-)_4R_1 + 4NaCl$$

$$4\overline{RCH_2N(CH_3)_3{}^+CI^-} + R_1(SO_3^-)(Na^+) \rightleftarrows \overline{[RCH_2N(CH_3)_3^+]_4}(SO_3^-)_4R_1 + 4NaCl$$

The dye anions ($R_1(SO_3^-)_4$) replaced exchangeable chloride anions which compensates positive electric charge of the tertiary amine or quaternary ammonium groups of the anion exchanger. The ion pairs are formed between these groups. Such interactions were revealed during the analysis of the ATR FT-IR spectra of the dye loaded anion exchangers, the absorption peaks at 1170–1047 nm and 1019 nm were attributed to the presence of $-SO_3^-$ and –S=O groups [20]. In the FT-IR spectrum of the weakly basic anion exchanger Purolite A 847 with loaded azo- (Lanasyn Navy M-DNL) and phthalocyanine (C.I. Acid Blue 249) dyes, the vibrations at 1042 nm and 1032 nm related to $-SO_3^-$ groups were detected by Kaušpėdienė et al. [35]. Although ion exchange is a significant mechanism in dyes sorption, some extent of physical adsorption can also occur. The attachment through hydrophobic π–π interactions between the aromatic rings of the dye and the matrix of the anion exchanger ('like attracts like') is considered. These interactions play a more extensive role in the case of the polyacrylic resins like Amberlite IRA 958, IRA 458 or IRA 67 compared to those of polystyrene-based materials. High affinity of dyes for the anion exchangers can also result from the formation of H-bondings, which can be created between nitrogen of the quaternary ammonium groups of strongly basic anion exchange resins and nitrogen of the —NH_2 group of dyes as well as oxygen of the carbonyl group of the resins and the —NH_2 group of dye. Also oxygen atom of the carbonyl group of anion exchangers and oxygen atom of the hydroxyl group or nitrogen atom of the azo group of dyes through the water molecules could interact. Possible interactions between the anions of C.I. Reactive Black 5 and the strongly basic anion exchangers of polyacrylic matrices are shown in Figure 6. Kaušpėdienė et al. [35, 36] also observed that more than one interaction was involved in dyes sorption on Purolite A 847 of polyacrylic matrix: ion exchange and nonelectrostatic interactions. The studies on the sorption of dyes and organic compounds on the ion exchangers reported so far showed that the size of the sorbate molecules has a considerable effect on the sorption degree. Organic dye molecules of different positions of sulfonic groups as well as the number of other anionic groups and their charges can interact in a different way with anion exchangers. Of significant importance in the removal of dye anions by an anion exchanger is the resin structure. Concerning the exchange of large molecular weight species like dyes, the macroporous property becomes important in providing an easier diffusion path for uptake compared with the gel structure. Amberlite IRA 958 as the anion exchanger with the macroporous structure possesses significant porosity in comparison with that of gel Amberlite IRA 458 or IRA 67. Therefore, sorption capacity of Amberlite IRA958 is much more bigger than that of Amberlite IRA 458 of the same functional groups in the case of C.I. Direct Blue 71 sorption. Taking into consideration the strongly basic anion exchangers Lewatit MonoPlus M 500 and Lewatit MonoPlus M 600 of polystyrene matrix and gel structure, a very low capacity is observed. Additionally, flat structure of the reactive and direct dyes can inhibit the interactions between the dyes and the anion exchangers. Also when the dye anions are too large, they are excluded because of resin structure ('sieve effect'). Besides, because large size dyes have the tendency to form aggregates in the solution rather than in the resin phase [25].

7.2. Influence of auxiliaries such as salts and surfactants

Auxiliaries such as inorganic electrolytes and surfactants are widely used in chemical treatment of fibers. They remain in wastewaters at a concentration close to the initial one. Depending on the class of dye, dyeing process requires application of other auxiliaries. Acid dyes are the most numerous group of those known. They occur in the form of sodium salts of coloured compounds containing 1–3 sulfonic or carboxylic groups. They belong to strong electrolytes, undergo complete dissociation in water into coloured anions. The condition of their binding with fibre is forming a sufficient number of the ammonium groups in wool which can be obtained by the addition of acids to dyeing baths, hence it is said 'they dye in the acid bath'. Disposal of the carboxylic acids to the textile sewages in Europe amounts from 15, 000 to 20, 000 tons/year [16].

Acid dyes are divided into three groups differing in dyeing conditions. The first group includes so-called well equalizing dyes that is providing uniform level dyeing of a relatively small molecular mass. They are used for dyeing in acid bath of pH 2–2.5 with the addition of strong acids, e.g. H_2SO_4.

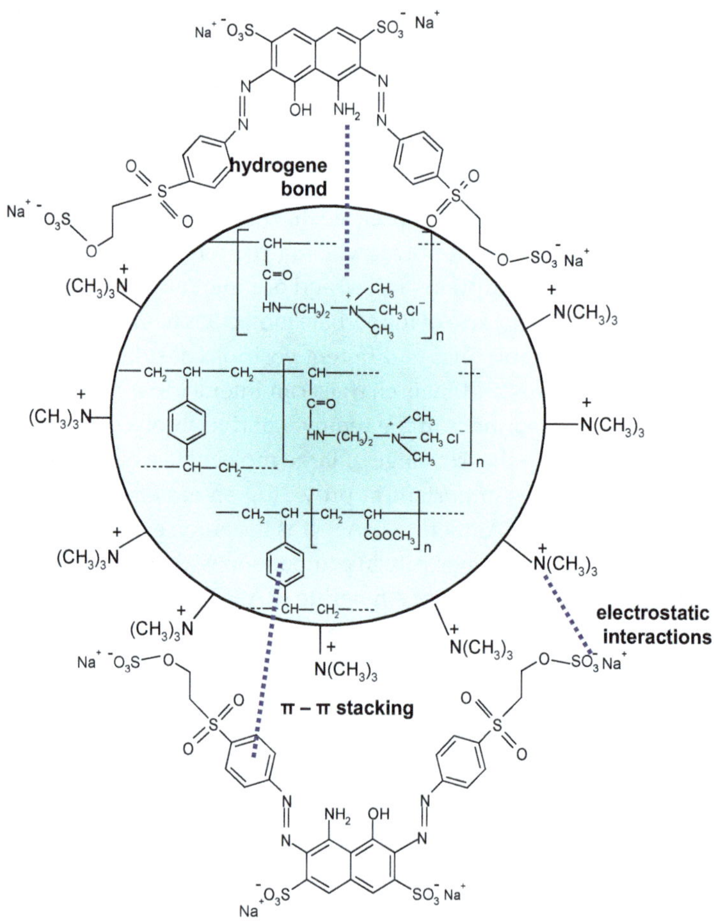

Figure 6. Mechanism of interactions between C.I. Reactive Black 5 and strongly basic anion exchanger of the polyacrylic matrix

The second group includes dyes of more developed molecules and greater affinity for wool. They are used for dyeing in the baths of pH 4.5–5 acidified with acetic acid. The third group has badly equalizing dyes of largely expended molecules and great affinity for wool. They are applied for dyeing in the baths of pH 5.5–6.5 in the presence of ammonium salts (sulfate, acetate), which decompose only at increased temperature acidifying the bath. Usually, sodium sulfate is added to the dyeing bath in the amount 10–20% in the proportion to the fibre and 1–2% of formic, acetic or sulfuric acid.

The influence of inorganic salts such as NaCl and Na_2SO_4 on the sorption of C.I. Acid Orange 7 from the solution of the initial concentration 100 mg/L was studied. The results indicate that the presence of these salts in the whole examination concentration of 1–25 g/L does not affect the dye adsorption significantly. In the case of the increasing amount of sodium sulfate in the solution from 1 to 25 g/L, the anion exchange capacities of the polyacrylic anion exchangers of the weakly basic functional groups Amberlite IRA 67 and of the intermediate base Amberlite IRA 478RF dropped from 9.98 mg/g to 9.2 mg/g and from 9.98 to 8.9 mg/g, respectively. The effect of Na_2SO_4 concentration on C.I. Acid Orange 7 removal using the strongly basic anion exchangers of the polyacrylic skeleton (Amberlite IRA 458 and Amberlite IRA 958) was insignificant. Similar observation of the effect of salts addition such as Na_2HPO_4 and NaH_2PO_4 was observed by Greluk and Hubicki [29]. In the presence of NaH_2PO_4 and Na_2HPO_4, the C.I. Acid Orange 7 retention by Amberlite IRA 958, in the system 200 mg/L dye and 0.1–2.0 g/L salts was not affected. Using weakly (Lewatit MonoPlus MP 62), intermediate (Lewatit MonoPlus MP 64) and strongly (Lewatit MonoPlus MP 500, Lewatit MonoPlus M 500, Lewatit MonoPlus M 600, Amberlite IRA 900 and Amberlite IRA 910) basic anion exchangers of the polystyrene–divinylbenzene matrix negligible influence of Na_2SO_4 addition in the range of 1–25 g/L on C.I. Acid Orange 7 uptake was observed. Quantitative removal of C.I. Acid Orange 7 of the initial concentration 200 mg/L from the system containing 0.1–2.0 g/L of NaH_2PO_4 and Na_2HPO_4 using strongly basic anion exchangers of type 1 Amberlite IRA 900 and of type 2 Amberlite IRA 910 was reported by Greluk [30].

The removal of C.I. Acid Orange 7 in the presence of sodium chloride even at relatively high concentration of 50 g/L was quantitative using all applied anion exchangers.

The removal of C.I. Acid Orange 7 from the solutions containing from 0.5 to 2.0 g/L of acetic acid was also studied. As shown in Figures 7 and 8, the amounts of the dye retained by the polystyrene and polyacrylic anion exchangers of various basicity decrease with the increasing amount of CH_3COOH in the system. The differences between the sorption capacities determined in the systems without the acetic acid and in the systems containing this acid do not exceed 15%. The same behaviour of the phenol–formaldehyde anion exchanger Amberlyst A 23 towards C.I. Acid Orange 7 was observed in the presence of the acid.

Direct dyes like the acid ones belong to strong electrolytes and are completely dissociated in water baths into coloured anions and sodium cations:

$$D^{Z-}(Na^+)_Z \leftrightarrow D^{Z-} + zNa^+$$

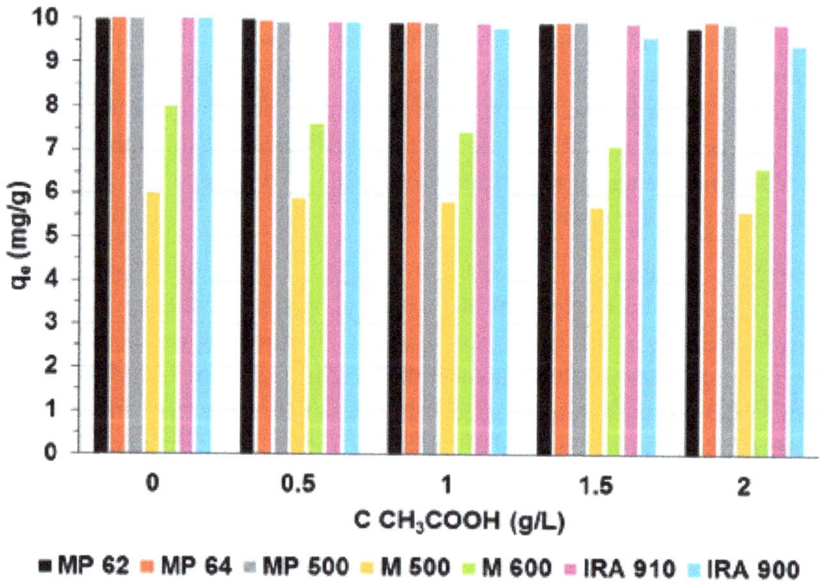

Figure 7. Influence of acetic acid concentration on C.I. Acid Orange 7 uptake by the anion exchangers of the polystyrene matrix

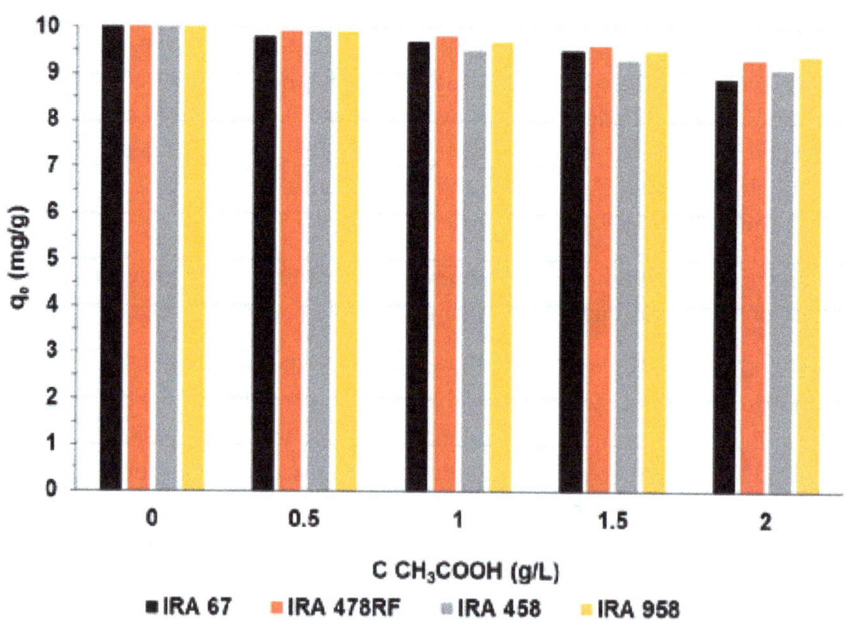

Figure 8. Influence of acetic acid concentration on C.I. Acid Orange 7 uptake by the anion exchangers of the polyacrylic matrix

where D^{z-} – the dye anion, z – the numer of sulfonic groups.

Flat structure and large molecular mass (usually 600-1000) of direct dyes make their tendency to form associated ions (colloidal electrolytes):

$$nD^{Z-} \leftrightarrow (D^{Z-})_n$$

where n – the association degree [14].

Association degree decreases with the increasing temperature. Alkalizing bath also promotes decomposition of associates; therefore, sodium carbonate is often added to the dyeing bath. Large negative charge of direct dye anions (2–4 sulfonic groups) causes that in the water bath they are repelled by the fibre surfaces of the negative electrokinetic potential dzeta. The addition of electrolyte, most frequently sodium sulfate or sodium chloride, decreases the negative potential dzeta facilitating the access of dye anions to the fibre surface. The dyeing bath contains (in the percentage of the dyeing product amount) from 0.5% to 2% Na_2CO_3 and 4–30% Na_2SO_4 depending on the method of dyeing and intensity of colour [14]. Removal of C.I. Direct Blue 71 from the systems containing 100 mg/L of dye and 1–25 g/L of NaCl and Na_2SO_4 on the weakly, intermediate and strongly basic anion exchangers was broadly described in the papers [20] and [25]. For the intermediate (Amberlite IRA 478RF) and strongly basic (Amberlite IRA 958 and IRA 458) anion exchangers of the polyacrylic matrix, the presence of NaCl and Na_2SO_4 in the whole examined concentration range of 1–25 g/L did not influence the adsorption capacities. The dye sorption was quantitative. The above mentioned anion exchangers have the same constitution of matrix, but different structure (gel or macroporous). Amberlite IRA 67 being of the same constitution of matrix and gel structure but of the tertiary amine functionalities exhibited insignificant drop of the sorption capacity with the increasing amount of electrolytes [20, 25]. It can be concluded that in the case of the polyacrylic anion exchange resins of different basicity, no significant influence of matrix structure was observed. Insignificant drop of the anion exchange capacities of Amberlyst A 23 towards C.I. Direct Blue 71 with the increasing amount of electrolytes was noticed, too.

The effect of the presence of NaCl and Na_2SO_4 on C.I. Direct Blue 71 sorption on the gelular polystyrene strongly basic anion exchangers of type I (Lewatit MonoPlus M 500) and type II (Lewatit MonoPlus M 600) was rather negligible. For Lewatit MonoPlus M 500, the increase in the amount of dye adsorbed at the equilibrium from 2.2 to 3.0 mg/g and from 2.2 to 2.8 mg/g was observed with the increasing NaCl and Na_2SO_4 concentration from 1 to 25 g/L, respectively. In the systems of the composition 100 mg/L – 1–25 g/L electrolyte – Lewatit MonoPlus M 600, the values of q_e ranged from 1.3 to 2.6 mg/g and from 1.3 to 2.4 mg/g with ranging concentrations of NaCl and Na_2SO_4, respectively. Gel structure of both anion exchangers hindered diffusion of C.I. Direct Blue 71 anions to the pores and interactions with the $-N^+(CH_3)_3$ or $-N^+(CH_3)_2C_2H_4OH$ functional groups. The type of the functional groups of these strongly basic anion exchangers is of no significance as far as the removal of C.I. Direct Blue 71 is concerned.

A remarkable increase of adsorption capacities was observed during the dye uptake by the polystyrene anion exchangers of macroporous structure and different basicity of functional groups. The dye uptake by the weakly basic anion exchanger Lewatit MonoPlus MP 62 was increased in the presence of salts, the values of q_e increase from 5.86 mg/g to 9.9 mg/g with the increasing concentration of both electrolytes. The sorption enhancement of Lewatit MonoPlus MP 64 was observed from 1.9 to 9.97 mg/g and from 1.9 to 9.95 mg/g with the increasing concentration of NaCl and Na_2SO_4, respectively. At the equilibrium, the sorption capacities of Lewatit MonoPlus MP500 increase from 1.89 to 9.98 mg/g and from 1.89 to 9.95 mg/g with the increasing amount of NaCl and Na_2SO_4 in the range of 1–25 g/L, respectively. For Amberlite IRA 900 of the $-N^+(CH_3)_3$ functional groups and Amberlite IRA 910 of the $-N^+(CH_3)_2C_2H_4OH$ functional groups, sorption capacities increased from 6.9 to 9.9 mg/g and from 5.9 to 9.2 mg/g with the increasing amount of NaCl in the solution. Similar dependences were noticed in the presence of Na_2SO_4, the addition of 25 g/L Na_2SO_4 enhanced the sorption yield of Amberlite IRA 900 and IRA 910 by about 36% and 30% compared with the systems without this salt.

Reactive dyes contain in their structure a system of atoms which can form covalent bonds with hydroxyl groups of cellulose. Hydroxyl groups of cellulose react with reactive systems in dye molecules according to the nucleophilic substitution mechanism (e.g. dyes with the symmetric triazine system) or nucleophilic attachment (residue of ethylsulfone sulfates). The alkaline medium (5–20 g/L Na_2CO_3) is the condition for the dye reaction with a fibre. To enhance it, the electrolyte in the form of sodium sulfate in the amount from 15 to 100 g/L depending on a dye is added to the bath [14]. The effect of sodium sulfate and sodium carbonate on the reactive dye adsorption on the anion exchangers was investigated in the salts concentration range 1–25 g/L with the constant initial C.I. Reactive Black 5 concentration of 100 mg/L.

Increasing amount of sodium sulfate and sodium carbonate in the system, insignificant decrease of the sorption capacities of Amberlite IRA 67, Amberlyst A 23, Lewatit MonoPlus MP 62 and Lewatit MonoPlus MP 64 were observed. Our previous studies revealed that in the 1000 mg/L C.I. Reactive Black 5 – 50-100 g/L Na_2SO_4 (or Na_2CO_3) system uptake of the reactive dye slightly increased by Amberlite IRA 67 [23]. C.I. Reactive Black 5 retention by the Amberlite IRA 478 RF, Amberlite IRA 458, Amberlite IRA 958 of the polyacrylic matrix and by Amberlite IRA 900 and IRA 910, Lewatit MonoPlus MP 500 of the polystyrene matrix was not affected in the presence of Na_2SO_4 and Na_2CO_3. It was reported by Greluk [30, 31] that the sorption capacities of Amberlites IRA 458 and IRA 958 devaluate about 2% in the presence of 100 mg/L of Na_2CO_3. The presence of NaCl in the range 25–100 g/L in the solution containing 200 mg/L of C.I. Reactive Black 5 caused significant decrease in the amount of the dye retained by Amberlite IRA 478RF according to Wawrzkiewicz [20]. The addition of Na_2SO_4 and Na_2CO_3 in the amount of 25 g/L enhanced adsorption capacities of Lewatit MonoPlus M 500 to C.I. Reactive Black 5 from 1.3 to 1.69 mg/g and from 1.3 to 1.72 mg/g, respectively. There was also noticed the increase of the amount of dye sorbed by Lewatit MonoPlus M 600 from the values 0.87 to 1.1 mg/g and from 0.87 to 1.2 mg/g with the increasing amount of sodium sulfate and sodium carbonate from 1 to 25 g/L, respectively.

Summing up, it could be stated that:

- Uptake of C.I. Acid Orange 7, C.I. Reactive Black 5 and C.I. Direct Blue 71 by some anion exchangers of different matrices and functionalities was not affected by the presence of inorganic electrolytes such as NaCl, Na_2SO_4 and Na_2CO_3
- Insignificant decrease of the sorption capacities was observed with the increasing amount of above mentioned salts as a consequence of competition in sorption between the dye anions and the chloride, sulfate and carbonate anions. Decrease of the sorption capacities in some cases is reflected in the affinity series of the anions towards the weakly basic anion exchangers:

$F^- < ClO_3^- < BrO_3^- < HCOO^- < IO_3^- < CH_3COO^- < H_2PO_4^- < HCO_3^- < Cl^- < CN^- < NO_2^- < Br^- < NO_3^- < HPO_4^{2-} < SO_3^{2-} < SO_4^{2-} < C_2O_4^{2-} < CrO_4^{2-} < MoO_4^{2-} < WO_4^{2-} < S_2O_3^{2-} < I^- < SCN^- < ClO_4^- <$ salicylate < citrate $< OH^-$

and towards the strongly basic anion exchangers:

$OH^- < F^- < ClO_3^- < BrO_3^- < HCOO^- < IO_3^- < CH_3COO^- < H_2PO_4^- < HCO_3^- < Cl^- < CN^- < NO_2^- < Br^- < NO_3^- < HPO_4^{2-} < SO_3^{2-} < SO_4^{2-} < C_2O_4^{2-} < CrO_4^{2-} < MoO_4^{2-} < WO_4^{2-} < S_2O_3^{2-} < I^- < SCN^- < ClO_4^- <$ salicylate < citrate

- Increasing salts concentration in the system, enhancement of the reactive and direct dyes occurred. The salting out effect reduces solubility of the dyes in the aqueous phase and promotes their sorption onto the hydrophobic part of sorbent. Kind of salts has a great effect on salting out. This phenomenon depends on the size of ions, their effective charge and ability of hydrates formation.
- The affinity of large organic anions such as dyes anions for the resins is influenced not only by the anion charge but also by the structure of the anion and its size.
- The type of functional groups of the anion exchangers, matrix structure (gel or macroporous) and composition play an important role in the dyes solution treatment in the presence of salts. The sieve effect is of significance when the sorption of the dyes anions is considered in the presence of chloride, sulfate, and carbonate anions.

Wastewater and after-process water from textile plants can contain surface-active substances. Surfactants are often applied as wetting, penetrating, dispersing and leveling agents in dyeing processes. They increase the solubility of dyes in water, in order to improve the dye uptake and dye fastness. Many articles published recently about the treatment of textile wastewaters focused on dyes removal and often neglected the influence of surfactant. Effect of the addition of such surfactants as sodium dodecyl sulfate (SDS) or cetyltrimethylammonium bromide (CTAB) on the dye removal by the used anion exchangers was studied from the systems containing 100 mg/L of dye in the presence of 0.1–1 g/L SDS or CTAB.

Investigating the uptake of C.I. Acid Orange 7, C.I. Reactive Black 5 and C.I. Direct Blue 71 by the applied weak, intermediate and strong base anion exchangers from the systems containing 100 mg/L of the dye in the presence of the cationic surfactant CTAB, major decrease in the sorption capacities with the increasing concentration of CTAB from 0.1 to 1 g/L was observed.

The attractive interactions between the dyes anions and positively charged head group of CTAB caused the formation of different types of aggregates (micelles) depending on the ratio of the dye to surfactant and decreasing concentrations of "free" dye anions in the aqueous phase at the same time. In case of anion exchangers' interaction with dyes, these observations were described by Greluk and Hubicki [29, 31], Greluk [30] and Wawrzkiewicz [15, 20, 23]. Sorption of basic and acid dyes on different low-cost sorbents like chemically treated wood shavings, iron humate as well as oxihumolite in the presence of different surfactants was explained in detail by Janoš [37] and Janoš et al. [38-40].

The anionic surfactant SDS affected the dye sorption in three ways: no impact on the dyes uptake, decrease in the dyes uptake or enhanced dyes uptake as shown in Figure 9. The first mentioned behavior was observed during the sorption of C.I. Acid Orange 7 (100 mg/L) from the solutions containing 0.1–2 g/L SDS not only for the polystyrene anion exchangers but also for the polyacrylic ones (Figure 9 a) and b)).

Decreasing of the anion exchange capacities with the increasing amounts of SDS can be explained as a competition of adsorption sites between these species compared with dyes anions causing reduction of the dye uptake. For example, the sorption of C.I. Reactive Black 5 (Figure 9 c) and d)) was slightly reduced with the increasing amount of SDS in the system for the most of weakly and strongly basic anion exchangers of the polystyrene matrix. Both C.I. Reactive Black 5 and SDS are negatively charged and the interactions between the anionic SDS and the dye anions must be repulsive, independent of the surfactant concentration. Competition between these species of adsorption sites caused reduction of the dye uptake.

For the polystyrene anion exchangers of macroporous structure, i.e. Lewatit MonoPlus MP 62, MP 64, MP 500 and Amberlite IRA 900 a diverse trend was found (Figure 9 e) and f)) in the case of C.I. Direct Blue 71 sorption. Increasing the concentration of SDS in the range 0.1–1 g/L, enhancement of the dye anions sorption was observed. The surfactant can be adsorbed on the active sites or on the hydrophobic parts of the matrix of the anion exchangers in the form of relatively small poor ordered surface aggregates and interact with the aromatic rings or anionic groups of the dye.

7.3. Effect of solution pH

The solution pH is a very important parameter during the adsorption process and is mainly influenced by two factors:

- Distribution of the dyes ionized species in the aqueous phase,
- Overall charge (functional groups and surface) of the adsorbent.

Weakly basic anion exchangers in the free base form ($RN(CH_3)_2$) function at low pH when the hydrogen ion concentration is sufficiently high to protonate the amine nitrogen atom. Protonation of the amine group through the donor lone pair of electrons on nitrogen atom occurred during equilibration with acid as described below [41]:

$$\underset{anhydrous}{\overline{RCH_2N(CH_3)_2}} + H_2O \xrightarrow{hydration} \underset{weak\ dissociation}{\overline{RCH_2NH(CH_3)_2^+ + OH^-}}$$

$$RCH_2NH(CH_3)_2^+ + OH^- + HCl \xrightarrow{acid\ addition} \underset{weakdissociation}{RCH_2NH(CH_3)_2^+Cl^- + H_2O}$$

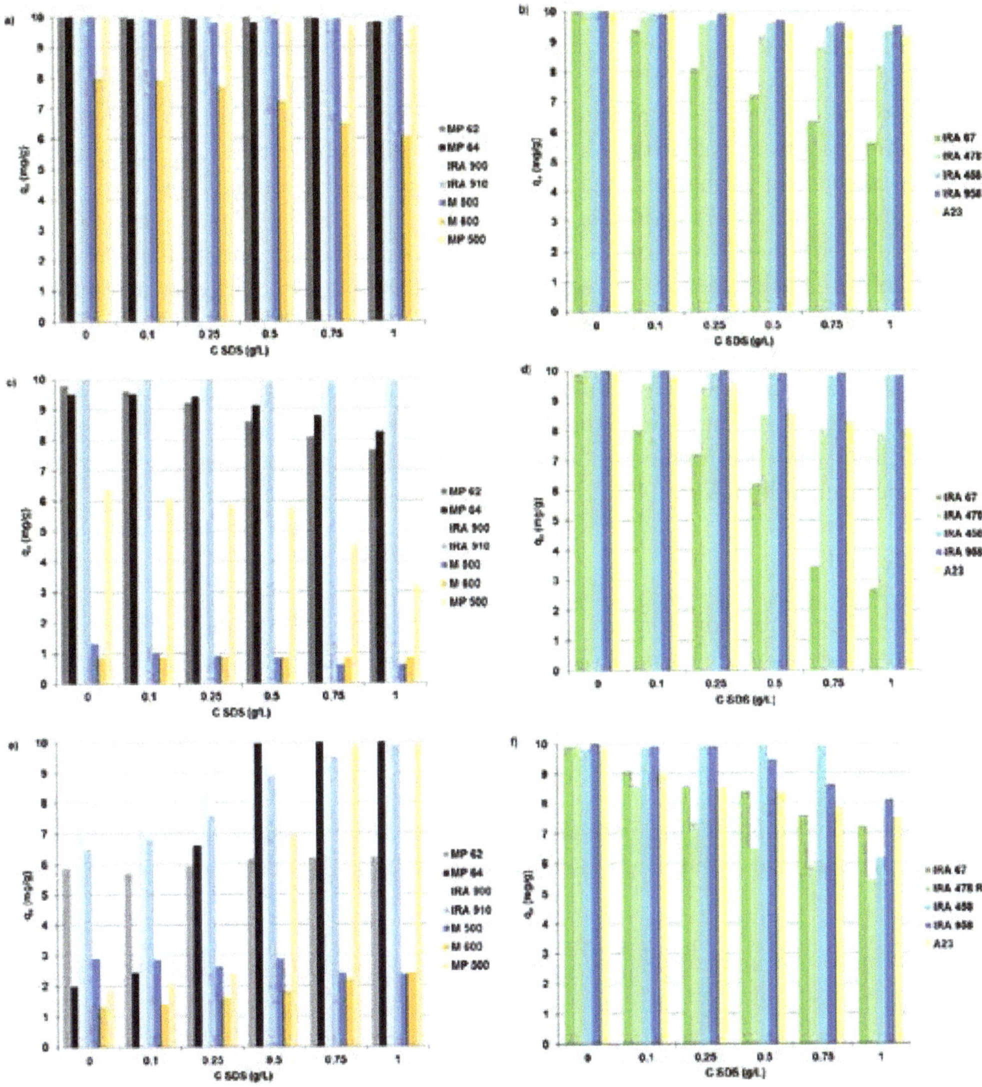

Figure 9. Influence of anionic surfactant sodium dodecyl sulfate (SDS) on uptake of C.I. Acid Orange 7 by the polystyrene anion exchangers (a) as well as the polyacrylic and phenol–formaldehyde anion exchangers (b), of C.I. Reactive Black 5 by the polystyrene anion exchangers (c) as well as the polyacrylic and phenol–formaldehyde anion exchangers (d) and of C.I. Direct Blue 71 by the polystyrene anion exchangers (e) as well as the polyacrylic and phenol–formaldehyde anion exchangers (f)

The resins were washed with 1 M HCl before the use taking the above into account. The relations between the initial pH of dyes solutions in the range of 1–12 and the sorption capacities of the anion exchangers were studied in the system containing 100 mg of dye per 1 L at 20°C. The decrease in the sorption capacities with the increasing initial solution pH was observed for the weakly basic anion exchangers because the capacity of the weak base anion exchange resin is a function of pH (it decreased with the increase of pH).

As the strongly basic anion exchangers function at any pH, there was observed no influence of solution pH of the dyes on their sorption on the strongly basic anion exchangers. As it was mentioned, formation of ion pairs due to the electrostatic attraction between the dye functional groups and the quaternary ammonium groups of strongly basic anion exchangers or the tertiary amine groups of weakly basic anion exchangers is probably a leading, though not the only one, mechanism of the acid, reactive and direct dyes retention. These dyes contain different groups such as –OH, $-SO_3Na$, –N=N–, $-NH_2$ that can participate in covalent, coulombic, hydrogen bonding or weak van der Waals forces. The occurrence of the double bond serves to enhance the interaction between the dyes and the anion exchangers macromolecule. The physical adsorption and π–π dispersion forces can arise from the aromatic nature of the resins and the dyes. A similar phenomenon was observed for the sorption of Acid Green 9 on the weak and strong base anion exchange resins [42], for the sorption of Sunset Yellow on the weakly basic anion exchanger Amberlite FPA 51 [43] and strongly basic Amberlite IRA 900 and IRA 910 [44] as well as for the sorption of Brilliant Yellow on Amberlite IRA 67, Amberlite IRA 458 and Amberlite IRA 958 [45].

7.4. Kinetic studies

To understand better the sorption process of dyes of various types of anion exchange resins, it is essential to determine the course of this process in time and the effect of different factors affecting their retention. The rate at which the dissolved dye is removed from the aqueous solution by solid sorbents is a significant factor for application in wastewater quality control, too. It is essential to evaluate the adsorption kinetics using theoretical models in order to design and control the sorption process units. Two common kinetic models, namely, the Lagergren pseudo first-order model (Equation 4) and the Ho and McKay pseudo second-order model (Equation 5) were fitted to the experimental data of dyes sorption on the anion exchangers:

$$\log(q_1 - q_t) = \log(q_1) - \frac{k_1}{2.303}t \tag{4}$$

$$\frac{t}{q_t} = \frac{1}{k_2 q_2^2} + \frac{1}{q_2}t \tag{5}$$

where q_1 and q_2 are the amounts of dye sorbed at equilibrium according to Equations (4) and (5), respectively (mg/g), q_t is the amount of dye sorbed at time t (mg/g), t is the time (min), k_1 is the constant rate of pseudo first-order adsorption (1/min), k_2 is the constant rate of the pseudo second-order adsorption (g/mg min) [46–50].

The values of k_2 and q_2 can be determined from the slope and intercept of the plot t/q_t vs. t, respectively. This dependence is defined as type 1 of the pseudo second-order expression or simply the pseudo second-order expression. Similarly, k_2 and q_2 can be calculated from the plots of $1/q_t$ versus $1/t$, $1/t$ versus $1/q_t$, q_t/t versus q_t and $1/q_2$-q_t versus t for type 2, type 3, type 4 and type 5 of the pseudo second-order expressions, respectively [51, 52].

The above mentioned equations were used due to the fact that the first one describes well the initial stage of dyes sorption, the second equation fits well experimental data in the whole range of process time for most adsorbate/adsorbent systems. Moreover, using these equations the dyes sorption process is considered as a chemical reaction (chemisorption) [46-52].

Kinetic behaviour of the anion exchangers towards C.I. Acid Orange 7, C.I. Reactive Black 5 and C.I. Direct Blue 71 was examined in the systems of different initial dyes concentrations ranging from 100 to 500 mg/L (in most cases) or even 1000 mg/L [2, 15, 20, 23, 25–31]. It was noticed that the amounts of dyes uptake increased with the contact time, and at some point in time reached an almost constant value where the amounts of dyes retained by the anion exchangers were in a state of dynamic equilibrium with the amounts of dyes desorbed from the anion exchangers. For the initial dyes concentration of 100 mg/L, the time required to reach equilibrium differed from a few to dozens minutes or even 12 h depending on the type of dye, composition of resins matrices as well as their structure, hydrophilic character of the skeletons and type of functional groups as can be seen from the data presented in papers [2, 15, 20, 23, 25–31]. C.I. Acid Orange 7 of the smaller molar weight saturated faster the available anion-exchanging sites compared with C.I. Reactive Black 5 and C.I. Direct Blue 71. The studies on the sorption of dyes and organic compounds on the ion exchangers reported so far showed that the size of the sorbate molecules has a considerable effect on the degree of fixation [53–55]. According to Dragan and Dinu [55] investigating the interactions of azo dyes such as Ponceau SS, Crocein Scarlet MOO, Congo Red and Direct Blue 1, differing in either the position of sulfonic groups or the number of anionic groups, with quaternized poly(dimethylaminoethyl methacrylate) it was stated that the number of sulfonic groups, position of the anionic charges and the whole structure of the dyes determined the dyes removal by the sorbent [55]. From the papers of Wojaczyńska and Kolarz [53, 54] about sorption of Methyl Orange, Acid Orange 10, Acid Red 44 and Direct Blue 1 on the divnylbenzene weak base anion exchangers of mono- and diethanolamine functional groups, it can be found that the copolymer gel heterogeneity has a marked effect on the degree of sorption and its course [54]. With the constant anion exchange capacity, the sorption properties decreased with an increase in the gel heterogeneity [53]. The dyes with a higher content of sulfonic groups were sorbed mainly by the formation of aggregates in the anion exchanger phase whereas Direct Blue 1 dye because of large size has the tendency to form aggregates in the solution rather than in the resin [54].

As follows from the data included in papers [2, 15, 20, 23, 25–31], generally the Lagergren equation (PFO) is not applied for description of sorption kinetics of C.I. Acid Orange 7, C.I. Reactive Black 5 and C.I. Direct Blue 71 on the chosen anion exchangers of different types. This results, among others, from the non-linear dependence $log(q_e - q_t)$ vs t, confirmed by low values of the determination R^2 and significant differences between the values of sorption capacity obtained experimentally and that calculated from the Lagergren equation. Moreover, the condition $log\ q_e = intercept$ is not satisfied which also indicates that the above equation is not applied [56–59]. Only in one case it was observed that the pseudo first-order expression better predicts the sorption kinetics than the pseudo second-order one. The closeness of the pseudo first-order equilibrium capacity to the experimentally determined equilibrium capacity indicates the usage of the pseudo first-order model to describe the kinetics of C.I. Direct Blue 71 uptake by Amberlite IRA 478RF. Several authors have also shown the applicability of the PFO kinetics in describing the sorption of dyes onto anion exchangers [53, 54] and different adsorbents [60–67]. Numerous applications of the Lagergren equation in sorption of dyes and inorganic ions have also been reported in the paper by Ho and McKay [49].

The linear form of the second-order equation proposed by Ho and McKay (PSO) can be applied for description of sorption of C.I. Acid Orange 7, C.I. Reactive Black 5 and C.I. Direct Blue 71 on ion exchangers of various types. In this case there are satisfied the following conditions: the dependence t/q_t vs t is linear, the determination coefficient reaches high values close to 1 and the calculated value of sorption capacity is largely consistent with the sorption capacity obtained experimentally. Comparison of quality of PFO and PSO equations fitting with the experimental data is presented for the system C.I. Reactive Black 5 – intermediate base anion exchanger Lewatit MonoPlus MP 64 in Figure 10.

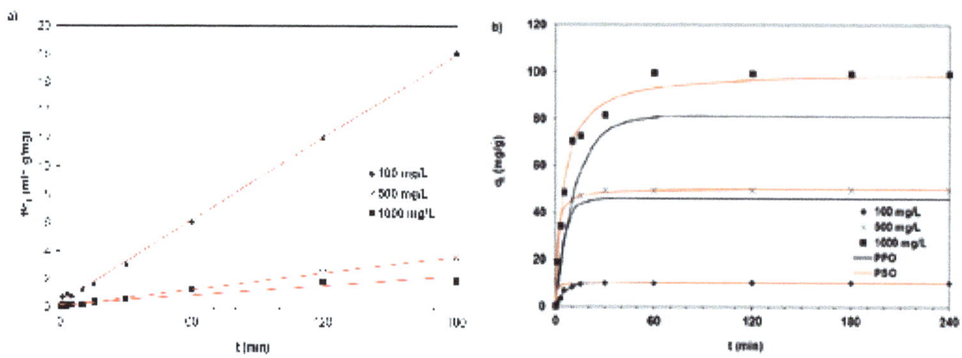

Figure 10. The pseudo second-order plot (a) and the fitting of the pseudo first-order and pseudo second-order equations (b) to the experimental data of C.I. Reactive Black 5 sorption on Lewatit MonoPlus MP 64 from the systems of different initial dye concentrations

In order to emphasize the influence of phase contact time on industrial effluents purification, suitable experiments were performed using the polyacrylic anion exchanger of the quaternary ammonium functionalities Amberlite IRA 958. Its effectiveness was confirmed in the batch experiments. The effluents from the textile industry containing different dyes and auxiliaries were shaken with 0.5 g of Amberlite IRA 958 from 1 to 144 h. The changes in absorbance values at the maximum wavelength in UV–vis spectra of the wastewater before and after sorption on Amberlite IRA 958 are presented in Figure 11. Analyzing the absorption values of untreated and purified wastewaters (Figure 11 a)), a significant colour reduction was observed after 12 h of phase contact time. The efficiency of decolourization exceeding 87% after only 1 h of contact time for the wastewater after the ozonation step is shown in Figure 11 b). The absorbance value at the maximum wavelength was reduced from 2.439 (before purification) to 0.1901 (after 12 h) in the case of wastewater containing Synazol Yellow KHL, SynazolBlue KBR and Synazol Red K3BS (Figure 11 c)). As shown in Figure 11 d), after 3 h of phase contact time the yield of decolourization was 88.3%, the increasing phase contact time to 144 h did not increase the adsorption efficiency significantly.

7.5. Desorption studies

The regeneration step is the key to the implementation of the anion exchange system on the commercial scale. Desorption studies help to evaluate the nature of adsorption process. Desorption experiments were performed using different regenerating agents such as 1 M NaCl, 1 M Na_2SO_4, 1 M Na_2CO_3, 1 M NaOH, 1 M HCl and even 1 M KSCN. As previously stated [2, 15, 20, 23, 25–31], the aqueous solutions mentioned above were ineffective for the dyes removal from the resin phase.

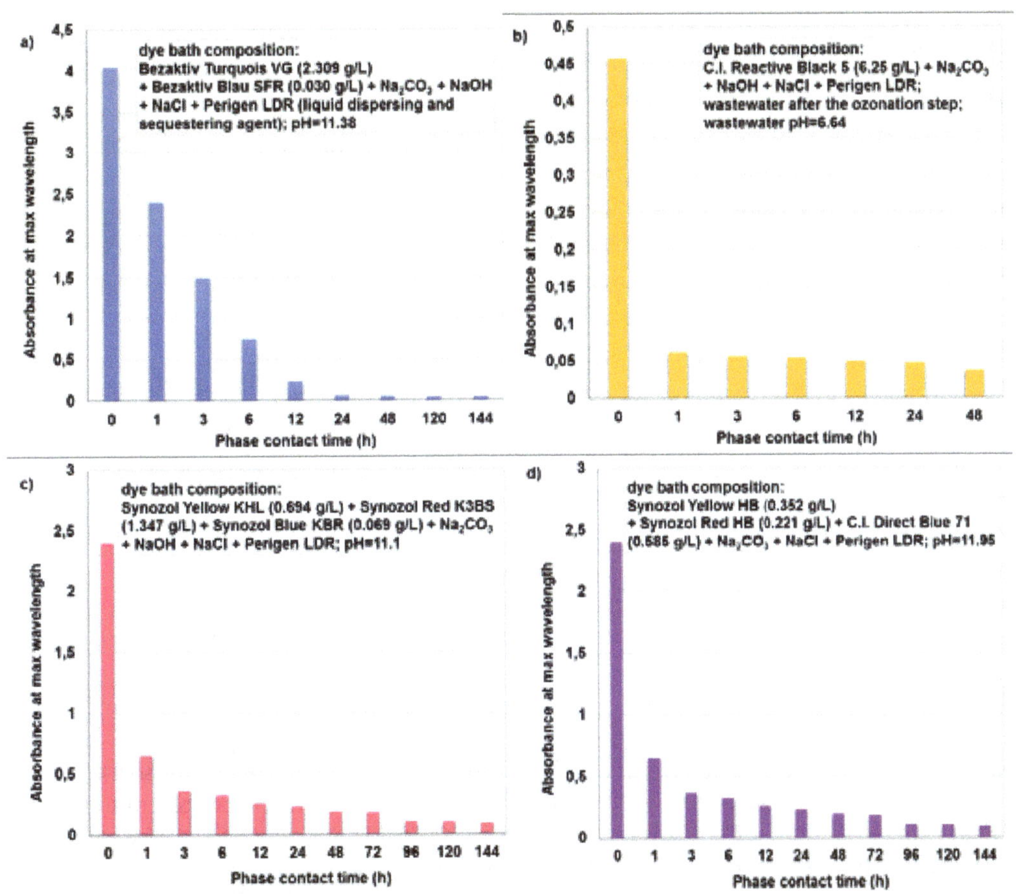

Figure 11. Influence of phase contact time on the purification of raw textile effluents of different compositions using Amberlite IRA 958: a) the absorbance values at max wavelength at 0 h and 1 h were recorded after ten times repeated dilution, b) and c) samples were not diluted before measurements, d) the absorbance values at max wavelength at 0 h was recorded after twice repeated dilution

Considering that the dye retention by the anion exchanger may not be only by ion exchanging but also by the hydrophobic interaction or hydrogen bonding, methanol was chosen for breaking these non-specific interactions. Regeneration of the anion exchangers using 10–90% methanol solutions was ineffective confirming that strong electrostatic attraction between the dyes and the anion exchange matrix is a predominant mechanism of adsorption. The mixtures of 90% methanol with 1 M KSCN, 1 M HCl or 1 M NaOH improved the dyes desorption from the anion exchangers in most cases. Greluk and Hubicki [28, 29, 31], Karcher et al. [68, 69], Liu et al. [70] and Wawrzkiewicz [20, 23, 25] confirmed that regeneration of the anion exchangers loaded with the acid, reactive and direct dyes was problematic and required usage of aggressive regenerants which could have negative impact on the cost of the process.

8. Conclusions

The possibility of the removal of C.I. Acid Orange 7, C.I. Reactive Black 5 and C.I. Direct Blue 71 using weakly (Amberlite IRA 67, Lewatit MonoPlus MP 62, Amberlyst A 23), intermediate (Lewatit MonoPlus MP 64 and Amberlite IRA 478RF) and strongly (Amberlite IRA 458, Amberlite IRA 958, Amberlite IRA 900, Amberlite IRA 910, Lewatit MonoPlus MP 500, Lewatit MonoPlus M 500 and Lewatit MonoPlus M 600) basic anion exchangers of the polyacrylic, polystyrene and phenol–formaldehyde skeletons and different matrix structure from aqueous solutions and wastewaters were discussed [2, 15, 20, 23, 25–28, 34–36]. Not only the number of sulfonic groups, position of the anionic charges, whole structure of the dyes and their molecular weight but also type of the anion exchangers determined the dyes removal by these sorbents.

Based on the values of the monolayer sorption capacities towards C.I. Acid Orange 7 (1370.4 mg/g), C.I. Reactive Black 5 (1655.2 mg/g) and C.I. Direct Blue 71 (1630.6 mg/g), it seems that the macroporous anion exchanger Amberlite IRA 958 can be a promising adsorbent for the textile wastewater treatment.

The affinity series of the dyes for Amberlite IRA 958 of quaternary ammonium functionality can be presented as follows:

C.I. Reactive Black 5 > C.I. Direct Blue 71 > C.I. Acid Orange 7.

Its advantageous behaviour for the acid, reactive and direct dyes resulted from combination of such properties as positive charge of functional groups and large pores of hydrophilic polyacrylic matrix. The experimental data indicated that the amounts of dyes adsorbed on the anion exchange resins are influenced by many factors such as initial dye concentration, phase contact time, solution pH, auxiliaries presence (NaCl, Na_2SO_4, Na_2CO_3, CH_3COOH and surfactants) and their concentrations. The reversibility of adsorption depends on whether the predominant mechanism is a strong binding bond such as ionic bonding or weak binding forces such as van der Waals interactions or H-bonding and occurred with high yield using sodium hydroxide or hydrochloric acid (or potassium thiocyanate) in methanol. The adsorption behaviour of Amberlite IRA 958 demonstrates that it can be a promising adsorbent for the textile wastewater treatment.

Author details

Monika Wawrzkiewicz* and Zbigniew Hubicki

*Address all correspondence to: m.wawrzkiewicz@op.pl

Maria Curie-Sklodowska University, Faculty of Chemistry, Department of Inorganic Chemistry, Lublin, Poland

References

[1] Hunger K, ed. Industrial Dyes – Chemistry, Properties and Applications, Weinheim: Wiley-VCH; 2003.

[2] Wawrzkiewicz M. Zastosowanie sorbentów różnego typu w procesie usuwania barwników z roztworów wodnych i ścieków przemysłowych. PrzemChem 2012;91(1):45–52.

[3] http://www.colour-index.com/ci-explained (accessed 9 January 2015).

[4] Word Dyes & Organic Pigments – Industry Study with Forecasts for 2013 & 2018, Freedonia Group INC, 2009 Cleveland, USA. http://www.freedoniagroup.com/brochure/25xx/2508smwe.pdf (accessed 9 January 2015).

[5] Pereira L, Alves M. Dyes – environmental impact and remediation. In: Malik A, Grohmann E. (eds.), Environmental Protection Strategies for Sustainable Development, Strategies for Sustainability, Dordrecht Heidelberg London New York: Springer; 2012. DOI 10.1007/978-94-007-1591-2_4, Springer (accessed 9 January 2015).

[6] Gupta VK, Suhas. Application of low-cost adsorbents for dye removal – a review. J Environ Manage 2009;90(8):2313–42.

[7] Elwakeel KZ. Removal of Reactive Black 5 from aqueous solutions using magnetic chitosan resins. J Hazard Mater 2009;167(1):383–92.

[8] Padmavathy S, Sandhya S, Swaminathan K, Subrahmanyam YV, Chakrabarti T, Kaul SN. Aerobic decolorization of reactive azo dyes in presence of various cosubstrates. ChemBiochemEngQuart 2003;17(2):147–51.

[9] Kumar Pandey A, Dubey V. Biodegradation of azo dye Reactive Red BL by Alcaligenes Sp. AA09. Int J EngSci 2012;1(12):54–60.

[10] Dotto GL, Vieira MLG, Esquerdo VM, Pinto LAA. Equilibrium and thermodynamics of azo dyes biosorption onto spirulina platensis. Brazil J Chem Eng 2013;30(1):13–21.

[11] Saranraj P. Bacterial biodegradation and decolourization of toxic textile azo dyes. African J Microbiol Res 2013;7(30):3885–3890.

[12] Majewska-Nowak K. Usuwanie barwników ze ścieków przemysłowych. Ochrona Środowiska 1986;488/4(30):17–22.

[13] Bilińska L, Bemska J, Biliński K, Ledakowicz S. Zintegrowana chemiczno-biologiczna oczyszczalnia ścieków włókienniczych. Inżynieria i Aparatura Chemiczna2012;51(4): 95–97.

[14] Mielnicki J. Zarys Chemicznej Obróbki Wyrobów Włókienniczych. Warshaw: Wydawnictwo Naukowo-Techniczne; 1991.

[15] Wawrzkiewicz M. Removal of C.I. Basic Blue 3 Dye by sorption onto cation exchange resin, functionalized and non-functionalized polymeric sorbents from aqueous solutions and wastewaters. Chem Eng J 2013;217:414–25.

[16] Hessel C, Allegre C, Maisseu M, Charbit F, Moulin P. Guidelines and legislation for dye house effluents. J Environ Manage 2007;83:171–80.

[17] Solecka M, Ledakowicz S. Biologiczne Procesy Oczyszczania Barwnych Ścieków Włókienniczych. Biotechnologia 2005;2(69):103–24.

[18] Kyzioł-Komosińska J, Rosik-Dulewska C, Dzieniszewska A, Pająk M. Wykorzystanie kompostu jako biosorbentu w usuwaniu barwników kwasowych ze ścieków przemysłu włókienniczego. ArchEnvironProtect 2011;37:3–14.

[19] Epolito WJ, Lee YH, Bottomley LA, Pavlostathis SG. Characterization of the textile anthraquinone dye Reactive Blue 4. Dyes Pigments 2005;67(1):35–41.

[20] Wawrzkiewicz M. Anion exchange resins for C.I. Direct Blue 71 removal from aqueous solutions and wastewaters: effects of basicity and matrix composition and structure. Industr Eng Chem Res 2014;53:11838–49.

[21] Bartkiewicz B, Umiejewska K. Oczyszczanie Ścieków Przemysłowych. Warshaw: PWN; 2010.

[22] Holme I. Ekologicol Aspects of Colour Chemistry: Developments in the Chemistry and Technology of Organic Dyes. Oxford: Society of Chemistry Industry; 1984.

[23] Wawrzkiewicz M, Hubicki Z. Remazol Black B removal from aqueous solutions and wastewater using weakly basic anion exchange resins. Central Eur J Chem 2011;9(5): 867–76.

[24] Babu BR, Parande AK, Raghu S, Kumar TP. Cotton textile processing: waste generation and effluent treatment. J Cotton Sci 2007;11:141–53.

[25] Wawrzkiewicz M. Comparison of the efficiency of Amberlite IRA 478RF for acid, reactive and direct dyes removal from aqueous media and wastewaters. Industr Eng Chem Res 2012;51(23):8069–78.

[26] Greluk M, Hubicki Z. Effect of basicity of anion exchangers and number of sulfonic groups of acid dyes on dyes adsorption on macroporous anion exchangers with styrenic polymer matrix. Chem Eng J 2013;215–216:731–9.

[27] Wawrzkiewicz M. Anion exchange resins as effective sorbents for acidic dye removal from aqueous solutions and wastewaters. Solvent Extract Ion Exch 2012;30:507–23.

[28] Greluk M, Hubicki Z. Kinetics, isotherm and thermodynamic studies of Reactive Black 5 removal by acid acryl resins. Chem Eng J 2010;162:919–926.

[29] Greluk M, Hubicki Z. Comparison of the gel anion exchangers for removal of Acid Orange 7 from aqueous solution. Chem Eng J 2011;170:184–93.

[30] Greluk M. Badania Procesu Sorpcji Barwników Kwasowych i Reaktywnych z Roztworów Wodnych na Anionitach Różnego Typu, PhD thesis, Maria Curie Sklodowska University Lublin; 2011.

[31] Greluk M, Hubicki Z. Evaluation of polystyrene anion exchange resin for removal of reactive dyes from aqueous solutions. Chem Eng Res Design 2013;91:1343–51.

[32] Langmuir I. The adsorption of gases on plane surface of glass, mica and platinum. J Am Chem Sci 1918;40:1361–403.

[33] Freundlich HMF. Über die adsorption in lösungen. ZeitschriftfürPhysikalischeChemie 1906;57:385–470.

[34] Suteu D, Bilba D, Coseri S. Macroporous polymeric ion exchangers as adsorbents for the removal of cationic dye Basic Blue 9 from aqueous solutions. J Appl Polymer Sci 2014;131(1):39620–39631.

[35] Kaušpėdienė D, Gefenienė A, Kazlauskienė E, Ragauskas R, Selskienė A. Simultaneous removal of azo and phthalocyanine dyes from aqueous solutions using weak base anion exchange. Water, Air, Soil Pollut 2013;224:1769–81.

[36] Kaušpėdienė D, Kazlauskienė E, Česũnienė R, Gefenienė A, Ragauskas R, Selskienė A. Removal of the phthalocyanine dye from acidic solutions using resins with the polystyrene divinylbenzene matrix. Chemija 2013;24(3):171–81.

[37] Janoš P. Sorption of basic dyes onto iron humate. Environ Sci Technol 2003;37:5792–8.

[38] Janoš P, Šmídová V. Effects of surfactants on the adsorptive removal of basic dyes from water using an organomineral sorbent – iron humate. J Colloid Interface Sci 2005;291:19–27.

[39] Janoš P, Coskun S, Pilařová V. Removal of basic (methylene blue) and acid (Egacid Orange) dyes from waters by sorption on chemically treated wood shavings. Biores Technol 2009;100:1450–3.

[40] Janoš P, Šedivy P, Ryznarová M, Grötschelová S. Sorption of basic and acid dyes from aqueous solutions onto oxihumolite. Chemosphere 2005;59:881–6.

[41] Harland CE. Ion Exchange. Theory and Practice. Cambridge: The Royal Society of Chemistry; 1994.

[42] Dulman V, Simion C, Bârsănescu A, Bunia I, Neagu V. Adsorption of anionic textile dye Acid Green 9 from aqueous solution onto weak or strong base anion exchangers. J Appl Polymer Sci 2009;113:615–627.

[43] Wawrzkiewicz M. Sorption of Sunset Yellow dye by weak base anion exchanger – kinetic and equilibrium studies. Environ Technol 2011;32:455–65.

[44] Wawrzkiewicz M, Hubicki Z. Kinetics of adsorption of sulphonated azo dyes on strong basic anion exchangers. Environ Technol 2009;30:1059–71.

[45] Leszczyńska M, Hubicki Z. Application of weakly and strongly basic anion exchangers for the removal of Brilliant Yellow from aqueous solutions. Desalin Water Treat 2009;2:156–161.

[46] Ho YS. Citation review of Lagergren kinetic rate equation on adsorption reactions. Scientometrics 2004;59:171–7.

[47] Ho YS, McKay G. A comparison of chemisorption kinetic models applied to pollutant removal on various sorbents. Process Safety Environ Protect 1998;76:332–40.

[48] Ho YS, McKay G. Sorption of dye from aqueous solution by peat. ChemEng J 1998;70:115–24.

[49] Ho YS, McKay G. A comparison of chemisorption kinetic models applied to pollutant removal on various sorbents. Trans IChemE 1998;76Part B:332–340.

[50] Ho YS. Second-order kinetic model for the sorption of cadmium onto tree fern: a comparison of linear and non-linear methods. Water Res 2006;40:119–25.

[51] Dizge B, Keskinler B, Barlas H. Sorption of Ni(II) ions from aqueous solution by Lewatit cation-exchange resin. J Hazard Mater 2009;167:915–26.

[52] Kumar KV. Linear and non-linear regression analysis for the sorption kinetics of methylene blue onto activated carbon. J Hazard Mater 2006;137B:1538–44.

[53] Wojaczyńska M, Kolarz B. Weak base polystyrene anion exchangers I. Relation between their structure and sorption of methyl orange. Die Angewandte Makromolekulare Chemie 1980;86(1):65–82.

[54] Wojaczyńska M, Kolarz B. Weak base polystyrene anion exchangers II. Relation between their structure and structure of sorbates. Die Angewandte Makromolekulare Chemie 1980;86(1):83–91.

[55] Dragan ES, Dinu IA. Interaction of dis-azo dyes with quaternized poly(dimethylaminoethyl methacrylate) as a function of dye structure and polycation charge density. J Appl Polymer Sci 2009;112(2):728–35.

[56] Fujiwara K, Ramesh A, Maki T, Hasegawa H, Ueda K. Adsorption of platinum(IV), palladium(II) and gold(III) from aqueous solutions onto L-lysine modified crosslinked chitosan resin. J Hazard Mater 2007;146(1–2):39–50.

[57] Özacor M, Şengil İA. A kinetic study of metal complex dye sorption onto pine sawdust. Process Biochem 2005;40:565–72.

[58] Yener J, Kopac T, Dogu G, Dogu T. Adsorption of Basic Yellow 28 from aqueous solutions with clinoptilolite and amberlite. J Colloid Interface Sci 2006;294:255–64.

[59] Wawrzkiewicz M, Hubicki Z. Kinetic studies of dyes sorption from aqueous solutions onto strongly basic anion-exchanger Lewatit MonoPlus M-600. Chem Eng J 2009;150:509–15.

[60] Ofomaja AE. Kinetic study and sorption mechanism of methylene blue and methyl violet onto Mansonia (Mansonia Altissima) wood sawdust. Chem Eng J 2008;143(1–3):85–95.

[61] Waranusantigul P, Pokethitiyook P, Kruatracchue M, Upatham ES. Kinetics of basic dye (methylene blue) biosorption by giant duckweed (Spirodela Polyrrhiza). Environ Pollut 2003;125(3):385–92.

[62] Mall ID, Upadhyay SN. Treatment of methyl violet bearing wastewater from paper mill effluent using low cost adsorbents. J Indian Pulp Paper Technol Assoc 1995;7(1): 51–7.

[63] Mittal A, Kurup L, Mittal J. Freundlich and Langmuir adsorption isotherms and kinetics for the removal of tartrazine from aqueous solutions using hen feathers. J Hazard Mater 2007;146(1–2):243–8.

[64] Mittal A, Mittal J, Kurup L. Batch and bulk removal of hazardous dye, indigo carmine from wastewater through adsorption. J Hazard Mater 2006;B137:591–602.

[65] Mittal A, Kaur D, Mittal J. Batch and bulk removal of triarylmethane dye, Fast Green FCF, from wastewater by adsorption over waste materials. J Hazard Mater 2009;463:568–77.

[66] Mittal A. Use of hen feathers as potential adsorbent for the removal of a hazardous dye, Brilliant Blue FCF from wastewater. J Hazard Mater 2006;B128(2–3):233–239.

[67] Ging R, Ding Y, Li M, Yang C, Liu H, Sun Y. Utilization of powdered peanut hull as biosorbent for removal of anionic dyes from aqueous solutions. Dyes Pigments 2005;64(3):187–92.

[68] Karcher S, Kornmüller A, Jekel M. Screening of commercial sorbents for the removal of reactive dyes. Dyes Pigments 2001;51(2–3):111–125.

[69] Karcher S, Kornmüller A, Jekel M. Anion exchange resins for the removal of reactive dyes from textile wastewaters. Water Res 2002;36(19):4717–24.

[70] Liu CH, Wu JS, Chiu HC, Suen SY, Chu KH. Removal of anionic reactive dyes from water using anion exchange membranes as adsorbers. Water Res 2007;41(7):1491–500.

5

Strengthening Dental Porcelains by Ion Exchange Process

Humberto Naoyuki Yoshimura and
Paulo Francisco Cesar

Abstract

Porcelains have been used in dentistry for many decades because of their excellent aesthetic qualities, besides other favorable characteristics. Despite these desirable characteristics, porcelain restorations may fail in the oral environment due to fracture. Studies on the clinical success rate of porcelain onlays, inlays and veneers have shown that their fracture rate is relatively high and is among the main reasons for failure of these restorations. The fracture of dental porcelains is a consequence of its brittle nature and low fracture toughness. Porcelains are also highly susceptible to weakening during their lifetime in the oral environment, because the sizes of defects tend to increase by the slow crack growth phenomenon. Therefore, in order to increase the lifetime of porcelain restorations, it is necessary to enhance their overall resistance to crack propagation. Among the methods proposed to strengthen glasses and ceramics, a potential method to improve the mechanical properties of dental porcelains is the chemical strengthening or tempering by the ion exchange process. In this chapter, the effects of chemical tempering on mechanical behavior of dental porcelains are reviewed. Dental porcelains are based on alkali-containing aluminosilicate glass compositions and can have leucite ($KAlSi_2O_6$) crystalline particles dispersed in the glassy matrix. The ion exchange process can be carried out by the paste method using KNO_3 salt at a temperature that is 80% of glass transition temperature (T_g) of porcelain during a short time (15 to 30 min). In this treatment, the small Na^+ ions in the glassy matrix are exchanged by larger K^+ ions from the salt, resulting in a K^+ concentration profile that results in a steep gradient of residual compressive stress by the ion stuffing effect at the surface region of the porcelain. No significant variations in strengthening have been observed when temperature and time varied around the above indicated values, since the increase in these parameters enhanced the stress relaxation process, which hinders the effect of higher ion interdiffusion. Although few porcelains with high leucite content have no strengthening response to ion exchange process, most dental porcelains can be strengthened and significant increases in fracture toughness (up to around 150%) have been reported. The same level of increase in flexural strength has been observed, but the variability of fracture stress also increases due to the relative small thickness of compressive layer and the decreasing resistance curve effect. The lower reliability is counterbalanced by significant increases of the resistance to slow crack growth phenomenon, leading to higher strength retention after long lifetimes even at low levels of fracture probability. Therefore, it is expected that the application of chemical tempering (strengthening by ion exchange) can improve the lifetime of dental porcelain restorations.

Keywords: Bioceramics, dental porcelain, ion exchange, chemical tempering, strength, toughening, lifetime

---

1. Introduction

Dental porcelains have been used in dental restorations due to their good qualities, including high color stability, high resistance to stain, good biocompatibility, low thermal conductivity, high wear resistance, and capacity to mimic dental structures [1–3]. Notwithstanding, disadvantages of these restorations include high susceptibility to fracture, risk of debonding, and microleakage [4–6]. For feldspathic porcelain onlays placed in posterior teeth after 6 years, the observed cumulative survival rate was ~60%, with bulk fracture in 16% of the restorations [7]. The reported clinical success rate for maxillary anterior porcelain veneers after 10 years was 64%, and main reasons for failure were fracture (11%) and large marginal defects (20%) [2]. Similar behavior was also observed for posterior feldspathic porcelain inlays, and marginal defects and fracture were 22% and 11% of the restorations, respectively, after an 8-year period of clinical assessment [8].

The high susceptibility to fracture of porcelain restorations is caused by their brittle nature, that is, their low capability to absorb strain energy due to an external loading before fast crack propagation occurs. The resistance to crack propagation can be quantified by the fracture toughness (K_{Ic}) which is given by [9,10]:

$$K_{Ic} = Y \cdot \sigma_f \cdot \sqrt{c} \tag{1}$$

where, Y is a geometrical constant, σ_f is fracture stress, and c is crack size that results in fracture. From Griffith's energy failure criterion, the term$\sigma_f c^{1/2}$ is constant, which implies that strength (σ_f) is not constant and varies inversely with the square root of critical flaw size ($c^{1/2}$). Furthermore, Equation 1 also shows that the porcelain strength is directly related with its fracture toughness, K_{Ic}.

Dental porcelains have relatively low values of K_{Ic} (around 0.6 to 1.2 $MPa.m^{1/2}$) [11,12], especially when compared to metals, which have K_{Ic} higher than around 30 $MPa.m^{1/2}$ [13]. As a consequence, porcelains have low strength values usually in the range of 40 to 120 MPa [14, 15]. Moreover, the strength of porcelain restorations can decrease during their lifetimes in the oral environment, since a weakening effect known as subcritical crack growth causes the flaw sizes to increase slowly over time [16,17]. Therefore, it is important to develop processing methods that can enhance the overall resistance to crack propagation, particularly strength and fracture toughness, in order to increase the lifetime of porcelain restorations.

Different methods have been proposed to strengthen dental porcelains, including addition of reinforcing phases, like ceramic fibers or phase transformable tetragonal zirconia particles, and incorporating a compressive surface layer, which can be achieved by thermal tempering, glazing with a glassy material having lower thermal expansion coefficient, or chemical tempering [18–20]. Among these, chemical strengthening by ion exchange is a promising method to significantly enhance the mechanical behavior of dental porcelain restorations.

2. Chemical tempering

Chemical tempering is a strengthening or toughening treatment by ion exchange process that introduces a residual compressive stress layer on the surface of glassy materials that hinders the crack propagation and increases the material resistance to fracture. In this treatment, alkali ions of the glass are removed and exchanged by other larger alkali ions from an external source at a temperature sufficiently high to promote ion interdiffusion. Figure 1 shows the sizes of different alkali ions. The most applied alkali ion pair for strengthening is the Na^+/K^+, but other pairs are also exchangeable, like Li^+ by Na^+ and K^+ by Rb^+, depending on the glass composition [21,22].

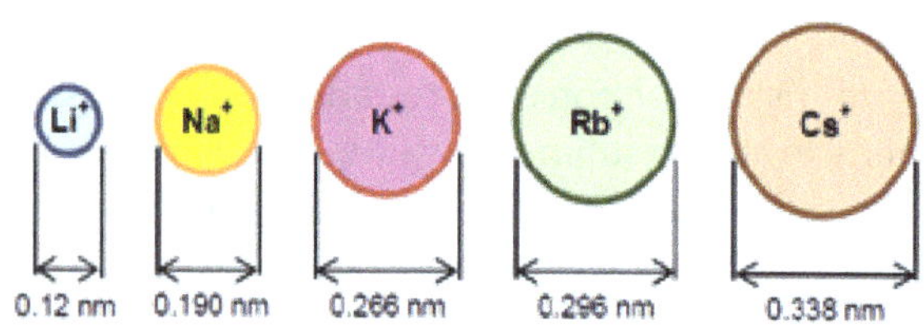

Figure 1. Pauling's calculated ionic diameters for alkali metals. Data from [21]

An usual practice is to make an ion exchange treatment in sodium-containing aluminosilicate glasses with a melt of KNO_3 salt, at a temperature between the melting point of salt and the glass transition temperature (T_g) of the glass. During the process, Na^+ ions diffuse out from the glass into the salt and simultaneously the diffusion of K^+ ions from the salt into the glass takes place, with equal and coupled counterdiffusing ion fluxes ($J_{Na+} = J_{K+}$) to maintain the eletro-neutrality (Figure 2) [23].

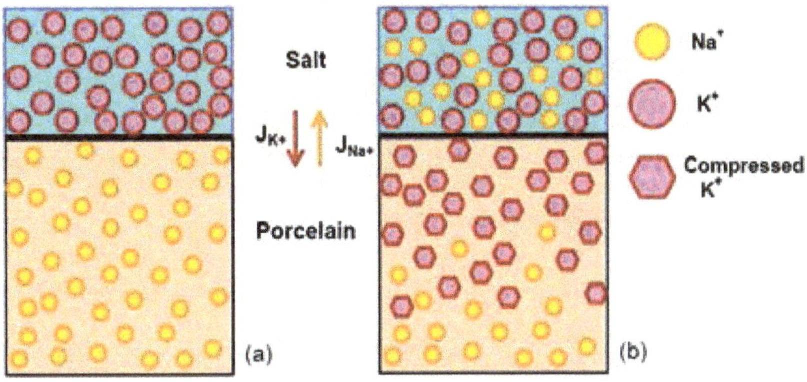

Figure 2. Schematics of (a) before and (b) after ion exchange process in dental glassy porcelain. J – ion flux

During the ion exchange process, concentration gradients of K^+ and Na^+ are formed at the region near the glass surface that can be described by the Fick's second law, given by [22–26]:

$$C_x = (C_S - C_0)\left[1 - erf\left(\frac{x}{2\sqrt{\tilde{D}\cdot t}}\right)\right] + C_0 \quad (2)$$

where, C_x is the ion concentration at depth x (from surface) after ion exchange time t, C_0 is the initial ion concentration in glass, erf(z) is the Gaussian error function, and $\check{D}$ is the interdiffusion coefficient given by:

$$\tilde{D} = \frac{D_{Na} \cdot D_K}{D_{Na}N_{Na} + D_K N_K} \quad (3)$$

where, N_i is fractional concentration of alkali ion i and D_i is its self-diffusion coefficient in mixed-alkali glass compositions, which increases exponentially with temperature by:

$$D = D_0 \exp\left(-\frac{Q_d}{kT}\right) \quad (4)$$

where, D_0 is preexponential factor, Q_d is activation energy for diffusion, k is Boltzmann's constant and T is temperature. Figure 3a shows examples of K^+ concentration profiles after ion exchange at different time or temperature.

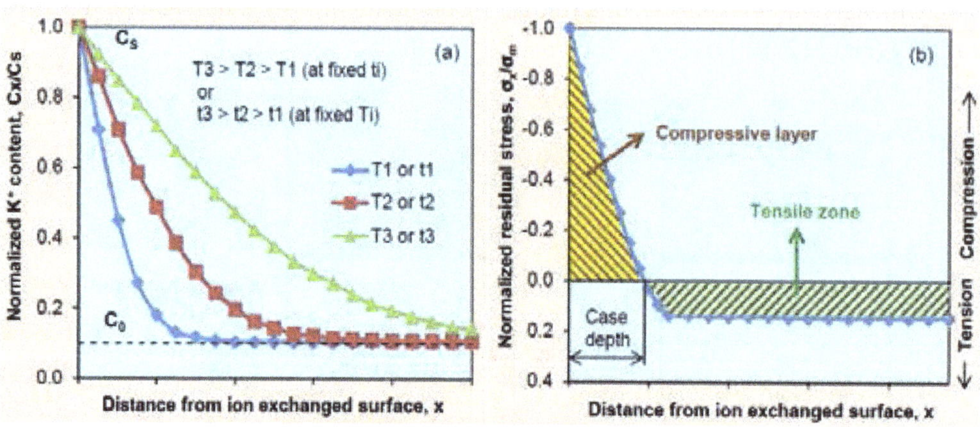

Figure 3. Normalized potassium concentration profiles (Equation 2) (a), and normalized residual stress profile (b) in dental glassy porcelain after ion exchange process

The bigger K^+ ions that replace Na^+ ions tend to cause material expansion (known as ion stuffing) in the exchanged surface layer, which is restricted by the non-ion exchanged glass region. This situation, when the ion exchange is carried out at a temperature lower than the T_g of glass, generates a residual compressive stress field parallel to the surface in the K-rich layer. This layer has a gradient of compressive stress similar to the potassium concentration gradient, that is, the compressive stress is high at the surface and decreases with the increase in distance from surface. In order to counterbalance the net stress state, a residual tensile stress field is generated below the compressive layer (Figure 3b).

The residual compressive stress layer adds a toughening contribution, K_{RC}, which hinders crack propagation, leading to an increase in the fracture toughness in ion exchanged glass, $K_{Ic,IE}$, by:

$$K_{Ic,IE} = K_0 + K_{RC} \tag{5}$$

where, K_0 is the fracture toughness of unreinforced glass. The higher K_{Ic} leads to a higher fracture stress, if a flaw size c is unaltered (Equation 1). Therefore, the generation of a surface compressive layer by ion exchange can result in the increase in fracture toughness and strength of glasses and porcelains.

The toughening effect depends on the thickness of compressive layer, known as case depth (Figure 3b), especially when the glass contains deep surface flaws. Because of the relative slow ion exchange rate, the case depths varying from few tens to hundreds of micrometers have been reported, depending on the ion exchange parameters, including time (up to hundreds of hours of treatment have been reported), temperature, salt composition, exchangeable ionic pair, and glass composition [22,26,27].

An effect that can lower the strengthening rate by the ion exchange process is the stress relaxation that can occur during this process, leading to a reduction in the magnitude of residual compressive stress [27]. Stress relaxation occurs by viscous flow of the glass and can be described by the Maxwell's model given by [28,29]:

$$\sigma = \sigma_0 \exp\left(\frac{G \cdot t}{\eta}\right) \tag{6}$$

where, σ is the remaining stress at time t, σ_0 is initial stress, and G and η are shear modulus and viscosity of glass, respectively. Since η decreases strongly with temperature, the rate of stress relaxation is more rapid with the increase in temperature.

The chemical tempering has been applied to strengthen cockpit windows for aircrafts, high speed train windshields, photocopier scanner glass, display windows in mobile personal electronic devices, compact disks for portable hard drives, high-end ophthalmic glasses, and glass items for drug delivery [22,26]. Advantageous characteristics of chemical tempering include: possibility to strengthen complex geometries and thin components (thickness of up to around 100 μm), which are difficult in thermal tempering; higher compressive stress level on the surface compared to thermal tempering; low level of internal residual tensile stress, with less fragmentation and explosion-like fracture propagation; and did not cause optical distortion. Disadvantageous characteristics include: limited to alkali-containing glasses; shallow depth of residual compressive stress layer (case depth); generation of corrosive alkali-containing salt residue; and high cost when long time of ion exchange is applied [22,26].

3. Dental porcelains

The feldspathic porcelain is a predominantly glassy material with variable crystalline content. Its basic structure has a network of silica with potassium, sodium, and other ions as network modifiers. In order to reduce the glass softening temperature and increase fluidity, metal oxide fluxes are added (CaO, K_2O, Na_2O), which decrease the softening temperature by reducing the amount of cross linking in the porcelain structure. The addition of alumina (between 8 and 20 wt%) is used for controlling the viscosity and decreasing the flow at high temperatures. When added, B_2O_3 in concentrations below 12 wt% also acts as a flux to form a less stable network of BO_4/SiO_4 [30–33].

Silica, soda, potash, and alumina are the constituents of mineral feldspar (Na_2O/ $K_2O.Al_2O_3.6SiO_2$), the main raw material used in the manufacturing of dental porcelain [30]. Silica and alumina account for most of the feldspar, about 70 and 17 wt%, respectively. Feldspar porcelains are relatively pure and colorless, so pigments are added to produce shades of natural tooth [34]. In the process of obtaining porcelain, feldspar is mixed with fluxes, then heated to temperatures between 1150 and 1530°C, and rapidly cooled in water. With the thermal shock, the glass breaks into fragments called frits. Opacifiers (TiO_2, ZrO_2) and pigments (Cr_2O_3, Fe_2O_3) are also added to this glass [34]. Heating feldspar to high temperatures leads to an incongruent melting resulting in the formation of leucite and a liquid glass. The leucite crystal is a mineral (potassium aluminum silicate) with high coefficient of thermal expansion compared to feldspathic glasses [30]. The porcelain structure after incongruent melting of feldspar has leucite ($K_2O.Al_2O_3.4SiO_2$) crystals involved in an aluminosilicate glassy matrix [32,33,35]. In general, the leucite ratio is governed by the K_2O content of the frit and the time and temperature of the heat treatment; thus, the desired leucite content can be achieved by controlling the appropriate time and the crystallization temperature range [36].

The frits based on leucite are used in Dentistry since the early 60s [31]. However, feldspar is not essential as a precursor for the formation of leucite, and many dental porcelains do not use feldspar as the raw material. These materials are called feldspar-free porcelain and are synthesized in the laboratory by controlled addition of leucite instead of mineral processing. It has been suggested that porcelains with a large amount of leucite dispersed in the glass matrix are called leucite-based instead of feldspathic porcelains [30].

Feldspathic porcelains are usually presented as a liquid-powder system, in which the liquid (water with dispersant) is mixed with porcelain powder to form a slurry or paste that is applied to the refractory die or metal framework. After production of the green body, it is taken into an electric furnace for the firing cycle (Figure 4a) [3,37]. Chemical reactions between the porcelain powder components are completed during the process of obtaining the frits. Therefore, the main purpose of firing is sintering of particles, although some chemical reactions may occur during prolonged firing times or during multiple firing [30]. According to the firing (sintering) temperature, the porcelain used for restorations and bridges can be classified as being of high fusion (850 to 1100°C) and low fusion (below 850°C) [30]. The firing procedure involves high heating rates (around 60°C/min) under vacuum and few minutes at the maximum temperature. During this procedure, sintering process transforms the porous green body (Figure 4b) in a translucent and dense solid, almost pore free (Figure 4c), by means of densification mechanisms with mass transport by viscous flow [3,37]. Vacuum firing is a resource used to reduce the porosity of these materials [31]. At the end of the firing procedure, the porcelain can have three distinct phases: a crystalline phase (leucite), the glass matrix, and pores.

Most dental porcelain developed for metal-ceramic restorations contain leucite ($KAlSi_2O_6$) as a main crystalline phase [38]. Leucite was first introduced so that the porcelain could reach a linear thermal expansion coefficient close to that of alloys used in metal-ceramic restorations. In this way, metal and porcelain have similar behavior when cooled together during the firing process, preventing the appearance of cracks in the porcelain [39]. Leucite is also the main crystalline constituent of most generations of porcelain for all-ceramic restorations [40]. In this case, leucite is not added with the aim of achieving thermal compatibility, but to increase the strength and fracture toughness of the material [32,41]. The amount, average crystal size, and crystal structure of leucite directly affect the thermal, optical, and mechanical properties of the final restoration [42]. One of the advantages having leucite as the crystalline phase is that the translucency of the porcelain is not lost, since the refractive index of this crystal is similar to that of the glassy matrix [43,44].

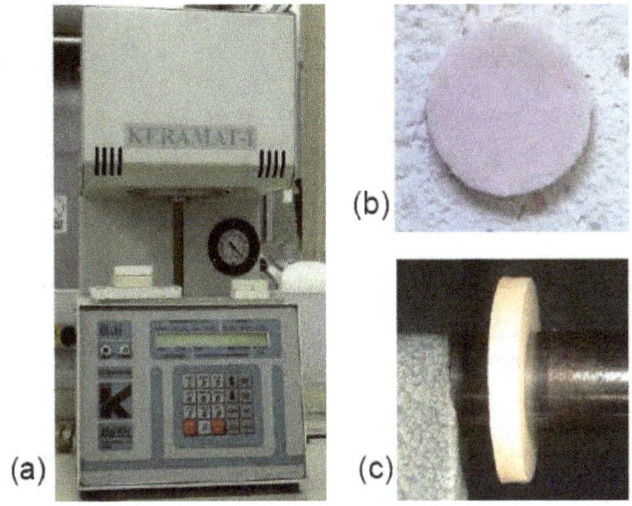

Figure 4. Dental porcelain furnace (a), green (b), and sintered (c) porcelain body [5]

The leucite crystal has a crystallographic polymorphic transformation (without change of composition) from tetragonal to cubic during heating. This transformation is displacive (martensitic) and accompanied by a marked change in the parameters of crystalline lattice with an increase of 1.2% by volume of the unit cell [38]. When a porcelain containing leucite is cooled from the firing temperature to room temperature, residual stresses arise in the material as a result of the large difference between the linear thermal expansion coefficients of the glass matrix (8.6 ppm °C^{-1}) and tetragonal leucite (22.3 ppm °C^{-1}). After cooling, tangential compressive stresses and radial tensile stresses appear in the glass matrix around the leucite particle and opposite stress fields in the tetragonal crystals [38,43]. The compressive stresses have a beneficial effect on the porcelains as they function as a mechanism for increasing the toughness, as opposed to the tensile stresses, which drive the cracks forward [45].

Figure 5 shows typical micrographs of dental porcelains containing leucite crystals. For porcelains with high leucite contents, the distribution of crystals usually is not homogeneous in the glassy matrix (Figure 5a), forming agglomerates of leucite particles (Figure 5b). In these micrographs, it is possible to see some circumferential cracks in the glass matrix surrounding leucite agglomerates, which reveal the radial tensile stresses generated during cooling [11]. The residual stress fields associated with the leucite crystals have significant effects on the fracture behavior of porcelain, since they change the propagation trajectory of a crack, driving it through the glassy matrix region with radial tensile stresses. The result is that a crack propagates bowing around leucite particles and agglomerates (Figure 5c). This effect is called crack deflection and is the main toughening mechanism caused by leucite crystals in dental porcelains. In fact, it has been observed that fracture toughness, K_{Ic}, increases with the increase of volume fraction of leucite [11].

The increase in porcelain's strength with the increase in leucite content has been observed experimentally. However, there is a tendency of this increase to achieve a maximum, because

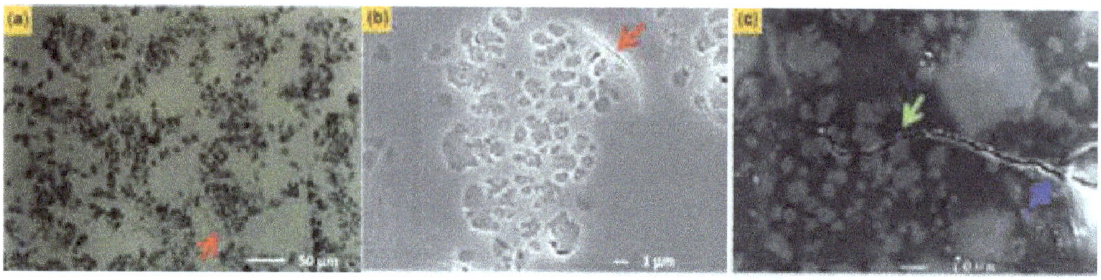

Figure 5. Micrographs (a – optical, b,c – scanning electron microscopy) of dental porcelains showing: (a) leucite ($KAlSi_2O_6$) crystals dispersed in glassy matrix; (b) agglomerate of leucite particles; (c) radial crack emanated from Vickers indentation corner (blue arrow) deflecting among leucite particles (green arrow). In (a,b), red arrow indicates circumferential crack around leucite agglomerate generated by chemical etching with HF solution (also used to reveal leucite particles)

for porcelains with high leucite volume fraction spontaneous microcracks can be generated around big leucite agglomerates during cooling after sintering. These flaws have large size (c value in Equation 1) and limit the porcelain's strength, despite the increase in fracture toughness, K_{Ic} [42,46]. Besides the quantity, the size, and distribution of the leucite particles also influences the mechanical properties of the porcelain [18,41].

4. Hardness and toughness of chemically strengthened dental porcelains

In this section, the effects of chemical strengthening using ion exchange by paste method on the hardness and fracture toughness of dental porcelains with different microstructures are shown. Table 1 shows descriptions of five porcelain powders used in this study, two recommended to be used as veneering materials for alumina cores (V and Cb) and three recommended for porcelain fused-to-metal restorations (C, D, and B), containing wide variation of leucite fraction (0 to 22 vol%).

Table 2 shows the chemical compositions of the porcelains measured by X-ray fluorescence spectroscopy (XRF 1500, Shimadzu), showing that all porcelains had aluminosilicate compositions with alkali and alkaline-earth metal oxides, besides other minor oxides. For porcelains containing leucite particles (C, D, and B), parts of SiO_2, Al_2O_3, and K_2O contents composed these particles. Considering the fraction and stoichiometry of leucite ($KAlSi_2O_6$ = $K_2O.Al_2O_3.4SiO_2$), the compositions of glassy matrix of these porcelains were calculated and are also shown in Table 2. Note that all porcelains had in the glassy matrix an initial K_2O content, and also potentially exchangeable Na^+ ions by K^+ ions from an external source.

Green specimens with bar shape (5 × 6 × 40 mm) were prepared by the vibration-condensation method, mixing the porcelain powder with distilled water and using a steel mold. Then, the specimens were vacuum sintered in a dental porcelain furnace (Keramat I, Knebel), following the firing schedules recommended by the manufacturers (sintering temperatures indicated in Table 1). After firing, the specimens were machined following the guidelines in ASTM C 1161 to the dimensions of 3 × 4 × 30 mm, and one of larger surfaces was mirror-polished using a semi-automatic polishing machine (Ecomet 3, Buehler), with diamond suspensions of 45, 15, 6, and 1 μm. For each material, 10 specimens were prepared.

Porcelain	Manufacturer / Brand Name	Manufacturer's Description	Leucite fraction (vol%)
V	VITA Zahnfabrik/Vitadur Alpha	Porcelain used with alumina frameworks. Sintering temperature: 960°C	0
Cb	Noritake/Cerabien	Porcelain used with alumina frameworks. Sintering temperature: 960°C	0
C	Dentsply/Ceramco Finesse	Leucite-based porcelain, used for metal-ceramic or all ceramic restorations, containing fine-grained leucite particles. Sintering temperature: 800°C	6
D	Ivoclar/d.Sign	Leucite-based porcelain, used for metal-ceramic or all ceramic restorations, containing leucite particles and crystals of fluorapatite. Sintering temperature: 875°C	15
B	Dentsply/Ceramco II	Leucite-based porcelain, used for metal-ceramic or all ceramic restorations, containing equiaxial leucite particles. Sintering temperature: 1000°C	22

Table 1. Description of five dental porcelains. Data from [47]

Oxide	V	Cb	C	D	B
SiO_2	75.9	82.9	70.1 (69.9)	67.6 (66.7)	72.0 (71.9)
Al_2O_3	10.0	5.6	6.4 (5.7)	9.0 (7.6)	9.5 (7.5)
K_2O	7.1	4.6	8.7 (8.6)	8.1 (7.7)	9.2 (8.8)
Na_2O	3.6	3.6	5.5 (5.8)	4.9 (5.8)	3.8 (4.9)
CaO	3.0	1.1	3.8 (4.0)	3.7 (4.4)	3.9 (5.0)
Others	0.3 ZrO_2	0.8 ZrO_2 0.7 MgO 0.3 CeO	5.0 (5.3) MgO 0.4 (0.5) Tb_4O_7	3.0 (3.5) ZnO 1.3 (1.6) ZrO_2 1.2 (1.4) BaO 0.6 (0.7) TiO_2 0.4 (0.5) P_2O_2	0.7 (0.9) BaO 0.6 (0.7) CeO
Traces (<0.2%)	Fe, Ni, Ti, Rb, Sr, Pb	Fe, Ni, Zn, Ti, Cr, Hf, V, I	Fe, Ni, Zr, Rb, Sr, Re, Cl	Fe, Ni, Cr, Hf	Fe, Ni, Rb, Sr, Cs, Tb, Cl

Table 2. Overall chemical composition (mol%) of dental porcelains. The calculated compositions of glassy matrix for porcelains containing leucite particles are given in parenthesis. Data from [48]

Later, the polished surfaces were subjected to an ion exchange treatment. First, the surface of the specimen was coated with a layer of a paste composed of distilled water mixed with potassium nitrate (KNO_3). The amount of paste placed on each specimen was controlled by mass measurement. Then, the specimens coated with the paste were heat treated in an electric furnace (FP32, Yamato). The ion exchange cycle was conducted at a heating rate of 5°C/min with a first step of drying at 150°C for 20 min to remove the water from the paste, followed by a step of melting of KNO_3 and ion exchange at 450°C for 30 min, and then cooled inside the furnace to room temperature. The KNO_3 paste residue was easily detached and removed with a wet piece of cotton and the ion exchange treatment did not affect the superficial appearance of the specimens.

The contents of K_2O and Na_2O on the polished surfaces before and after ion exchange were determined by energy dispersive spectroscopy (EDS, Noram) coupled in a scanning electron microscopy (SEM, JSM 6300, Jeol). This analysis showed that the KNO_3 paste method caused the decrease of Na_2O content with the increase of K_2O content in all porcelains (Figure 6), indicating that the Na^+ ions from the glassy matrix were successfully exchanged by K^+ ions from the paste. The porcelains Cb and B (with 0 and 22% of leucite, respectively) had the highest relative increase in K_2O content of around 35%.

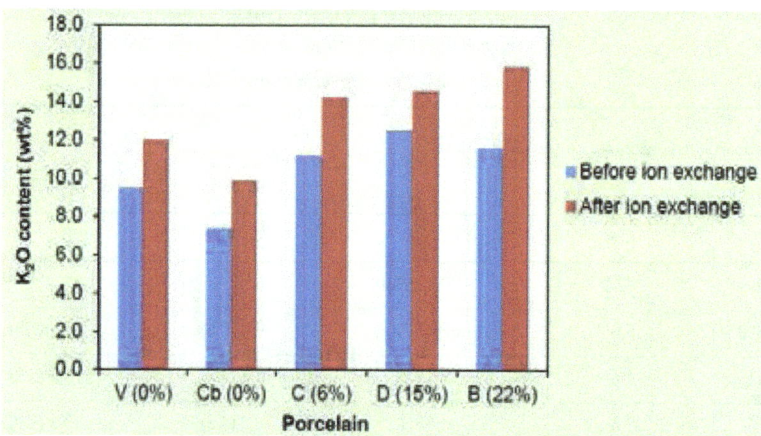

Figure 6. K_2O content on the surface of dental porcelains before and after ion exchange process (value in parenthesis is the volume fraction of leucite crystal). Data from [47]

Vickers hardness and fracture toughness by indentation fracture (IF) method were evaluated on the polished surface before and after ion exchange. These properties were measured using a Vickers microhardness tester (MVK-H-3, Mitutoyo) with load of 9.8 N and dwell time of 20 s. The diagonal of hardness impression and the length of radial crack emanated from the corner of hardness impression were measured using an optical microscope (Zeiss) under magnification of 200 times, within 30 s after indentation to minimize the slow crack growth phenomenon [16]. Vickers hardness, HV, and fracture toughness, K_{Ic}, were calculated according to the following equations [16,49]:

$$HV = \frac{1.8544 \cdot P}{(2a)^2} \tag{7}$$

$$K_{Ic} = 0.016\left(\frac{E}{H}\right)^{1/2}\left(\frac{P}{c^{3/2}}\right) \tag{8}$$

where, P is the indentation load, a and c is the half-size of diagonal of the indentation or radial/median crack, respectively (Figure 7), E is the elastic modulus, and H is the material's hardness [defined as $H = P/(2a^2)$]. The elastic modulus of each porcelain was determined by the ultrasonic pulse-echo method [50].

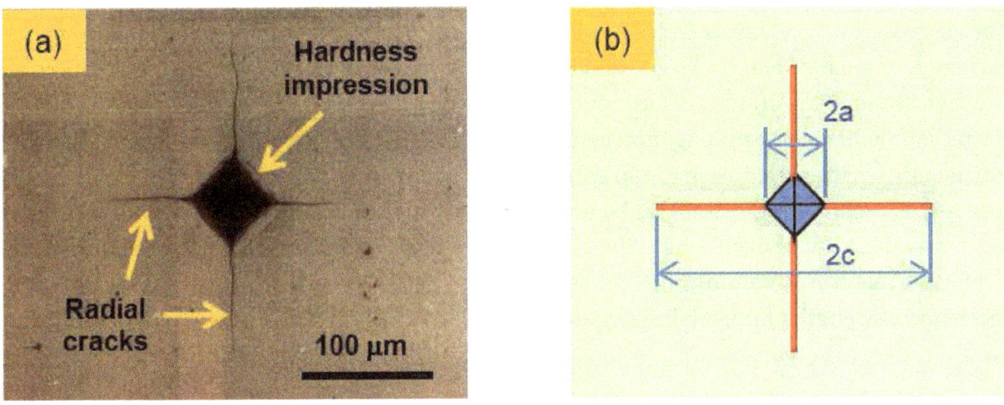

Figure 7. Optical micrograph of a Vickers hardness impression and the radial/median cracks generated on the corners of impression on the polished surface of a dental porcelain (a), and schematic image showing the dimensions a and c used to calculate the hardness, HV, and fracture toughness, K_{Ic}, by indentation fracture (IF) method

Figure 8a shows that ion exchange process increased significantly the fracture toughness, K_{Ic}, of most of the tested dental porcelains, with the increase in this property achieving up to around 150% (variation from 0.61 to 1.56 MPa.m$^{1/2}$ in porcelain C). However, there was also one porcelain (B with highest leucite content) that had no positive response to this toughening treatment. In general, the increase in K_{Ic} was higher for the porcelains with lower leucite content. The variation in Vickers hardness, HV, followed similar tendency as K_{Ic} (Figure 8b), but with lower relative increases that achieved up to around 70% (variation from 7.6 to 12.7 GPa in porcelain Cb).

Another work also observed beneficial effects of applying ion exchange (with K-containing compound at 450°C for 30 min) to increase the fracture toughness of dental porcelains. For eight dental porcelains, K_{Ic} increased between 39% and 116%, but no significant differences in hardness values were observed [51].

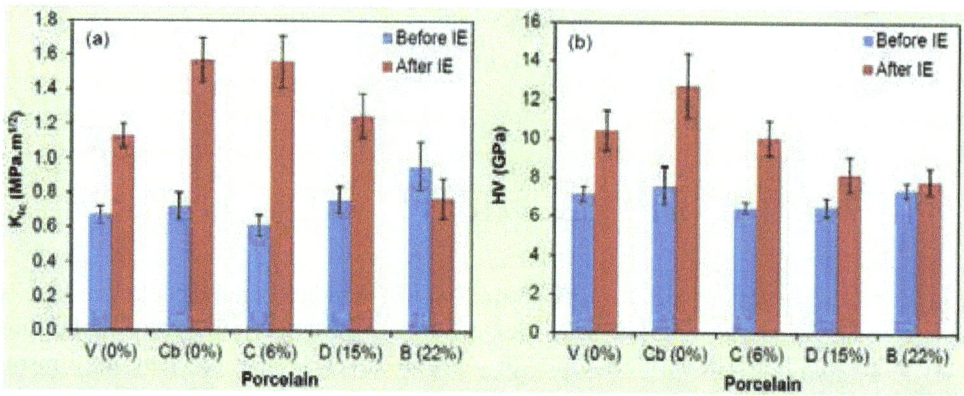

Figure 8. Fracture toughness, K_{Ic} (a), and Vickers hardness, *HV* (b), of dental porcelains before and after ion exchange (IE) process (in parenthesis the vol% of leucite crystal). Data from [47]

The increase in K_{Ic} value is indicative of the operation of ion stuffing mechanism by ion exchange of smaller Na^+ by larger K^+ in the glassy matrix, which introduced residual compressive stress fields on the surface region of the porcelain. The compressive stresses hinder the radial/median crack propagation generated by the Vickers indentation (Figure 9), decreasing the ratio c/a (ratio between the sizes of radial/median crack and indentation diagonal) and increasing fracture toughness. This behavior is highly desirable since the compressive surface layer may decrease or even inhibit the generation of large and deep cracks on the surface of a dental porcelain restoration during mastication. Since surface cracks are deleterious to the mechanical strength of porcelains, decreasing its size results in lower strength degradation.

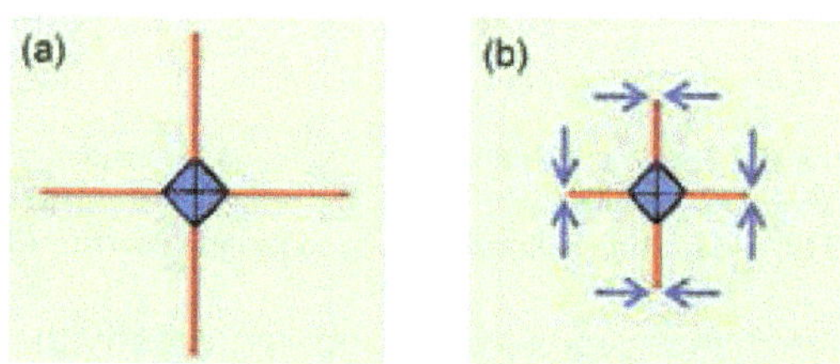

Figure 9. Schematic images of radial/median cracks generated at the corners of Vickers impression: (a) before ion exchange, without residual stresses and (b) after ion exchange, indicating the shortening of the cracks due to the residual compressive stress fields (indicated by arrows)

The residual stress (MPa) introduced by ion exchange was calculated according to the following equation [47]:

$$\sigma_r = \frac{K_{Ic,b} - K_{Ic,a}}{2\left(\frac{c}{\pi}\right)^{1/2}} \tag{9}$$

where, $K_{Ic,b}$ and $K_{Ic,a}$ are the fracture toughness measured before and after ion exchange, respectively. Figure 10 shows the calculated residual compressive stress values. It can be seen that ion exchange by paste method can generate significant compressive stresses, up to around 90 MPa, on the surface of dental porcelains.

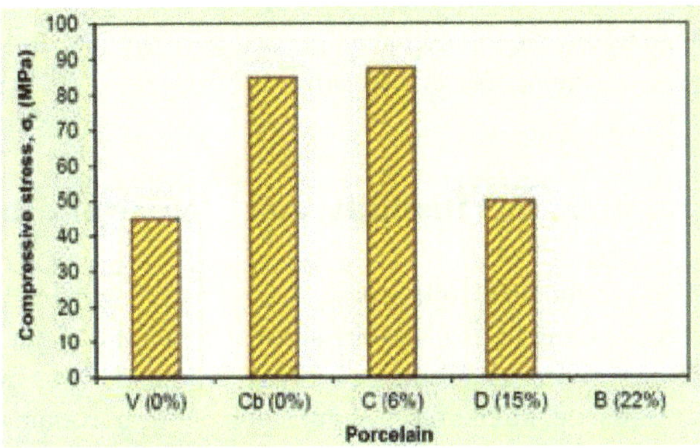

Figure 10. Residual compressive stress generated on the surface of dental porcelains by ion exchange process. Data from [47]

For porcelain B, which had the highest leucite fraction (22%), however, there was no introduction of residual compressive stress, although significant increase in K_2O content after ion exchange have been detected (Figure 6). A possible explanation could be the occurrence of stress relaxation caused by the viscoplasticity of glassy matrix during the heat treatment of ion exchange process. The temperature used for ion exchange (450°C), however, seemed to be sufficiently lower than the glass transition temperatures, T_g, for all porcelains, as indicated by their annealing point (Table 3). This point is defined as the temperature at which the glass viscosity is 10^{14} Pa.s and most of the internal stresses are reduced within about 15 min [52]. The annealing point was determined by calculating the viscosity curve as a function of temperature from the chemical composition using the program SciGlass (SciGlass v.7.7, MDL Information Systems) [53].

Porcelain	V	Cb	C	D	B
T (°C) at $\eta = 10^{14}$ Pa.s	708	673	564	641	659

Table 3. Calculated annealing point (temperature) of dental porcelains

The overall and glassy matrix chemical compositions (Table 2) of porcelain B, compared to the other porcelains, did not justify the ineffectiveness of ion exchange to improve the mechanical properties in the porcelain B. The tendency that the relative increases in fracture toughness and hardness decreases with the increase in leucite content (Figure 8) suggests that the beneficial effects of ion exchange are counterbalanced by the toughening effect of leucite particles. It is possible that the ion exchanged K^+ ions could preferentially occupy the sites under the tensile residual stressed regions around leucite particles and agglomerates, which could be energetically more favorable causing less increase in residual stresses. The chemistry of the glassy matrix also affected the response to ion exchange treatment, as can be seen from the results of both completely glassy porcelains (V and Cb, Figure 8). Both porcelains had the same initial Na_2O content (3.6 mol%, Table 2), but in porcelain Cb higher residual compressive stress was generated (Figure 10). It is difficult to predict the interactive effects of different ions present in the glassy matrix in the ion exchange process between Na^+ and K^+ ions, since even low concentrations of some elements can have strong effects [26].

5. Ion exchange on strength, reliability, and lifetime of dental porcelains

In this section, the effects of chemical toughening using ion exchange by the paste method on the strength and lifetime of dental porcelains are shown. In this study, a feldspathic porcelain (Ultropaline Super Transparent, Jen Dental – UST) recommended for metal-ceramic or all ceramic restorations was used. The chemical composition of this porcelain is shown in Table 4.

SiO_2	Al_2O_3	K_2O	Na_2O	CaO	MgO	Traces
70.2 (69.9)	10.4 (9.6)	10.6 (10.7)	4.1 (4.6)	3.1 (3.5)	1.3 (1.5)	Ti, Fe, Zr, Ni

Table 4. Chemical composition (mol%) measured by XRF spectroscopy and calculated glassy matrix composition (in parenthesis) of Ultropaline Super Transparent (UST) dental porcelain (containing 12 vol% leucite particles)

Disc-shaped green specimens were prepared by vibration-condensation method and then sintered at 930°C following the firing schedule recommended by the manufacturer. After firing, the specimens were machined and mirror-polished with diamond suspensions down to 1 μm. For each test condition at least 10 specimens (12.5 mm in diameter and 1 mm in thickness) were prepared.

For the chemical tempering, a paste was prepared by mixing 10 g of KNO_3 powder (Merck) with 4 mL of deionized water. Porcelain discs, containing 0.4 g of this paste on the polished surface, were subjected to the ion exchange treatment in an electric furnace (FP-32, Yamato) with a heating rate of 5°C/min at 470°C (or other specified temperature) during 15 min, after an intermediate step at 150°C for 20 min for drying. After the treatment, the paste residue was easily removed with sprayed water.

The porcelain's strength was determined in biaxial flexural mode, which is an adequate loading condition for thin specimens, like the dental restorations. The biaxial flexure strength (σ_f) was determined using a piston-on-three-balls loading device in a universal mechanical testing machine (Syntech 5G, MTS) at a constant stress rate of 10 MPa/s (or other specified rate), with the specimen immersed in artificial saliva (Table 5) heated to 37°C (Figure 11). Flexural test performed in artificial saliva at 37°C is more severe (strength tends to be lower) than in usual laboratory environment, but these conditions are more clinically relevant, since they are closer to the oral environmental conditions.

KH_2PO_4 (2.5 mM)	Na_2HPO_4 (2.4 mM)	$KHCO_3$ (1.5 mM)	NaCl (1.0 mM)	$MgCl_2$ (0.15 mM)	$CaCl_2$ (1.5 mM)	Citric acid (0.002 mM)
100 mL	100 mL	100 mL	100 mL	100 mL	100 mL	6 mL

Table 5. Composition of artificial saliva [5,54]

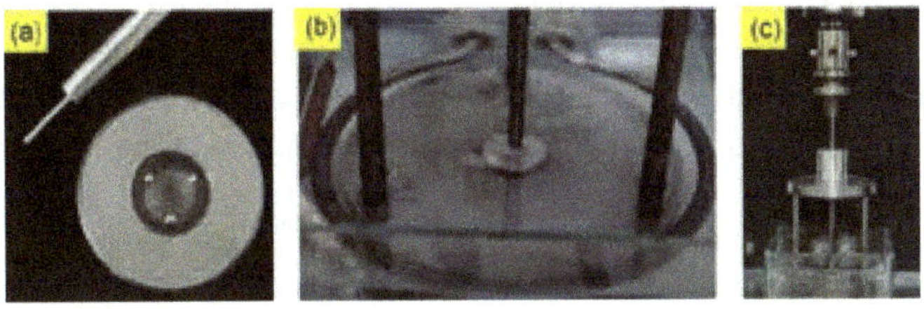

Figure 11. Images of biaxial flexural device: (a) loading piston and three-ball support; (b) disc specimen positioned in the flexural device and immersed in artificial saliva with heating element; (c) general view of the biaxial flexural loading device [5]

The biaxial flexural strength (σ_f) was calculated using the following equation [5,55]:

$$\sigma_f = \frac{0.2387 \cdot F}{w^2}\left\{(1+\upsilon)\left[1-2\ln\left(\frac{B}{A}\right)\right]+(1-\upsilon)\left[\left(\frac{A}{C}\right)^2-\frac{1}{2}\left(\frac{B}{C}\right)^2\right]\right\} \tag{10}$$

where, F is the load at fracture, w is the specimen thickness, A is the radius of the support circle (4 mm), B is the radius of the piston (0.85 mm), C is the radius of the specimen, and ν is the Poisson's ratio (determined by ultrasonic pulse-echo method [56]).

Figure 12 shows the effects of ion exchange temperature on biaxial flexural strength, σ_f, and fracture toughness, K_{Ic}, determined by the indentation fracture method. It can be seen that ion exchanged specimens had significantly higher σ_f values (around 130 MPa) compared to the control sample – without chemical tempering (57 MPa). This substantial increase of around 130% in strength was directly related to the increase in material's resistance to crack propagation, that is, fracture toughness. This property increased from 1.3 MPa.m$^{1/2}$ in control sample to around 2.8 MPa.m$^{1/2}$ in chemically tempered samples (relative increase of around 120%) [57]. The ion exchange by the paste method although short in time results in significantly strengthening and toughening of dental porcelains. Other works also showed increases between 20 to 83% in flexural strength by applying this method for feldspathic porcelains [58,59], although there are also reports showing no significant increases in some dental porcelains [60], particularly those with high K_2O content [61].

No effects of ion exchange temperature on σ_f and K_{Ic} were observed when this treatment was carried out between 430 and 510°C (Figure 12), which corresponded to a range between 75 and 89% of the glass transition temperature (T_g) of UST porcelain. This temperature (T_g), determined by differential thermal analysis, DTA (404S, Netzsch), at a heating rate of 5°C/min in air, was 575°C [57]. At this temperature range, one could expect a significant increase in the ion exchanged K^+ ions with the increase in temperature, since the kinetic of ion exchange by interdiffusion with Na^+ ions of the porcelain is exponentially dependent on the temperature (Equations 2–4). In fact, it was determined by XRF spectroscopy that the sodium content in the KNO_3 paste residue increased with the increase in temperature of ion exchange treatment (Figure 13). The increase in K_2O content and reduction of Na_2O content in porcelain were also confirmed by EDS analysis in SEM. In this case, it seems that the increase in ion exchange rate was counterbalanced by the stress relaxation with the increase in temperature (Equation 6), inhibiting further increase in residual compressive stress and increases in toughness and strength. Therefore, an appropriate temperature for making ion exchange in dental porcelains with K^+ ions exchanged by Na^+ ions by paste method seems to be around 80% of glass transition temperature (T_g), or around 100°C lower than T_g (for UST porcelain at 470°C).

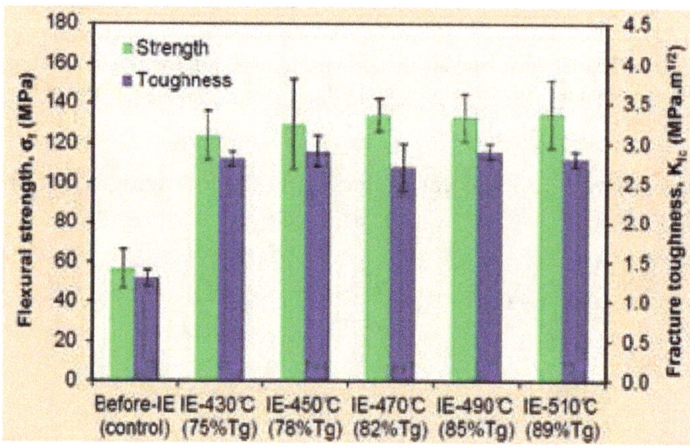

Figure 12. Biaxial flexural strength, σ_f, and fracture toughness, K_{Ic}, of UST dental porcelain before and after ion exchange (IE) at different temperature. In parenthesis is the temperature of IE relative to the glass transition temperature (T_g = 575°C). Data from [57]

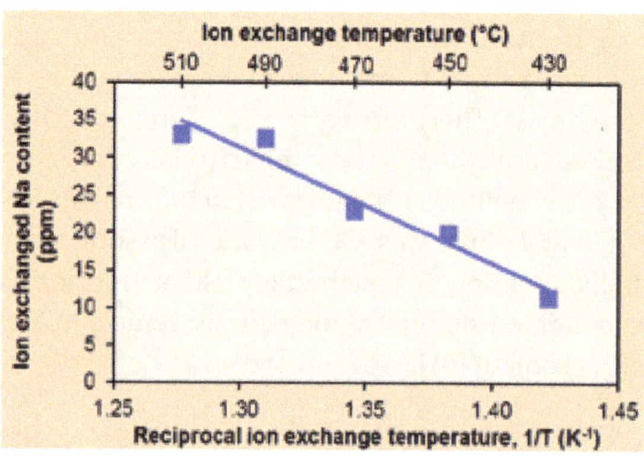

Figure 13. Sodium (Na) content in the KNO_3 paste residue after ion exchange treatment at different temperature for UST porcelain. Data from [57]

Small variations in flexural strength with varying ion exchange temperatures were also reported in another work. In general no significant differences were observed in strength when ion exchange with K-containing compound was applied to three dental porcelains between 300 and 600°C (for 30 min), although maximum values were observed at 450°C. Besides, small variations in strength with varying ion exchange time (10 to 90 min at 450°C) were also observed [62].

In order to evaluate the effects of the size of surface flaws on the strength of ion exchanged porcelain, the polished surface of specimens was indented with Vickers impression with increasing load from 2 to 49 N. In Figure 14a, it can be seen that the biaxial flexural strength, σ_f, of UST porcelain, without (control) and with ion exchange treatment (KNO_3, 470°C, 15 min), decreased with the increase in indentation load, since this increase causes the increase in radial/median crack size, c (Figure 7b). However, the decrease in σ_f was more accentuated in the ion exchanged specimens. These results showed that the beneficial effects of ion exchange are more pronounced for small surface cracks, and less effective for larger and deeper flaws. This behavior is related to the gradient of K^+ ion content introduced in the surface region, which results in a gradual decrease of residual compressive stress to the interior of porcelain. The positive effects of ion exchange disappear for flaws deeper than the thickness of compressive stress layer.

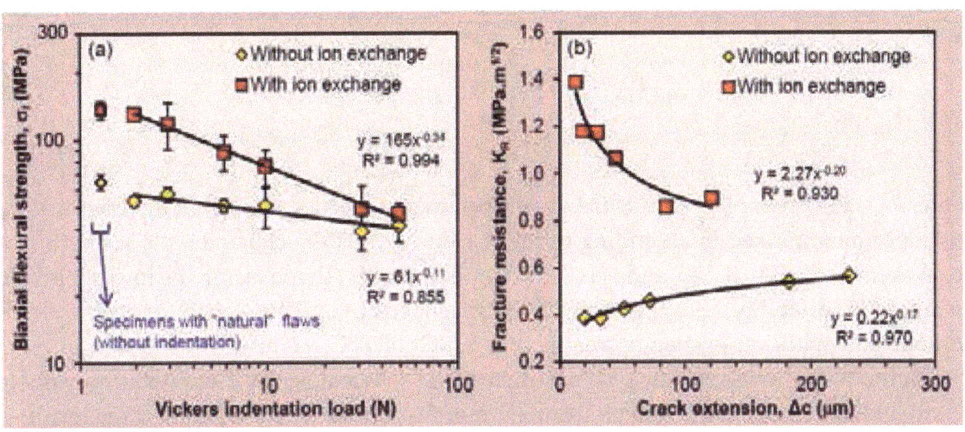

Figure 14. Biaxial flexural strength, σ_f, after a Vickers indentation at different loads (a) and fracture resistance, K_R, as a function of crack extension, Δc (b), for UST dental porcelain with and without ion exchanges. Data from [63,64]

The fracture resistance, K_R, or K_{Ic} as a function of crack extension, Δc, can be evaluated using the following equation [64–66]:

$$K_R = k \cdot (\Delta c)^q \quad (11)$$

Where, the parameters k and q are determined using the data obtained by power law fits on the results of biaxial flexural strength, σ_f, as a function of indentation load, P, in Figure 14a.

The calculated fracture resistance, K_R, values as a function of crack extension, Δc, are shown in Figure 14b. For the non-treated porcelain, K_R increased with the increase in crack size, the so-called rising R-curve behavior (crack growth resistance curve). This behavior is observed in leucite-containing porcelains and is caused by the friction between rough crack surfaces caused by crack deflection around leucite agglomerates [11]. Since this mechanical grip acts in the crack wake, the shielding effect at the crack tip is intensified with the increase in crack extension [67]. Rising R-curve effect is a desirable material behavior since it is necessary additional energy to propagate the crack, besides that needed at the crack tip [65,66]. For the ion exchanged porcelain, an opposite result was observed, with the decrease in K_R as the crack size increased. The decreasing residual compressive stress from the ion exchanged surface cancelled the rising R-curve effect of porcelain microstructure and in addition, resulted in a decreasing R-curve effect. This behavior makes the strength of ion exchanged porcelain more sensitive to the size of flaws.

The variability of strength in ceramic materials is closely related to its flaw population, since fracture is a probabilistic event due to a random-like distribution of the strength-limiting flaws [68]. The strength variability of ceramic materials can be evaluated using Weibull statistic, which is based on the weakest-link theory, where the more severe flaw results in fracture propagation and determine the strength [69]. The Weibull two-parameter distribution is given by [14]:

$$P_{f,i} = 1 - \exp\left[-\left(\frac{\sigma_{f,i}}{\sigma_0} \right)^m \right] \tag{12}$$

where, $P_{f,i}$ is the probability of fracture of ith specimen, i is order number of $\sigma_{f,i}$ (fracture stress of ith specimen, ranked in ascending order of values), σ_0 is the characteristic strength (scale factor, defined for $P_f = 63.2\%$), and m is the Weibull modulus (shape factor; the lower this value, the higher is variability). For this analysis, 30 specimens of UST porcelain for each condition (without and with ion exchange treatment – KNO_3, 470°C, 15 min) were tested in biaxial flexural mode at a stress rate of 1 MPa/s (in artificial saliva at 37°C). The Weibull parameters (σ_0 and m) were calculated based on the maximum likelihood method [70] and the results are shown in Figure 15.

It can be seen in Figure 15 that the ion exchange treatment increased more than 100% the characteristic strength, σ_0, but it also caused larger variability in fracture stress, reducing around 50% the Weibull modulus, m. Higher variability means lower reliability in the strength of porcelain, which was caused by the decreasing R-curve behavior in ion exchanged porcelain (Figure 14b), since shallow cracks were significantly toughened by high residual compressive stress level near the surface, but deeper surface cracks were less shielded by the decreasing resistance to crack growth. Therefore, strongest specimens in non-treated porcelain were strengthened more than the weakest ones. Clinically, it is more relevant to consider the fracture stress in an acceptable fracture probability, for example, at $P_f = 5\%$. Although strengthening is not so high as compared to the σ_0 value ($P_f = 63.2\%$), even at this low level of $P_{f,5\%}$ a significant

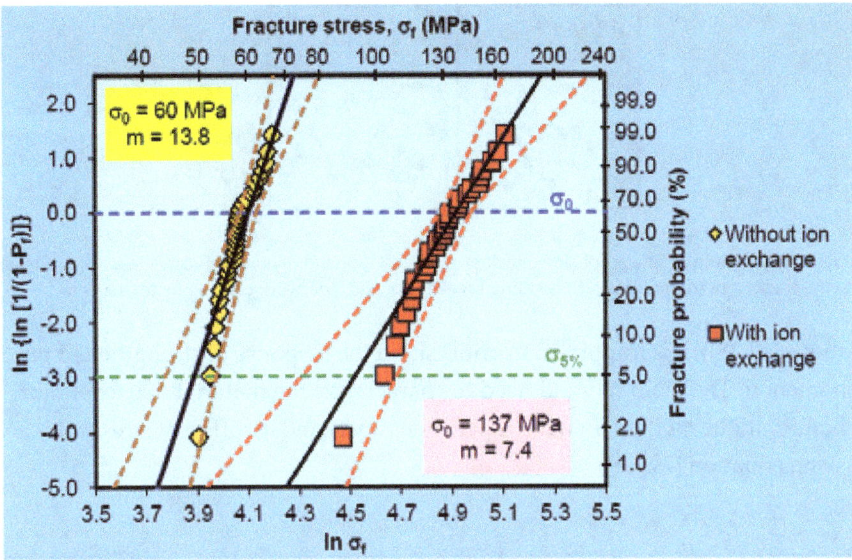

Figure 15. Weibull plot of biaxial flexural strength data, σ_f, for UST dental porcelain with and without ion exchange. Dotted lines are 95% confidence interval; m is Weibull modulus, σ_0 is characteristic strength, and $\sigma_{5\%}$ is fracture stress at 5% fracture probability. Data from [71]

strengthening effect (at 95% confidence interval) is observed (Figure 15). In practice, the strengthening effect could be even higher than for P_f = 5%, since clinical studies of inlays constructed with feldspathic porcelains have shown high fracture rates, up to 48% in evaluation periods of up to 3 years [72,73]. Therefore, although ion exchange reduces the mechanical reliability of dental porcelain, the strengthening effect is still significant even for low levels of fracture probability.

Another relevant factor related to the lifetime of a porcelain restoration is the mechanical degradation over time. When a ceramic material is subjected to a stress level lower than the fracture stress, the flaws (cracks) can growth slowly in a stable manner up to the time at which loading comes to a halt, reducing the material's strength due to the increase in crack size, or when a flaw achieves a critical size, given by the Griffith-Irwin fracture criterion (Equation 1), that results in (fast) fracture. This phenomenon is known as slow (or subcritical) crack growth, SGC, and silicate glasses, like porcelains, usually are highly susceptible to this type of degradation [74–76]. SCG occurs mainly by a stress corrosion mechanism (Figure 16), in which water molecules diffuse and are adsorbed at the crack tip, and then cause bonding rupture of glass network yielding Si–OH groups on each fracture surface, resulting in crack growth, by [76,77]:

$$\text{Si}-\text{O}-\text{Si}+\text{H}_2\text{O}\rightarrow 2\text{Si}-\text{OH} \tag{13}$$

Since the oral environment is aggressive to porcelain restorations and has many characteristics that favor SCG (water from saliva and dentin tubules, masticatory stresses, temperature and

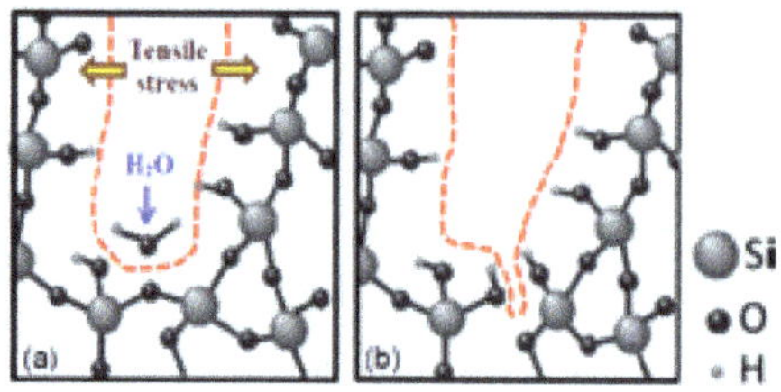

Figure 16. Schematic molecular images of slow crack growth (SCG) phenomenon in silicate glass: (a) diffusion of water molecule to the crack front (dotted line); (b) bonding breakage of SiO_4^{4-} network causing increase in crack size

pH variations [67,78]), it is important to understand the response of strengthened porcelain to this phenomenon. The main method used to characterize the material's susceptibility to SCG is the dynamic fatigue method, in which specimens are tested in different stress rates and using the following equation [79–81]:

$$\log \sigma_f = \frac{1}{n+1} \log \dot{\sigma} + \log \sigma_{f0} \tag{14}$$

where, σ_f is the flexural strength, $\dot{\sigma}$ is the stress rate, σ_{f0} the scaling parameter, and n is the slow crack growth, SCG, susceptibility coefficient (the higher is n value the lower is the susceptibility). Figure 17 shows the results of this test for UST porcelain, non-treated and ion exchanged (KNO_3, 470°C, 15 min). It can be seen that ion exchanged porcelain had significantly higher n value (relative increase of around 50%), showing more resistance to SCG degradation. This effect is substantial and can impact the lifetime of a restoration.

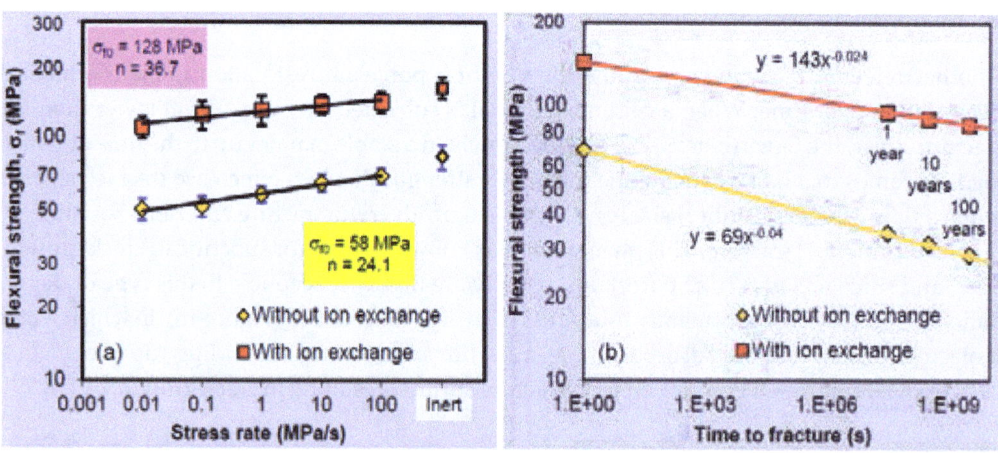

Figure 17. Biaxial flexural strength, σ_f, as a function of stress rate (a) and predicted flexural strength as a function of time to fracture (b) for UST dental porcelain, with and without ion exchange, in artificial saliva at 37°C. In (a), σ_{f0} is the scaling parameter and n is the slow crack growth, SCG, susceptibility coefficient. Inert strength was determined at 100 MPa/s in air with a drop of silicone oil on the tensile surface to inhibit the occurrence of SCG. In (b), the slope of fitted curve is related with n. Data from [63,71]

Using the results of the dynamic fatigue test, it is possible to extrapolate the strength decrease after long lifetimes, as shown in Figure 17b. For both conditions, the average strength decreases over time, and after 10 years the expected remaining strength drops to around 30 MPa for the non-treated porcelain, but it still remains high (around 90 MPa) for the ion exchanged porcelain [71]. The increase in the stress corrosion coefficient, n, is a significant effect, since the difference in strength between ion exchanged porcelain and non-treated one increases over time. Therefore, besides increasing the strength the compressive layer generated by ion exchange process also decreases the rate of strength degradation by slow crack growth phenomenon.

Using the results of Weibull distribution (Figure 15) and dynamic fatigue test (Figure 17a) it is possible to construct the strength-probability-time (SPT) diagram [79,82,83], as shown in Figure 18. This diagram makes possible the estimation of a fracture stress at any time during the lifetime of a dental restoration at any fracture probability level. For example, it is possible to verify that UST porcelain after ion exchange has at least twice the fracture stress than non-treated porcelain even at a fracture probability as low as 1% during long lifetimes (e.g., 100 years). Note that the difference in fracture stress increases over time, at any level of fracture probability.

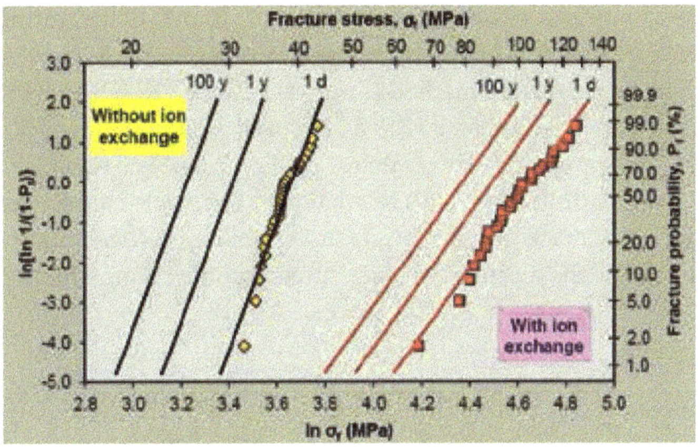

Figure 18. SPT (strength-probability-time) diagram for 1 day (1 d), 1 year (1 y), and 100 years (100 y) for UST dental porcelain, with and without ion exchange. Data from [63,71]

6. Two-step tempering processes

There are some two-step tempering methods proposed to overcome some limitations of the conventional (one step) ion exchange process. It was demonstrated that the application of ion exchange by the paste method after a thermal tempering treatment in a feldspathic porcelain resulted in an increase in the Weibull modulus (m = 14.6) compared to only thermal tempered condition (m = 8.7), with small decrease in average strength value [84]. Similar results were observed after applying a thermal tempering followed by a short-time chemical tempering in a silica-soda-lime glass dielectric, in which the reliability of thermal shock resistance was enhanced, with no change in the strength [85].

In order to increase the case depth, a two-step ion exchange was applied to a feldspathic porcelain. The specimens coated with a slurry containing 10 mol% LiCl and 90 mol% NaCl were heat treated at 750°C for 30 min and then cooled and heat treated at 450°C for 30 min. The first step was conducted above the melting temperature of chloride mixture and the second step below the glass transition temperature (T_g) of porcelain. The intent of first step was to exchange Na^+ ions and some K^+ ions in porcelain by smaller Li^+ ions from chloride mixture, considering that the high diffusivity at 750°C could result in a deeper exchanged layer. In the second step, some of Li^+ ions would be re-exchanged by bigger Na^+ ions, introducing a deep compressive layer. The determined thickness of ion exchanged layer was at least 140 μm [86].

Another two-step ion exchange method was proposed to introduce an engineered stress profile (ESP), in which a designed ion exchange stress profile with steep increase of the compression stress from the surface achieves a maximum at a predetermined depth and then decreases towards the interior of the glass. In the ESP glass, surface cracks growth in a stable manner up to the maximum compressive stress and are subsequently arrested, resulting in an uncommon surface crack pattern with a set of arrested cracks, before unstable fracture occurs. This behavior results in significant lower variability of fracture stress. The ESP is generated by a short second ion exchange process carried out for partial removal of the stuffing ion introduced in the first extended treatment [87,88]. This method was applied to a leucite-reinforced glass-ceramic prepared by heat-pressing method (Empress), using a KNO_3 bath in first step during 11 h at 450°C followed by a second step with a bath of 70 mol% KNO_3 and 30 mol% $NaNO_3$ at 400°C for 30 min (both temperature lower than T_g of Empress). The two-step method resulted in increase in Weibull modulus (m = 11.7) in relation to the single-step ion exchange (m = 5.1) [89], and very high slow crack growth (SCG) susceptibility coefficient (n = 107) [90]. Figure 19 shows schematically the residual surface stress profiles from different ion exchange methods.

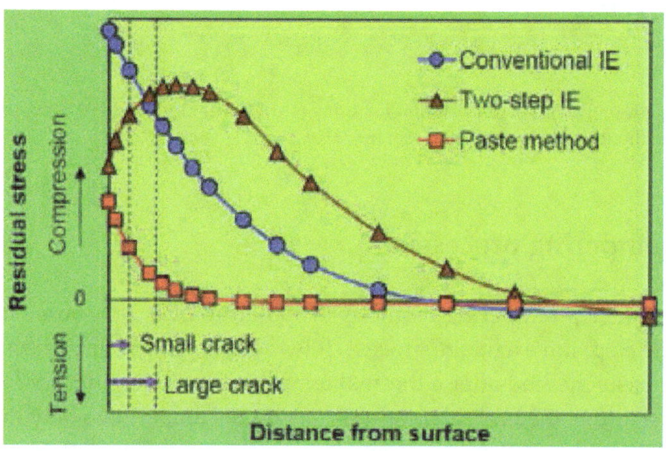

Figure 19. Residual surface stress profiles expected from different ion exchange (IE) methods: conventional extended (one-step) IE; two-step IE; and paste method (short IE treatment). Vertical dotted lines represent the depths of small and large surface crack. Data from [71]

7. Concluding remarks

Dental porcelains have been used for many decades because of their excellent aesthetic qualities, including the possibility to mimic natural teeth, besides other favorable characteristics. However, the low strength and fracture toughness, and high susceptibility to mechanical degradation by slow crack growth result in relatively low lifetimes in the oral environment. The introduction of a compressive surface layer by ion exchange is a promising process to enhance the mechanical behavior of dental porcelains. Exchanging small ions (e.g., Na^+) by larger ones (e.g., K^+) in the glassy matrix at temperatures lower than its glass transition temperature (T_g) can result in a steep gradient of residual compressive stress, which significantly increases the fracture toughness (K_{Ic}) in the porcelain surface. This increase leads to significant strengthening effects, even using the paste method, which is carried out in short times (less than 1 h). However, the strengthening is more pronounced for shallow surface flaws, because of the limited thickness of compressive layer and decreasing resistance curve (R-curve) behavior, which increases the variability of fracture stress and decreases the mechanical reliability (decreases the Weibull modulus, m). On the other hand, strengthening by ion exchange also significantly increases the resistance to slow crack growth phenomenon (stable crack propagation at low stresses intensified by water corrosion). This results in lower strength degradation over time and counterbalances the negative effect of decreasing reliability, leading to higher strength retention after long lifetimes (decades) even at a level of low fracture probability (e.g., 5%). The two-step ion exchange method has the advantage of increasing the material reliability, but at the expense of longer periods of treatment (above 10 h). In conclusion, it is expected that the application of chemical tempering (strengthening by ion exchange) can improve the lifetime of dental porcelain restorations.

Acknowledgements

The authors acknowledge the Brazilian agencies FAPESP, CAPES, and CNPq for the financial support.

Author details

Humberto Naoyuki Yoshimura[1*] and Paulo Francisco Cesar[2]

*Address all correspondence to: humberto.yoshimura@ufabc.edu.br

1 Center for Engineering, Modelling and Applied Social Sciences, Universidade Federal do ABC, Santo André, SP, Brazil

2 School of Dentistry, Universidade de São Paulo, São Paulo, SP, Brazil

References

[1] Braga RR, Ballester RY, Daronch M. Influence of time and adhesive system on the extrusion shear strength between feldspathic porcelain and bovine dentin. Dental Materials 2000;16(4) 303–310.

[2] Peumans M, DeMunck J, Fieuws S, Lambrechts P, Vanherle G, Van Meerbeek B. A prospective ten-year clinical trial of porcelain veneers. The Journal of Dentistry 2004;6(1) 65–76.

[3] Fredericci C, Yoshimura HN, Molisani AL, Pinto MM, Cesar PF. Effect of temperature and heating rate on the sintering of leucite-based dental porcelains. Ceramics International 2011;37(3) 1073–1078.

[4] Burke FJT, Watts DC. Fracture resistance of teeth restored with dentin-bonded crowns. Quintessence International 1994;25(5) 335–340.

[5] Pinto MM. Influence of storage pH on subcritical crack growth of dental porcelains with different microstructures. Master thesis. University of São Paulo; 2006.

[6] Sasahara RMC, Ribeiro FC, Cesar PF, Yoshimura HN. Influence of the finishing technique on surface roughness of dental porcelains with different microstructures. Operative Dentistry 2006;31(5) 577–583.

[7] Smales RJ, Etemadi S. Survival of ceramic onlays placed with and without metal reinforcement. Journal of Prosthetic Dentistry 2004;91(6) 548–553.

[8] Hayashi M, Tsuchitani Y, Kawamura Y, Miura M, Takeshige F, Ebisu S. Eight-year clinical evaluation of fired ceramic inlays. Operative Dentistry 2000;25(6) 473–481.

[9] Yoshimura HN, Molisani AL, Narita NE, Manholetti JLA, Cavenaghi JM. Mechanical properties and microstructure of zinc oxide varistor ceramics. Materials Science Forum 2006;530–531, 408–413.

[10] Yoshimura HN, Gonzaga CC, Cesar PF, Miranda Jr WG. Relationship between elastic and mechanical properties of dental ceramics and their index of brittleness. Ceramics International 2012;38(6) 4715–4722.

[11] Yoshimura HN, Cesar PF, Miranda Jr WG, Gonzaga CC, Okada CY, Goldenstein H. Fracture toughness of dental porcelains evaluated by IF, SCF, and SEPB methods. Journal of the American Ceramic Society 2005;88(6) 1680–1683.

[12] Gonzaga CC, Okada CY, Cesar PF, Miranda Jr WG, Yoshimura HN. Effect of processing induced particle alignment on the fracture toughness and fracture behavior of multiphase dental ceramics. Dental Materials 2009;25(11) 1293–1301.

[13] Meyers MA, Chawla KK. Mechanical Behavior of Materials. 2 ed. Cambridge: Cambridge University Press; 2009.

[14] Cesar PF, Yoshimura HN, Miranda Jr WG, Miyazaki CL, Muta LM. Relationship between fracture toughness and flexural strength in dental porcelains. Journal of Biomedical Materials Research: Part B – Applied Biomaterials 2006;78(2) 265–273.

[15] Gonzaga CC. Subcritical crack growth in dental ceramics: effect of the material (microstructure) and the test method. Doctoral thesis. University of São Paulo; 2007.

[16] Cesar PF, Soki FN, Yoshimura HN, Gonzaga CC, Styopkin V. Influence of leucite content on slow crack growth of dental porcelains. Dental Materials 2008;24(8) 1114–1122.

[17] Borba M, Araújo MD, Fukushima KA, Yoshimura HN, Cesar PF, Griggs JA, Della Bona Á. Effect of the microstructure on the lifetime of dental ceramics. Dental Materials 2011;27(7) 710–721.

[18] Cesar PF. Fracture toughness and microstructure of dental porcelains. Doctoral thesis. University of São Paulo; 2002.

[19] Medeiros IS, Luz LA, Yoshimura HN, Cesar PF, Hernandes AC. $Al_2O_3/GdAlO_3$ fiber for dental porcelain reinforcement. Journal of the Mechanical Behavior of Biomedical Materials 2009;2(5) 471–477.

[20] Araújo, MD, Miranda RBP, Fredericci C, Yoshimura HN, Cesar PF. Effect of fiber addition on slow crack growth of a dental porcelain. Journal of the Mechanical Behavior of Biomedical Materials 2015;44 85-95.

[21] Kistler SS. Stresses in glass produced by nonuniform exchange of monovalent ions. Journal of the American Ceramic Society 1962;45(2) 59–68.

[22] Varshneya AK. Chemical strengthening of glass: lessons learned and yet to be learned. International Journal of Applied Glass Science 2010;1(2) 131–142.

[23] Barsoum MW. Fundamentals of Ceramics. 2 ed. Bristol: IOP; 2003.

[24] Callister Jr WD. Materials Science and Engineering: an introduction. 7 ed. New York: John Wiley & Sons; 2007.

[25] Morris DJ, Myers SB, Cook RF. Indentation crack initiation in ion-exchanged aluminosilicate glass. Journal of Materials Science 2004;39(7) 2399–2410.

[26] Gy R. Ion exchange for glass strengthening. Materials Science and Engineering B 2008;149(2) 159–165.

[27] Sane AY, Cooper AR. Stress buildup and relaxation during ion exchange strengthening of glass. Journal of the American Ceramic Society 1987;70(2) 86–89.

[28] Hodge IM. Adam-Gibbs formulation of enthalpy relaxation near the glass transition. Journal of Research of the National Institute of Standards and Technology 1997;102(2) 195–205.

[29] Kese KO. Relaxation and Nanomechanical Studies of the Vickers residual stress field in glass. Doctoral thesis. Royal Institute of Technology, Stockholm; 2004.

[30] Mackert Jr JR. Dental ceramics. In: Anusavice KJ (ed.) Science of Dental Materials. 10 ed. Rio de Janeiro: Guanabara Koogan; 1998. p345–366.

[31] Jones DW. Development of dental ceramics: an historical perspective. Dental Clinics of North America 1985;29(4) 621-644.

[32] Denry IL. Recent advances in ceramics for dentistry. Critical Reviews in Oral Biology & Medicine 1996;7(2) 134–143.

[33] Braga RR. Influence of time and adhesive system on the bond strength of feldspathic porcelain and bovine dentin. Doctoral thesis. University of São Paulo; 1999.

[34] McLean JW. The science and art of dental ceramics, volume I: the nature of dental ceramics and their clinical use. Chicago: Quintessence; 1979. p23–51.

[35] Ban S, Matsuo K, Mizutani N, Tanikawa H, Kaikawa K, Hasegawa J. Effect of cooling condition on leucite crystals in dental porcelain. Dental Materials Journal 1999;18(2) 137–143.

[36] Ong JL, Farley DW, Norling BK. Quantification of leucite concentration using X-ray diffraction. Dental Materials 2000;16(1) 20–25.

[37] Fredericci C, Yoshimura HN, Molisani AL, Bellinati CE, Alcântara RQM, Cesar PF. Sintering of commercial dental leucite-based porcelains. Materials Science Forum 2008;591–593 692–696.

[38] Denry IL, Mackert JR, Holloway JA, Rosenstiel SF. Effect of cubic leucite stabilization on the flexural strength of feldspathic dental porcelain. Journal of Dental Research 1996;75(12) 1928–1935.

[39] McLean JW. The science and art of dental ceramics. Operative Dentistry 1991;16(4) 149–156.

[40] Piché PW, O`Brien WJ, Groh CL, Boenke KM. Leucite content of dental selected dental porcelains. Journal of Biomedical Materials Research Part A 1994;28(5) 603–609.

[41] Mackert Jr JR, Russell CM. Leucite crystallization during processing of a heat-pressed dental ceramic. International Journal of Prosthodontic 1996;9(3) 261–265.

[42] Kon M, Kawano F, Asaoka K, Matsumoto N. Effect of leucite crystal on the strength of glassy porcelain. Dental Materials Journal 1994;13(2) 138–147.

[43] Lee HH, Kon M, Asaoka K. Influence of modification of Na_2O in a glass matrix on the strength of leucite-containing porcelains. Dental Materials Journal 1997;16(2) 134–143.

[44] Yoshimura HN, Pinto MM, Lima E, Cesar PF. Optical properties of dental bioceramics evaluated by Kubelka-Munk model. In: Bose S (ed.) Biomaterials Science – Processing, Properties and Applications III. Hoboken: John Wiley & Sons; 2013. p71–79.

[45] Seghi RRS, Sorensen JA. Relative flexural strength of six new ceramics materials. International Journal of Prosthodontic 1995;8(3) 239–246.

[46] Cesar PF, Yoshimura HN, Miranda Jr WG, Okada CY. Correlation between fracture toughness and leucite content in dental porcelains. Journal of Dentistry 2005;33(9) 721–729.

[47] Cesar PF, Gonzaga CC, Miranda Jr WG, Yoshimura HN. Effect of ion exchange on hardness and fracture toughness of dental porcelains. Journal of Biomedical Materials Research: Part B – Applied Biomaterials 2007;83(2) 538–545.

[48] Yoshimura HN, Cesar PF, Soki FN, Gonzaga CC. Stress intensity factor threshold in dental porcelains. Journal of Materials Science: Materials in Medicine 2008;19(5) 1945–1951.

[49] Anstis GR, Chantikul P, Lawn BR, Marshall DB. A critical evaluation of indentation techniques for measuring fracture toughness: I direct crack measurements. Journal of the American Ceramic Society 1981;64(9) 533–538.

[50] Yoshimura HN, Molisani AL, Narita NE, Cesar PF, Goldenstein H. Porosity dependence of elastic constants in aluminum nitride ceramics. Materials Research 2007;10(2) 127–133.

[51] Seghi RR, Denry I, Brajevic F. Effects of ion exchange on hardness and fracture toughness of dental ceramics. International Journal of Prosthodontic 1992;5(4) 309–314.

[52] Barsoum MW. Fundamentals of ceramics. Bristol: IOP; 2003.

[53] Mazurin OV. Glass properties: compilation, evaluation, and prediction. Journal of Non-Crystalline Solids 2005;351(12–13) 1103–1112.

[54] ten Cate JM, Duijsters PPE. Alternating demineralization and remineralization of artificial enamel lesions. Caries Research 1982;16(3) 201–210.

[55] ASTM-F394-78. Standard test method for biaxial flexure strength (modulus of rupture) of ceramic substrates. West Conshohocken: American Society for Testing Materials; 1996.

[56] Yoshimura HN, Molisani AL, Siqueira GR, Camargo AC, Narita NE, Cesar PF, Goldenstein H. Effect of porosity on mechanical properties of a high purity alumina. Cerâmica 2005;51(319) 239–251.

[57] Rosa V, Fredericci C, Moreira MF, Yoshimura HN, Cesar PF. Effect of ion exchange temperature on mechanical properties of a dental porcelain. Ceramics International 2010;36(6) 1977–1981.

[58] Seghi RR, Crispin BC, Mito W. The effect of ion exchange on the flexural strength of feldspathic porcelains. International Journal of Prosthodontic 1990;3(2) 130–134.

[59] Giordano RA, Campbell S, Pober R. Flexural strength of feldspathic porcelain treated with ion exchange, overglaze, and polishing. The Journal of Prosthetic Dentistry 1994;71(5) 468–472.

[60] Mirkelam MS, Pamuk S, Balkaya MC, Akgünör G. Effect of Tuf-Coat on feldspathic porcelain materials. Journal of Oral Rehabilitation 2005;32(1) 39–45.

[61] Piddock V, Qualtrough AJ, Brough I. An investigation of an ion strengthening paste for dental porcelains. International Journal of Prosthodontic 1991;4(2) 132–137.

[62] White SN, Seghi RR. The effect of ion strengthening time/temperature kinetics on the flexural strength of feldspathic porcelains. Dental Materials 1992;8(5) 320–323.

[63] Rosa V. Effect of ion exchange on Weibull parameters, subcritical crack growth and R-curve behavior of a dental porcelain. Master thesis. University of São Paulo; 2007.

[64] Cesar PF, Rosa V, Pinto MM, Yoshimura HN, Xu LR. Effect of ion exchange on R-curve behavior of a dental porcelain. Journal of Materials Science 2011;46(1) 117–122.

[65] Krause Jr RF. Rising fracture toughness from bending strength of indented alumina beams. Journal of the American Ceramic Society 1988;71(5) 338–343.

[66] Fischer H, Rentzsch W, Marx R. R-curve behavior of dental ceramic materials. Journal of Dental Research 2002;8(81) 547–551.

[67] Pinto MM, Cesar PF, Rosa V, Yoshimura HN. Influence of pH on slow crack growth of dental porcelains. Dental Materials 2008;24(6) 814–823.

[68] Lamon J. Ceramics reliability: statistical analysis of multiaxial failure using the Weibull approach and the multiaxial elemental strength model. Journal of the American Ceramic Society 1990;73(8) 2204–2212.

[69] Quinn GD. Fractography of ceramic and glasses. Washington: NIST; 2007.

[70] ASTM C 1239-07: Standard practice for reporting uniaxial strength data and estimating Weibull distribution parameters for advanced ceramics. West Conshohocken: ASTM International; 2007.

[71] Rosa V, Yoshimura HN, Pinto MM, Fredericci C, Cesar PF. Effect of ion exchange on strength and slow crack growth of a dental porcelain. Dental Materials 2009;25(6) 736–743.

[72] Isidor F, Brondum K. A clinical evaluation of porcelain inlays. Journal of Prosthetic Dentistry 1995;74(2) 140–144.

[73] Qualtrough AJE, Wilson NHF. A 3-year clinical evaluation of a porcelain system. Journal of Dentistry 1996;24(5) 317–323.

[74] Gonzaga CC, Yoshimura HN, Cesar PF, Miranda Jr WG. Subcritical crack growth in porcelains, glass-ceramics and glass-infiltrated alumina composite for dental restorations. Journal of Materials Science: Materials in Medicine 2009;20(5) 1017–1024.

[75] Yoshimura HN, Gonzaga CC, Cesar PF, Miranda Jr WG. Subcritical crack growth velocities (v-K curves) of dental bioceramics. Materials Science Forum 2012;727–728 1211–1216.

[76] Freiman SW, Wiederhorn SM, Mecholsky Jr. JJ. Environmentally enhanced fracture of glass: a historical perspective. Journal of the American Ceramic Society 2009;92(7) 1371–1382.

[77] Rosa V, Cesar PF, Pereira CFS, Pinto MM, Yoshimura HN. Effect of test environment and microstructure on the flexural strength of dental porcelains. Journal of Prosthodontics 2011;20(4) 275–279.

[78] Yoshimura HN, Pinto MM, Gonzaga CC, Cesar PF. Effects of artificial saliva storage on flexural strength and lifetime prediction of a dental porcelain. Cerâmica 2009;55(334) 190–198.

[79] Gonzaga CC, Cesar PF, Miranda Jr WG, Yoshimura HN. Slow crack growth and reliability of dental ceramics. Dental Materials 2011;27(4) 394–406.

[80] ASTM C 1368-06: Standard test method for determination of slow crack growth parameters of advanced ceramics by constant stress-rate flexural testing at ambient temperature. West Conshohocken: ASTM International; 2006.

[81] Gonzaga CC, Cesar PF, Miranda Jr WG, Yoshimura HN. Determination of the slow crack growth susceptibility coefficient of dental ceramics using different methods. Journal of Biomedical Materials Research: Part B – Applied Biomaterials 2011;99(2) 247–257.

[82] Davidge RW, McLaren JR, Tappin G. Strength-probability-time (SPT) relationships in ceramics. Journal of Materials Science 1973;8(12) 1699–1705.

[83] Wachtman JB, Cannon WR, Matthewson MJ. Mechanical properties of ceramics. 2 ed. New York: John Wiley & Sons; 2009.

[84] Anusavice KJ, Shen C, Vermost B, Chow B. Strengthening of porcelain by ion exchange subsequent to thermal tempering. Dental Materials 1992;8(3) 149–152.

[85] Yoshimura HN, Guedes S. Glass dielectric prepared by two-step toughening. Glass International 2010;33(6) 22–26.

[86] Holloway JA, Denry IL, Rosenstiel SF. Surface layer characterization after dual ion exchange of a leucite-reinforced dental porcelain International Journal of Prosthodontic 1997;10(2) 136–141.

[87] Green DJ, Tandon R, Sglavo VM. Crack arrest and multiple cracking in glass through the use of designed residual stress profiles. Science 1999;283(5406) 1295–1297.

[88] Sglavo VM, Green DJ. Flaw insensitive ion-exchanged glass: II, production and mechanical performance. Journal of the American Ceramic Society 2001;84(8) 1832–1238.

[89] Fischer H, Brehme M, Marx R. Effect of ion exchange of glazed dental glass ceramics on strenght parameters. Journal of Biomedical Material Research Part A 2005;72(2) 175–179.

[90] Fischer H, Marx R. Suppression of subcritical crack growth in a leucite reinforced dental glass by ion exchange. Journal of Biomedical Material Research Part A 2003;66(4) 885–889.

6

Effect of Extremely High Specific Flow Rates on the Ion-Exchange Resin Sorption Characteristics

Miroslav Kukučka and Nikoleta Kukučka

Abstract

Behavior of the strongly basic, macroporous ion-exchange resin Amberlite IRA 958-Cl is circumstantially explained in the book chapter. Effects of different specific flow rate (SFR) and determination of its optimum value, as well as effects of the empty-bed contact time (EBCT) values on the removal of NOM, arsenic, sulfate, electrical conductivity, bicarbonate and chlorine from groundwater using strongly basic ion exchange resin (SBIX) is examined in this chapter. Determination of the resin's sorption characteristics is also part of the investigation. A new approach of pseudo equilibrium adsorption capacity is presented. Investigations of determination of optimum value of water flow rate and resin's sorption characteristics were conducted with native groundwater and native groundwater with addition of oxidizing agent. Sodium hypochorite was added to the raw water with the aim of oxidizing NOM and As(III) to As(V). The intention was to find out whether the resin might be used beyond the range of operating conditions recommended by the manufacturer. Results will provide a better understanding of treatment of groundwater with similar physico-chemical composition which is important when designing a water treatment plants for settlements in areas with such groundwater. Also, prolonging the resin working cycle can ultimately lead to water treatment plants cost reduction. Namely, when the results for the optimum SFR are obtained for a pilot plant system for concrete groundwater of distinct physico-chemical characteristics, it is simple to design a unique ion-exchange water treatment system of any capacity. The obtained results make a sound basis for designing an appropriate plant for the removal of NOM from groundwater of the region of the town of Zrenjanin. The tested resin can also be used in the process of pretreatment of the same water, since it allows the removal of about 50 % of the naturally occurring arsenic.

Keywords: Specific flow rate, resinsorption characteristics, arsenic, natural organic matter

---

1. Introduction

Natural organic matter (NOM) and arsenic present a major problem when found in drinking water. There have been many published scientific papers on the methods for their removal over the years [1-9]. Groundwater investigated in this chapter is one of the most arsenic and NOM contaminated water in Europe [10], and nevertheless population which inhabits these areas are using this groundwater as a drinking water without any treatment process.

Behavior of the strongly basic, macroporous ion-exchange resin Amberlite IRA 958-Cl [11] will be circumstantially explained in the book chapter. Effects of different specific flow rate (SFR) and determination of its optimum value as well as the effects of the empty-bed contact time (EBCT) values on the removal of NOM, arsenic, sulfate, electrical conductivity, bicarbonate, and chlorine from groundwater using strongly basic ion-exchange resin (SBIX) will be examined in this chapter. Determination of the resin's sorption characteristics is also part of the investigation. A new approach of pseudo equilibrium adsorption capacity will be presented.

Investigations of determination of optimum value of water flow rate and resin's sorption characteristics were conducted with native groundwater and native groundwater with addition of oxidizing agent. Sodium hypochorite was added to the raw water with the aim of oxidizing NOM and As(III) to As(V). The addition of sodium hypochlorite to water yields hypochloric acid, and the liberated atomic oxygen acts as an efficient disinfection agent and a very strong oxidant. Apart from oxidizing the present NOM and arsenic(III), the liberated nascent oxygen can also oxidize the present inorganic species such as nitrites, iron(II), and the like. This is important because of resin affinity for certain ions and their competition for resin binding sites which can affect NOM and arsenic ion-exchange and adsorption.

The intention was to find out whether the resin might be used beyond the range of operating conditions recommended by the manufacturer. Results will provide a better understanding of treatment of groundwater with similar physicochemical composition which is important when designing a water treatment plants for settlements in areas with such groundwater. Also, prolonging the resin working cycle can ultimately lead to water treatment plants cost reduction.

1.1. Composition and characteristic of dissolved natural organic matter and arsenic

Natural organic matter is frequently found dissolved in groundwater. In most cases, the presence of NOM gives these waters a characteristic yellow color. NOM consists usually of humic substances originated from the geological formations of the location of groundwater source [12]. Their molecules have a supramolecular structure, formed by condensation of smaller molecules resulting from the degradation of organic matter [13]. Humic matter has a very stable structure, a proof of this being the fact that they have been present in the Earth's crust for thousands of years [13, 14].

Piccolo and Stevenson showed that humic acids exhibit affinity to metal ions from the soil, forming thus the complexes of different stability constants and other characteristics [15].

Humin of the investigated groundwater contains humic acids (HAs) and fulvic acids (FAs). Humins are macromolecular polymers whose structure and characteristics are determined by their origin and process of humification. Like HAs, FAs too, are naturally present in water, soil, and turf. They are formed by chemical and microbiological degradation of plant material (humification). There are opinions that FAs are formed after the constitution of HAs. The FAs are richer in oxygen and poorer in carbon than HAs. Like HAs, FAs contain a number of reactive functional groups including carboxylic, hydroxylic, phenolic, quinonic, and semiquinonic [16]. Molar masses of FAs are smaller than those of HAs. FAs contain more constitutive groups that are structurally similar to carbohydrates, originated from polysaccharides. In the determination of total carbon content (TOC) in groundwater, HAs and FAs make the source of total dissolved organic carbon (C_T). Previous investigations showed that there exists labile (unstable) and non-labile components of NOM [17]. By taking advantage of the fact that labile NOM are sensitive to permanganate, it is possible to determine their fraction by standard method as the permanganate consumption (COD). The fraction of labile organic carbon is designated as C_L, and the fraction of non-labile carbon (C_{NL}) is calculated as the difference between the TOC and C_L. Anderson and Schoenau showed that the FA fraction depends essentially on the content of C_L, and that it is independent of C_{NL} [18]. Labile components of organic matter consist of cell biopolymers such as carbohydrates, amino acids, peptides, amino sugars and fats. Contents of these components in FA fraction is smaller, and because of that it is biologically and chemically more resistant. The HA skeleton contains strongly condensed aromatic structures surrounded by the side chains of aliphatic components. The FA fractions are mainly composed of carbohydrates, microbiological metabolites, and "younger" materials that are not significantly related to the mineral fraction. This explains the strong correlation between C_L and FA content, which is not observed for HAs. The strong correlation between C_{NL} and HAs indicates that C_{NL} consists of the fraction of organic carbon which is stabilized due to the chemical and physical associations with the mineral matrix. The groundwater that contains FAs has a characteristic yellow color.

Arsenic is a very toxic metalloid, occurring in nature in several oxidation states (-3, 0, 3, 5). In natural waters, it is present in its inorganic forms of oxyanions, primary and secondary arsenite, as well as in primary, secondary, and tertiary arsenate [19]. The water pH and redox potential influence dominantly the oxidation state of arsenic in natural waters. At the pH < 6.9 dominates $H_2AsO_4^-$, whereas at higher pH it is $HAsO_4^{2-}$. In strongly acidic media, under oxidation conditions the molecular form H_3AsO_4 is dominant, whereas under alkaline conditions, it is AsO_4^{3-}. Under reduction conditions, at the water pH < 9.2 dominates the amphoteric neutral molecular species H_3AsO_3 [20]. Arsenic is a constituent component of many ores and by the process of their dissolution reaches groundwater. Apart from natural arsenic,

arsenic in groundwater may be due to anthropogenic activities. Namely, many insecticides and fungicides, wood-protecting agents, chemicals used in semiconductor production, and additives to various alloys and glasses contain arsenic compounds [21]. It is known that prolonged exposure of the human organism to small doses or short exposure to high doses of arsenic causes skin disease and serious disturbance of internal respiratory and digestive organs, blood circulation, and of the nervous system. Also, Ng et al. proved that arsenic is carcinogenic [22].

The EU Drinking Water Directive from 1998 recommended that the maximum tolerable concentration (MTC) of arsenic in drinking water should be 10 μg/L [23]. Three years later, in its revision of the corresponding standard, the US EPA adopted also the same MTC value of 10 μg/L [24]. Because of the geological and mineralogical origin of arsenic in groundwater, its removal is a technologically very complicated and demanding operation.

1.2. Principles of NOM and arsenic removal using strongly basic ion-exchanging resins

Ion exchange is a process used to remove ionic components dissolved in water, the process being based on their high affinity to bind to the ion-exchange resin. As a consequence, ions are captured and exchanged for the resin's ions. When the resin capacity for exchanging particular ions is reached (exhaustion), it is usually regenerated, so that it can be used in repeated cycles. It is reported that the application of special macroporous strongly basic anionic resins is suitable for the removal of NOM from natural waters [25]. These resins are very effective in removing the water coloring matter (humins) and function in the regime of regeneration with sodium chloride [26, 27]. The main principle of NOM removal by these macroporous resins is based on the finding that more than 90% of NOM in groundwater are in fact weak acids represented mostly by molecules and also by dissociated anions and often cations of alkaline earth elements. Molecules of humic acids are large and complex, composed of benzene rings and other interconnected chemical structures. Every NOM molecule contains a number of carboxylic groups. Like carbonic acid, humic acids do not fully dissociate at a neutral pH. Depending on the number of carboxylic functional groups in their molecules, NOM can act as polyvalent species. These NOM properties make, in fact, the basis for their removal from the anionic resin by a saturated solution of sodium chloride. Strong base anionic macroporous exchange resins efficiently exchange chloride for all the anions present in raw water. The anions from water interact with the resin functional groups in dependence of their relative affinity, as well as the rate and dynamics of the ionic exchange. Under ideal conditions, when the resin is saturated, sulfate and anions of organic acids occur adsorb on the surface, while the other anions, such as nitrate, chloride and bicarbonate, are sorbed in the interior of the resin beads. In the initial cycle of ionic exchange, the resin plays the role of a dealkalizer and removes bicarbonates from the raw water. Also, the beginning of the working cycle is characterized by a decrease in the pH of the effluent, which often falls below pH 6. In the further course of the work cycle, the resin adsorbs anions that have a higher affinity than bicarbonate, which is released into the effluent. This is accompanied by a pH increase, which may attain a value that is higher than that of the raw water. In the further course of the working regime, chloride and nitrate are released and their places occupy anions of organic acids and sulfate.

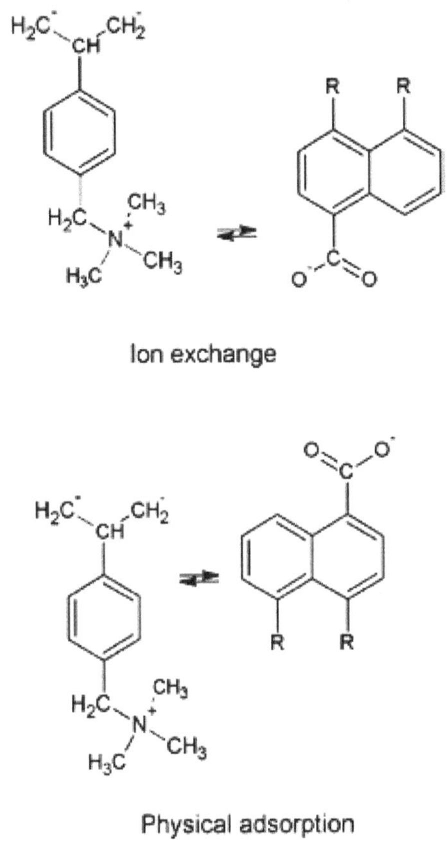

Figure 1. Possible mechanism of reversible sorption of organic molecules on ion exchange resin

Group of authors showed that in a complex solution containing humic acids and arsenic, complex compounds are formed between arsenic and carboxylic groups of humic acids [28]. By investigating this interaction, Warwick et al. found that the deprotonated functional groups of humic acids and arsenic form associates, depending on the pH, ionic strength, and arsenic concentration [29].

In the process of removal of organic ions, such as those of NOM using anion exchange resin, two sorption mechanisms may be involved, and these are ion exchange and physical adsorption (Figure 1). Ionic exchange includes the transfer of the ions from the ionic exchanger and electrostatic interaction between the ions of functional groups, presented as quaternary ammonium ion-exchanging groups and participation of the carboxylic groups. This interaction is of donor-acceptor type. Quaternary ammonium groups [-N$(CH_3)^{3+}$] are specific active groups which determine the intensity of the ionic exchange. Physical adsorption takes place as the van der Waals interaction between the non-polar (hydrophobic) groups present in the NOM molecules and central structure of the polymer resin to which substituents are attached [30, 31].

The adsorption of arsenic ion on the surface of strongly basic ion exchanging resin may be presented as follows:

$$R - Cl + H_2AsO_4^- \rightarrow R - H_2AsO_4 + Cl^- \quad (1)$$

$$2R - Cl + HAsO_4^{2-} \rightarrow R_2 - HAsO_4 + 2Cl^- \quad (2)$$

Since the affinity of the acrylic resin with quaternary ammonium groups is higher to doubly charged ions than to the singly charged ones, the efficiency of the ionic exchange is higher at higher pH values, when doubly charged arsenate ions are dominant [32].

Removal of As(III) by ionic exchange is less effective, since at the pH < 9 it occurs in water in the form of the molecule H_3AsO_3 [20]. On the other hand, As(V) is present in the form of the anions $H_2AsO_4^-$ and $HAsO_4^{2-}$, which makes the ionic exchange more probable. It is just the specific phenomenon of the occurrence of arsenous acid as an uncharged species that cannot be removed from water which allowed the development of procedures for distinguishing between arsenite and arsenate [33, 34]. Therefore, in order to remove arsenite from water, it has to be oxidized to arsenate. The regeneration of the ion-exchange resin proceeds according to the following reaction:

$$RH_2AsO_4^- + Cl^- \rightarrow RCl + H_2AsO_4^- \quad (3)$$

Regeneration can be done with HCl and NaCl. The use of HCl yields arsenic acid, H_3AsO_4, which has no influence on the equilibrium of ion exchange and the regeneration is more efficient [32]. The strongly basic resin that has been pretreated by anions, e.g. chloride, is capable of removing a wide spectrum of anions from water, depending on their relative affinities, shown by the following order [35]:

$$CrO_4^{2-} >> SeO_4^{2-} >> SO_4^{2-} >> HSO_4^- > NO_3^- > Br^- > HAsO_4^{2-} > SeO_3^{2-} > HSO_3^{3-} > NO_2^- > Cl^-$$

2. Materials and methods

2.1. Specificity of the investigated groundwater and geological survey of aquifer origin

In the period of the Middle Miocene to the Quaternary, the space of the Central and South-Eastern Europe was covered by the Pannonian Sea, formed in the spacious depression between the Alpine, Carpathian, and Dinaridian mountain ranges, where very thick sediment series were deposited. The northern part of Serbia-Vojvodina is situated in the south-east part of this sedimentation zone on an area of 21,506 km^2. The basic mass of these sediments make the Mesozoic clastites and carbonates along with crystalline schists and granitoids of the Paleozoic and Proterozoic eras. The town of Zrenjanin is located in the eastern part of the Vojvodina - called the Middle Banat, and the specificity of the location stems primarily from the geological and hydrological characteristics of the terrain. The water-bearing formations in this region were formed in the final phase of the existence of the Pannonian Sea. On the territory of the town of Zrenjanin, in the geological profile to a depth of 400 m, there are several water-bearing strata, separated by strata layers of pure or sandy clay or clayey sands. The thickness of the water-bearing strata is from several meters to about 50 m, most often 10-20 m. The groundwater that can be used as drinking water occurs to a depth of about 400 m. In the upper part, to a depth of about 60 m, an unconfined aquifer (phreatic aquifer) is formed, whereas below it the aquifer is under the pressure (artesian aquifer). The groundwater of the artesian aquifer, to which belongs the investigated water, is by their chemical composition significantly different from the waters of the phreatic aquifer. The investigated water belongs to the type of bicarbonate alkaline waters, with the bicarbonate share of about 85-90% equivalents. The total mineralization is most often lower compared to that of the phreatic water and is in the range 250-1500 mg/L. Content of sodium chloride increases generally with increase in the depth of the artesian aquifer. Local deviations are usually a consequence of tectonic disturbances. The ammonium content is usually below 2 mg/L, although there have been registered very high contents (up to 10 mg/L) at the locations south of Zrenjanin. The iron content often exceeds the maximum tolerable concentration (MTC); it is usually around 1 mg/L and decrease with depth. If compared to the contents in phreatic water, contents of total iron and manganese show a decrease. The artesian water is softer than phreatic water, and its hardness rarely exceeds the limit of 10° dH. Very soft waters (2-5° dH) have been found north of Zrenjanin. Artesian waters are often weakly alkaline, and their temperature is in the range of 15-20°C. In the Zranjanin region, these waters are of the very characteristic yellow color, have a specific taste, and in many settlements are used as drinking water. People that are accustomed to yellow water are reluctant to change their habit and this represents a problem. Namely, yellow color is due to the presence of undesirable organic (humic) matter. Content of this matter, expressed via the COD, determined by the permanganate method, is in the range of 20-150 mg/L, and in the extreme cases exceeds even 200 mg/L [36]. Up to now, water supply in the Middle Banat region has been based on the exploitation of artesian groundwater. Waters of this aquifer, in addition to marked humic matter load, contain significant amounts of orthophosphate, sodium, and arsenic. The measured arsenic contents are in the range from 0.040 to 0.380 mg/L, so that the Zrenjanin region is in this respect one of the most endangered regions in Europe. The size of the area of the arsenic occurrence suggests the supposition that it is of geological origin. However, there exists the possibility that it has been brought to Vojvodina by the Tisa River, whose water dissolves arsenic minerals of the Carpathian, where there are large deposits of arsenic ores [37]. In some Zrenjanin locations, many domestic and foreign corporations searched for practical groundwater treatment models, but without satisfactory results. A likely reason for this is the methodological research inadequacy for the actual problem. Pilot tests were conducted without a clear project assignment and a defined method of removal of many pollutants for the given case [38]. Groundwater from the Zrenjanin region as a resource outside the natural water cycle is of extremely complex chemical composition, which demands a complex technological treatment. Values of measured physicochemical parameters are given in Table 1.

Parameter	MTC	Groundwater
Color, permanent (°Co-Pt)	5	60
pH	6.8-8.5	8.06
COD (mgO_2/L)	2.0	7.73
TOC (mg/L)	/	10.64
Electrical conductivity (μS/cm)	1000	820
Chloride (mg/L)	200	6.97
Bicarbonate (mg/L)	/	645
Sulfate (mg/L)	250	25.6
Arsenic (mg/L)	0.01	0.2658

Table 1. Selected physicochemical parameters of the investigated groundwater

2.2. Strongly basic macroporous ion-exchange resin

To study the removal of dissolved NOM, arsenic bicarbonate, pH value, electrical conductivity, chlorine, and sulfate, an acrylic SBIX type Amberlite IRA 958-Cl, manufactured by Rohm and Haas Company, whose characteristics are given in Table 2 [11] was used. This type of resin has a rigid polymer porous network in which persist an intrinsic porous structure even after its drying. The resin large pores permit access to the interior exchange sites of the bead. They are also referred to as macroreticular or fixed-pore resins. Macroporous resins are manufactured by a process that leaves a network of pathways throughout the bead. This sponge-like structure allows the active portion of the bead to contain a high level of divinylbenzene crosslinked without affecting the exchange kinetics. Unfortunately, it also means that the resin has a lower capacity because the beads contain less exchange sites.

Properties	
Matrix	Crosslinked acrylic macroreticular structure
Functional groups	Quaternary ammonium
Physical form	White opaque beads
Ionic form as shipped	Chloride
Total exchange capacity	≥ 0.8 eq/L (Cl^- form)
Moisture holding capacity	66-72 % (Cl^- form)
Specific gravity	1.05-1.08 (Cl^- form)
Bulk density	655-730 g/L (Cl^- form)
Effective size	470-570 μm
Mean diameter	700-900 μm

Uniformity coefficient	≤ 1.8	
Suggested operating conditions		
Maximum operating temperature	80°C	
Minimum bed depth	600 mm	
Service flow rate	8-30 BV/h	
Regenerants	NaOH	NaCl
Flow rate (BV/h)	2-6	2-6
Concentration (%)	2	10
Level (g/l)	6-40	160-300
Minimum contact time	30 minutes	
Rinse water requirements	5 to 10 BV	
Rinse flow rate	Same as regenerant for first bed displacement, than same as service flow rate	

Table 2. Physicochemical characteristics of the Amberlite IRA 958-Cl macroporous resin and operating conditions recommended by the manufacturer

2.3. Determination of optimal specific flow rate of strongly basic ion-exchange resin

To study the influence of the different SFRs in the process of removal of NOM and arsenic from groundwater, a specially designed pilot plant system 1 was made (Figure 2). The pressure and the amount of water were ensured using the waterworks facilities of the Melenci settlement explained in Section 2.1. The mean operating water pressure and temperature were 3 bar and 18°C, respectively. The composite pressure vessels used for plant construction were Structural type Q-0844 and Q-0635, of total volume of 33.6 L and 14.4 L, respectively. The vessel was filled with 25 L and the later with 10 L of resin. The cross-section area of the tank Q-0844 was 0.034 m^2, and the resin bed height was 0.735 m. The tank Q-0635 of the cross-section of 0.020 m^2 had the bed height of 0.5 m. The bigger tank was used for smaller SFRs, from 2 to 80 BV/h, and the smaller one for the SFRs ranging from 100 to 300 BV/h. The control valve (CV) was a Fleck 3150, injector type, with manual control. The pilot plant was equipped with four Franck Plastic flow meters (R), for the flow ranges of 50-500 L/h, 100-1000 L/h, 200-2000 L/h, and 600-6000 L/h, respectively. The plant was also equipped with a Prominent dosing pump (DP), with a Zenner pulse water meter (F) for dosing sodium hypochlorite, and with a vessel for its mixing with water (MT), to achieve a homogeneous solution with a constant chlorine concentration of 0.5 mg/L, determined by the colorimetric method [39] and a micro filter (M). Chlorinated water for the determination of the residual chlorine was sampled via the tap (S). The first part of the study was concerned with the investigation of the effect of different SFR and the determination of its optimum value, as well as of the EBCT value for the removal of NOM and arsenic from groundwater without addition of sodium hypochlorite as oxidizing agent. Sodium hypochorite was added to the raw water with the aim of oxidizing NOM and As(III) to As(V) in the second part of the study. Special attention was paid to those SFRs that were smaller than 6 BV/h and higher than 30 BV/h, in order to determine the ion-exchange capabilities and sorption possibilities of using the resin beyond the prescribed range of operating conditions in the treatment of the given groundwater. The first part of the series of measurements was carried out using the resin from the original package of 25 L. The resin was backwashed by standard procedure for 20 min, i.e. to the end of foam formation and appearance of an unpleasant odor. The other two experiments in the series of investigations using either chlorinated or non-chlorinated water were carried out on the resin regenerated according to the procedure described in Section 2.2.1. After the mentioned preparation of the resin, groundwater was passed through the bed in a down flow direction. The EBCT and SFR in these investigations were from 0.5 to 0.0033 h and from 2 to 300 BV/h, respectively. During the experiment, samples of both raw and treated water were taken regularly. Three series of samples of both chlorinated and non-chlorinated water were used to determine the following physicochemical parameters: pH, electrical conductivity, sulfates, bicarbonates, COD, TOC, and total arsenic. All the results represent mean values of three repetitions. Each series of measurements encompassed 30 samples of the effluent and one sample of groundwater. In the first part of the study, involving non-chlorinated groundwater, the overall contact time of water and resin at different SFR values was in average 38.5 h, whereas the average volume of the water passing through the bed was 17,187.50 L. This volume corresponded to 687.5 volumes of the resin in the bed. In the second part of the study, dealing with chlorinated water, the overall contact time of water and resin was in average 34.1 h, at different SFR values. The average volume of groundwater that passed through the bed was 16,650 L, which was equivalent to 666 volumes of the resin.

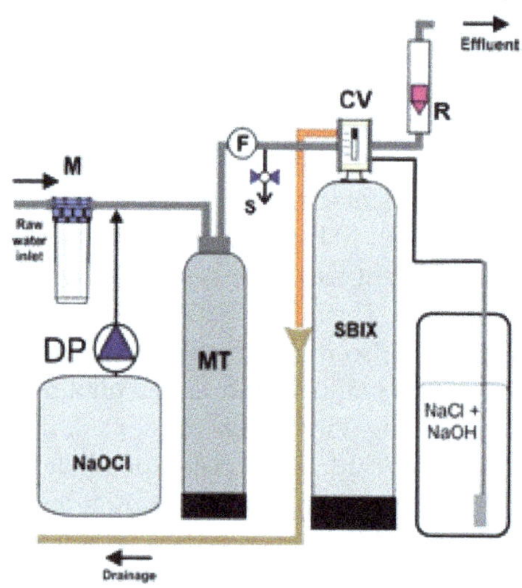

Figure 2. Scheme of pilot plant 1

2.3.1. *Regeneration of SBIX*

The resin regeneration was carried out in three phases. In the first phase, by opening the control valve, the resin was counterwashed with raw water for 10 min. In the second phase, at a SFR of 4 BV/h, the regenerant (solution of NaCl (10%) and NaOH (2%)) from the regenerant vessel was injected with the aid of an injector inbuilt in the control valve. The overall contact time of the resin with the regenerant was 120 min in a regime encompassing successive repetition of 10-min passing the regenerant and 10-min of stopped flow. This regeneration method ensured a better dynamic equilibrium between the adsorbed and desorbed humic acid molecules, i.e. NOM and arsenic. In the beginning of the regeneration, the effluent was light yellow, and then became darker, and at the end of the first 10 min, it was of a typical brown color. The effluent began to regain a lighter color after about 80 min of regeneration, to become yellow at about the 100th min, and this color remained to the end of the regeneration. The third phase consisted of the fast rinsing of the resin with raw water, whereby the resin was dispersed in the whole working volume, and this was accompanied by the measurement of the electrical conductivity. When the effluent electrical conductivity dropped to that of the raw water, further washing was stopped. At the same time, the measured pH of the effluent became slightly alkaline, like the raw water. Measurements of the electrical conductivity and pH were necessary to check whether the resin had been freed of the regenerants. After the resin regeneration, the experiments were repeated both with chlorinated and non-chlorinated water, and it was subsequently regenerated two times.

2.3.2. *Calculation of flow parameters*

In order to study the removal process, it was necessary to calculate the flow parameters and the resins exposure time to the groundwater flowing through pilot plant 1. Specific flow rate was calculated using equation 4, where Q is the flow rate of chlorinated or non-chlorinated groundwater (L/h) and V_{SBIX} is the volume of the resin bed (L). The SFR is expressed in the units of the bed volume per hour (BV/h), and it represents a universal quantity which can be applied to characterize any ion exchanger. Namely, when the results for the optimum SFR are obtained for a pilot plant system for concrete groundwater of distinct physicochemical characteristics, it is simple to design a unique ion-exchange water treatment system of any capacity. The empty bed contact time (EBCT) was expressed and calculated in the way shown in equation 5. The EBCT is used as a measure of the duration time of the contact between the resin granules and water flowing through the bed. The increase in the EBCT value presents increased time available for the adsorption of dissolved matter on the resin beads.

$$SFR = \frac{Q}{V_{SBIX}} \tag{4}$$

$$EBCT = \frac{V_{SBIX}}{Q} \tag{5}$$

2.4. Strongly basic ion-exchange resin sorption characteristics determination

The objective of this research was to experimentally examine the sorption efficiency of NOM and arsenic on Amberlite IRA 958-Cl resin. The aim was to compare experimentally obtained amounts of NOM and arsenic sorbed on the resin as indicators of the resin sorption capacity with the empirical sorption capacity based on the official data of the resin manufacturer. The investigations were conducted using native groundwater without (first part of investigation) and with addition of sodium hypochlorite (second part of investigation). For this purpose, a pilot plant 2 was designed and manufactured as shown in Figure 3. The main part of pilot plant 2 was the transparent container, manufactured by Atlas, filled with 1.4 L (970 g) of the resin Amberlite IRA 958-Cl. It was equipped with a flow meter (IHTM) for the flow range from 5 to 100 l/h, with a pump (Prominent) and pulse water meter (Zenner) for dosing sodium hypochlorite solution, and a vessel for mixing it with water, to obtain a homogeneous solution with a constant concentration of the residual chlorine. Samples of chlorinated water were taken from the tap. Concentration of the residual chlorine after the mixing vessel was measured by the colorimetric DPD method [39] and was kept constant as in the investigation of optimal flow parameters, at a level of 0.5 mg/L. The dosing system was not used in the first part of investigation, but only in the second one. In the first experiment in the series of measurements, both with native and chlorinated groundwater, a original packing of resin was used, which was backwashed with groundwater for 20 min, that is to the visually observed end of foaming and appearance of an unpleasant odor. The other two experiments in the series of investigation of chlorinated and non-chlorinated water were done using regenerated resin regenerated as described in Section 2.3.1. After the described resin preparation, the water was passed through it in a down flow direction. Samples of both the influent and effluent were taken regularly. Three series of samples of both chlorinated and non-chlorinated water were examined for the following parameters: COD, TOC, and total arsenic. All the measurement results are expressed as mean values of the determined parameters. The characteristics of the NOM and arsenic and of the resin applied as well as the chemical composition of groundwater and applied analytical methods and equipment for the determination of TOC, COD, and arsenic are identical to those described by the authors [40]. According to the resin manufacturer [11], the average sorption capacity of organic matter is 10-15 g of $KMnO_4$/L resin. The capacity expressed as COD is between 2.5 and 3.75 g O_2/L resin. The COD sorption capacity calculated per mass unit of SBIX, taking into account the filling density of 720 g/L, ranges from 3.47 to 5.21 mg O_2/g resin.

2.4.1. Mathematical expressions of volume and sorption parameters

The efficiency of the removal of NOM and arsenic was followed by determination of COD, TOC, and arsenic concentrations in the effluent as a function of the overall water volume that passed through the SBIX bed, expressed as bed volume (BV). Bed volume is a dimensionless quantity that expresses the water volume as the number of the volume of resin bed (equation 6), where V_{out} is the effluent volume that passed through until the moment of sampling (L) and V_{SBIX} is a constant, standing for the volume of SBIX, and it was 1.4 L in all series of experiments. The BV is a quantity that is not characteristic for only one concrete ion-exchange resin, but is universally applicable to any ion exchanger of any capacity, for the same treated water. The results obtained for BV can be used to calculate the SBIX volume, if one knows the volume of water and vice versa.

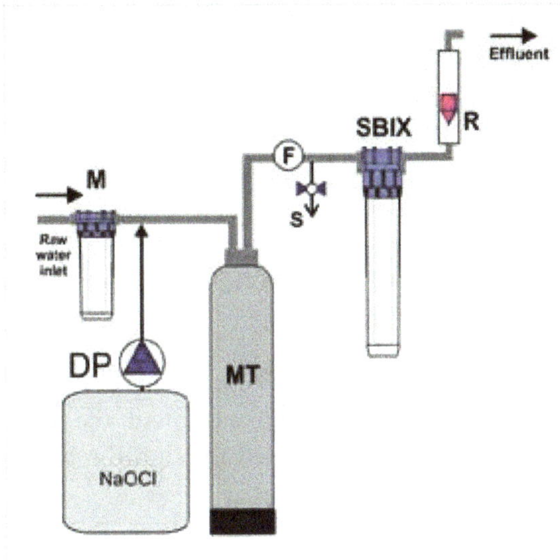

Figure 3. Scheme of pilot plant 2

$$BV = \frac{V_{out}}{V_{SBIX}} \tag{6}$$

NOM and arsenic content in the total volume of investigated chlorinated or non-chlorinated water during the whole experiment are calculated in the following way:

$$COD_{total} = V_{total} \cdot CODC_{in} \tag{7}$$

$$TOC_{total} = V_{total} \cdot TOCC_{in} \tag{8}$$

$$As_{total} = V_{total} \cdot AsC_{in} \tag{9}$$

where COD_{total} (mg O_2) and TOC_{total} (mg TOC) stand for the differently expressed total amounts of NOM; As_{total} (mg) is the total amount of arsenic - the two quantities representing in fact the resin load; V_{total} (L) is the total volume of water treated in the experiment, whereas $CODC_{in}$ (mg O_2/L) and $TOCC_{in}$ (mg O_2/L) designate the NOM concentration in the influent, and AsC_{in} (mg/L) is the corresponding arsenic concentration.

The overall load of the resin with organic matter and arsenic is calculated as shown in equations 10, 11, and 12, where COD_{load} (mgO_2/g) and TOC_{load} (mg TOC/g) represent the overall amounts of NOM; As_{load} (mg/g) the total amount of arsenic to which the resin was exposed in the experiment, and m represents the mass of the resin (g).

$$COD_{load} = \frac{COD_{total}}{m} \quad (10)$$

$$TOC_{load} = \frac{TOC_{total}}{m} \quad (11)$$

$$As_{load} = \frac{As_{total}}{m} \quad (12)$$

The amounts of NOM and arsenic that came in contact with the resin due to the volume of chlorinated or non-chlorinated water that passed through the bed in the real time of the experiment are calculated in the following way:

$$COD_{in} = V_{out} \cdot CODC_{in} \quad (13)$$

$$TOC_{in} = V_{out} \cdot TOCC_{in} \quad (14)$$

$$As_{in} = V_{out} \cdot AsC_{in} \quad (15)$$

where COD_{in} (mg O_2) and TOC_{in} (mg TOC) stand for the amount of the NOM; As_{in} (mg As) is the amount of arsenic, and V_{out} (L) is the volume of water that passed to the moment of sampling.

Contents of NOM and arsenic in the instantaneous sample, i.e. in the effluent after the flow of a given volume of chlorinated or non-chlorinated water, are calculated as follows:

$$COD_{out} = V_{out} \cdot CODC_{out} \quad (16)$$

$$TOC_{out} = V_{out} \cdot TOCC_{out} \quad (17)$$

$$As_{out} = V_{out} \cdot AsC_{out} \quad (18)$$

where COD_{out} (mg O_2) and TOC_{out} (mg TOC) are the differently expressed amounts of NOM; As_{out} (mg As) is the amount of arsenic that was not sorbed onto the resin and is found in the effluent from the pilot unit, whereas $CODC_{out}$ (mgO_2/L) and $TOCC_{out}$ (mg TOC/L) are the concentrations of NOM in the effluent determined by the two different methods. Finally, AsC_{out} (mg/L) represents the arsenic concentration in the effluent.

The amounts of NOM and arsenic that in a given time instant of sampling, after collecting the effluent volume V_{out} (L), were sorbed onto the SBIX are calculated in the following way:

$$COD_{ads} = COD_{in} - COD_{out} \tag{19}$$

$$TOC_{ads} = TOC_{in} - TOC_{out} \tag{20}$$

$$As_{ads} = As_{in} - As_{out} \tag{21}$$

where COD_{ads} (mg O_2) and TOC_{ads} (mg TOC) are the adsorbed amounts of dissolved organic matter, and As_{ads} (mg) the amount of arsenic adsorbed.

The sorption efficiency can be presented by the following expressions:

$$E_{COD} = \frac{\left(CODC_{in} - CODC_{out}\right)}{CODC_{in}} \cdot 100(\%) \tag{22}$$

$$E_{TOC} = \frac{\left(TOCC_{in} - TOCC_{out}\right)}{TOCC_{in}} \cdot 100(\%) \tag{23}$$

$$E_{As} = \frac{\left(AsC_{in} - AsC_{out}\right)}{AsC_{in}} \cdot 100(\%) \tag{24}$$

where E_{COD} and E_{TOC} are the efficiencies of sorption of NOM, and E_{As} is the efficiency of arsenic sorption, all expressed as the corresponding percentages.

2.5. Calculation of equilibrium adsorption capacity of SBIX toward NOM

Effects of NOM adsorption by measuring of effluent NOM concentrations (C) in mg/L, during the experiments were shown in the dependence of cumulative time ($\Sigma\tau$) in hours, i.e. number of effluent's bed volumes (BV). Shapes of these curves are the primary overview of adsorption process kinetics toward starting, breakpoint, and pseudo-equilibrium stages. The big challenge and most important scope of this work was pathway to calculation of mass of the adsorbent (m) in g, in the mass transfer zone (MTZ) [1, 1x] corresponding to each experimental sampling, i.e. measuring point. Elucidation of the m values was significant operation for reaching the final target, which represents the determination of Amberlite IRA-958 pseudo-adsorption capacity toward NOM. It was supposed that the mass of exhausted resin is proportional to effluent NOM concentrations in the moment of sampling time. Based on the obtained percent equation data, all masses of the resin corresponding to effluent NOM concentrations were calculated using the mathematical expression (25).

$$m = \frac{C}{C_0} \cdot M \tag{25}$$

Adsorbed amount of NOM (α) in mg O_2, at each sampling point was calculated as shown:

$$\alpha = (C_0 - C) \cdot V \tag{26}$$

where V is the volume of the effluent between two samplings (L).

Regarding the fact that samples have been taken randomly, calculation of the adsorption process rate as represented in equation 27 was an essential step:

$$VADS = \alpha / \tau \tag{27}$$

where *VADS* (mg/h) is adsorption rate and τ (h) is time between two sampling points.

Obtaining *VADS* data was a real contribution to equalize the adsorption process during the contact time of the complete experiment. Adsorption capacity of Amberlite IRA-958 in the time of experiment duration - specific adsorption capacity (*ADC*) in mg/gh - was calculated using equation 28:

$$ADC = VADS / m \tag{28}$$

The mutual relation of *ADC* values at measured *C* values was described as exponential regression plot. General equation form of the curve was represented (equation 29).

$$ADC = \beta \cdot e^{-\delta \cdot C} \tag{29}$$

β ($L\ g^{-1}\ h^{-1}$) and δ are empiric adsorption coefficients obtained from the experimental data and calculated from the plot ln *ADC* vs. *C* (equation 30) as intercept and slope, respectively.

$$\ln ADC = \ln \beta - \delta \cdot C \tag{30}$$

The total *ADC* (ADC_a) in mg/gh, during the experiment, was calculated by integration of the surface above the curve using the following equation 31:

$$ADC_a = \left[\left(C_{\max} - C_{\min} \right) \cdot ADC_{\max} \right] - \int_{C_{\min}}^{C_{\max}} \beta \cdot e^{-\delta \cdot C} dC \tag{31}$$

where C_{max} and C_{min} are the highest and lowest measured values of C, and ADC_{max} is the maximal value of ADC.

Value of adsorption capacity of the resin (Q_{pe}) in mg O_2 of NOM per gram of the resin at pseudo equilibrium was calculated as a function of overall experiment time ($\Sigma\tau_{max}$) in h, using following equation 32:

$$Q_{pe} = ADC_a \cdot \Sigma\tau_{\max} \tag{32}$$

Investigated adsorption process is flow and time dependent. There is no equilibrium at every sampling point, but only the adsorption capacity could be recorded in the appointed time. Dynamic adsorption kinetics is different to batch experiments adsorption kinetics [41-46]. The difference is that time of adsorption in batch processes is infinitely and equilibrium could be detected with differences depending of time and temperature. In dynamic fixed bed adsorption, process time is defined and finite when sampling. During the adsorption and sampling as well, lasting irreversible adsorption is a consequence of constant addition of adsorbate which is dominant with regard to reversible equilibrium.

3. Results and discussion

The following results present data obtained during the determination of optimal specific flow rate, sorption characteristics, and kinetics of strongly basic ion-exchange resin.

3.1. Results obtained during determination of optimal specific flow rate of strongly basic ion-exchange resin

During the changes of SFR and EBCT, significant data of measured parameters were obtained. Dependence of measured parameters and flow rate values in experiments with non-chlorinated and chlorinated water are shown in Figures 4-7. Figure 4 presents the dependence of pH value and bicarbonate content in the effluent and SFR values (A) and influence of SFR on electrical conductivity and chloride concentrations (B) in both non-chlorinated and chlorinated water.

In the initial sorption contact of groundwater with the resin, an abrupt decrease in the effluent pH effluent was noted. This observation was especially pronounced in the case of chlorinated water. A high affinity of the resin to bicarbonate ions in the initial phase of ion exchange lead to the increase in the effluent acidity. The minima of the pH and bicarbonate concentration were observed at the lower SFR values compared to those measured in non-chlorinated water. These results are the consequence of the presence of the nascent

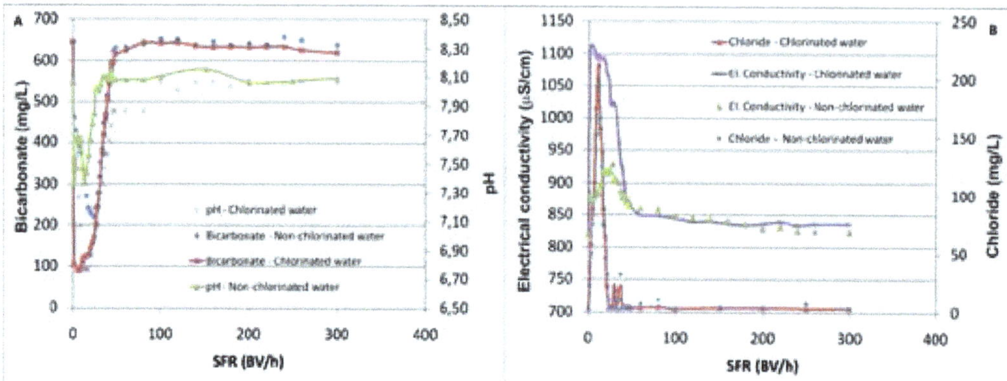

Figure 4. Influence of SFR on the bicarbonate and pH (A), and electrical conductivity and chloride (B) content

oxygen in chlorinated water, which accelerates ionic exchange of chloride for bicarbonate on the resin. The initial intensive exchange of these ions is gradually counterbalanced by the adsorption of the anions of humic acids taking place to the SFR of 50 BV/h, since the sites at which the dynamic equilibrium of sorption and desorption of bicarbonate is shifted to the interior of the resin granules, while humic acid binds to the functional groups on their outer surface. At the end of the investigated range of SFR, the HCO_3^- concentration in the effluent remained almost constant and asymptotically reached the concentration in the groundwater. At higher SFR, effluent pH value approached the pH value of the raw water as a consequence of the change in the bicarbonate concentration. After the value of SFR 50 BV/h, the resin is in the chloride form. Figure 4(A) also shows that in the initial part of the working cycle of resin with chlorinated water, there has been an immediate decrease of bicarbonates in the effluent already at SFR 2-6 BV/h after which the contents of bicarbonates slowly rises to SFR 50 BV/h, where the bicarbonates till the very end of the working cycle reach the concentration similar to that in raw water. The enormous increase in the chloride concentration at small SFR takes place simultaneously with the marked decrease in the content of bicarbonate in the effluent Figure 4(B). This chloride concentration rise, a singly charged small and high diffusion rate, results in the increase in the measured conductivity. A higher increase in the chloride concentration was observed in the case of chlorinated water, which is understandable taking into consideration the release of chloride from hypochloric acid. The differences in the peaks of chloride concentration and electrical conductivity in the effluent for chlorinated and non-chlorinated water are certainly a consequence of the generation of nascent oxygen, which also activates the resin's functional groups and thus contributes to a more effective ion exchange.

The effect of nascent oxygen on the oxidation of As(III) to As(V) is evident from the fact that in a wide range of SFR, from 2 to 300 BV/h, the total arsenic concentration in the effluent of chlorinated water was constantly below the MTC value [47] (Figure 5(A)). On the other hand, in the case of non-chlorinated water, the arsenic concentrations in the effluent at the SFR<100 BV/h were by 10 times and at 100<SFR<300 BV/h even by 17 times

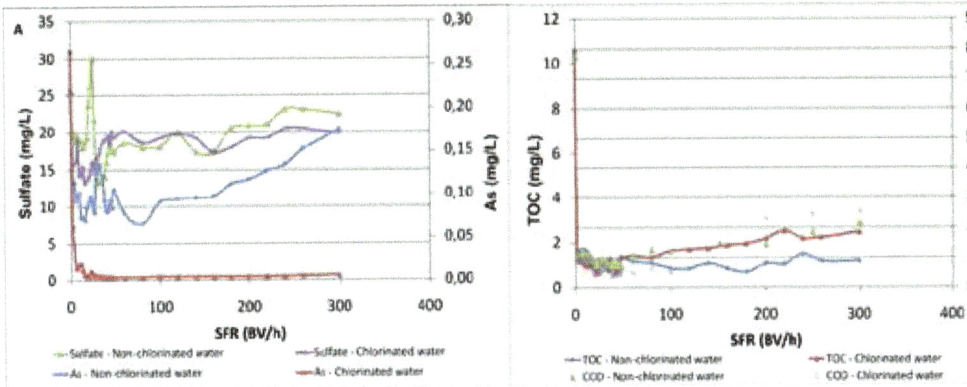

Figure 5. Changes in the sulfate and arsenic concentration (A) as well as NOM concentration (B) of the effluent with SFR

higher. Also, it is evident that the presence of atomic oxygen in chlorinated water hinders desorption of sulfate from the resin and changes the position of sulfate anions in the Clifford series [35]. This effect was especially pronounced at smaller SFR values of about 25 BV/h, when desorption of sulfate ion was observed. At 170<SFR<300 BV/h, the sulfate removal was more efficient in the case of chlorinated water. At low SFR, the effect of the oxidant on the removal of NOM is more pronounced. In the effluent of the water treated with oxidant, this is evidenced as the lower values of both COD and TOC (Figure 5(B)). At the SFR between 2 and 50 BV/h, there is a competition between humic acids, arsenic, and sulfate with bicarbonate and chloride, which is especially pronounced in the presence of nascent oxygen. With the increase in SFR of chlorinated water, the measured NOM contents in the effluent were higher, which means that NOM removal from non-chlorinated water was higher under these flow conditions. Evidently, the resin's functional groups exhibit a lower affinity toward the newly formed, smaller, organic molecules resulting from the NOM degradation than to the original NOM molecules from the groundwater. Thus, the NOM concentration measured as COD in chlorinated effluent reaches tolerable level of 2 mg O_2/L [47] at 180 BV/h. In the case of non-chlorinated water, this value was not reached until 275 BV/h, suggesting that in such a medium the resin shows a higher affinity to NOM. The effect of contact time of the resin with chlorinated and non-chlorinated water on the NOM removal is presented in Figure 6(A). Changes of arsenic content in the effluent as a function of the contact time of chlorinated and non-chlorinated water are presented in Figure 6(B). As can be seen, more intensive changes in the NOM concentration take place at shorter contact times. To an EBCT value of about 0.05 h, it comes out that the adsorption-desorption processes are especially pronounced in chlorinated water. At longer contact times, the COD and TOC values of the effluent become apparently constant. It is evident that in the shortest time of contact of water with the resin, the arsenic removal was most efficient in the presence of nascent oxygen. This is a consequence of the oxidation-reduction processes by which is generated As(V), which exhibits remarkable adsorption affinity to the resin.

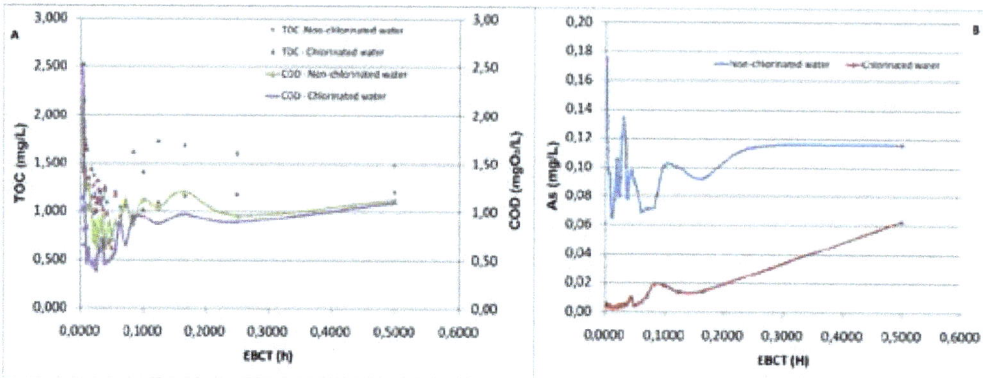

Figure 6. Influence of EBCT on the NOM (A) and arsenic (B) removal

3.2. Results obtained during strongly basic ion-exchange resin sorption characteristics determination

During the investigation of optimal specific flow rate of strongly basic ion-exchange resin, it has come to the conclusion that optimal flow rate of the Amberlite IRA-958 was 40 L/h, i.e. SFR of 30 BV/h ± 5% when used on described groundwater. This research was conducted in order to determine resin sorption characteristics based on the experimental data and data given by the manufacturer [11]. Based on the data given by the manufacturer, one gram of the resin can adsorb from 3.47 to 5.21 mgO_2 NOM expressed as COD. Investigated groundwater contains 7.73 mgO_2/L NOM, expressed as COD. The calculation shows that one gram of SBIX may adsorb the NOM from 0.45 L to 0.67 L of such water. Taking into account the mass of the resin of 970 g, it can be calculated that the overall volume of the groundwater from 436.5 L to 650 L may be effectively treated with the used the SBIX. If this is expressed via the BV values, the expected adsorption capacity of the resin for NOM is in the range from 312 BV to 464 BV.

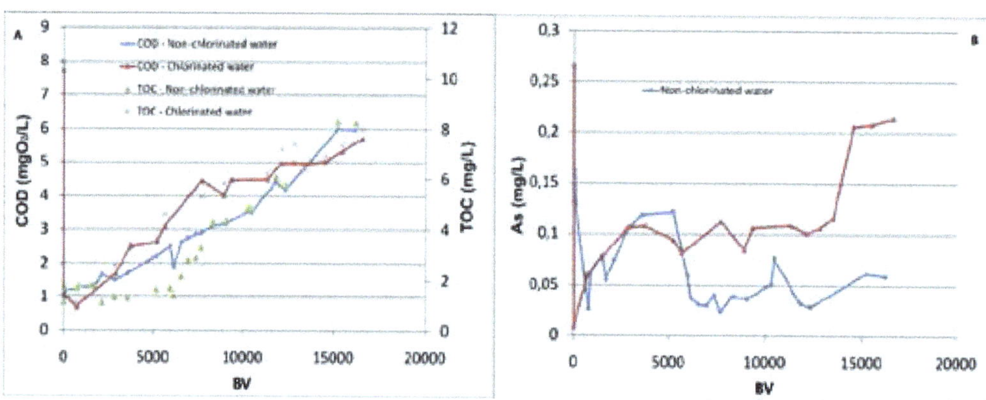

Figure 7. Changes of NOM (A) and arsenic (B) content in the effluent as a function of the overall volume of raw water passed through the SBIX

In the initial part of the sorption process, at small volumes of raw water of about 2,900 BV, as well as at the large volumes of over 13,500 BV, the NOM removal from chlorinated water is more efficient. In the first part of the operation cycle, the nascent oxygen generated from hypochloric acid causes NOM degradation and more efficient sorption of the degradation products of HAs and FAs compared to the native humins present in non-chlorinated water. These processes take place in the environment of the ion competition, as described by authors [40]. At increased volumes of groundwater passing through the system, surface adsorption is a dominant process, by which organic matter accumulates on the outer surface of the resin pearls. In this process, native HAs from non-chlorinated water are more efficiently bonded to the resin functional groups than the smaller molecules formed by the degradation of humin under the influence of nascent oxygen. Due to the larger number of the newly formed molecules in the effluent of chlorinated water at a smaller overall volume of water passing through the system, the resin's saturation and COD limit of 2 of mgO_2/L [47] is attained faster, at about 3,000 BV. At the same time, the adsorption of larger molecules on the resin that was in contact with non-chlorinated water is slower, and the COD limit appears at 4,000 BV. From the analysis of TOC values of the effluent from chlorinated water, it comes out that the changes of this parameter with the volume of treated water are similar to the changes observed for the COD. However, the TOC changes in the effluent of non-chlorinated water show that up to the value of about 8,000 BV, the NOM removal expressed via TOC is more efficient than if expressed via COD, and after that these values become proportionally equivalent. This difference can be explained by the fact that in the absence of the oxidant in water, the COD determination by the permanganate method gives lower results. Namely, permanganate is not a sufficiently strong oxidizing agent to oxidize entirely the humic matter, but only its more easily oxidizable part - fulvic acids. On the other hand, the TOC method encompasses all the dissolved NOM, and this is the reason for the discrepancy between the COD and TOC results presented in Figure 7(A). In the initial phase of resin saturation, the NOM molecules that are more easily oxidized (that is C_L) are preferentially adsorbed on the resin. With the increase in the amount of adsorbed NOM, proportionally to the volume of treated water above 8,000 BV, the resin affinity to sorption of organic matter becomes lower and larger amounts of easily oxidized NOM remain in the effluent, so that the changes in COD and TOC are proportional, which is evident from Figure 7(A). Namely, the appearance of COD and TOC curves above 8,000 BV becomes identical, since the COD value of, for example, 3 and 6 mgO_2/L is proportional to the TOC values of 4 and 8 mg/L. As can be seen from Figure 7(B), the arsenic concentrations in the effluent from chlorinated water are below the MTC of 0.01 mg/L only up to about 700 BV. By monitoring the effect of the further groundwater flow, it can be seen that arsenic removal is approximately equally efficient up to about 5,000 BV, from both chlorinated and non-chlorinated water. At the BV values above about 7,000, the effect of nascent oxygen on arsenic removal represents a complex process. In the investigated groundwater of complex composition, a number of redox reactions can take place in parallel. The reactions of HA degradation dominate over the reactions of As(III) oxidation. With the increase in the volume of treated water exceeding 5,000 BV, the amount of arsenic removed was twice larger when the treated water contained hypochlorite. In the presence of the oxidizing agent, the processes of NOM sorption in competition with arsenate binding to SBIX are more pronounced. The increased concentrations of arsenic in the effluent from chlorinated water appear as a consequence of arsenic desorption. Desorption process was especially pronounced at the volumes exceeding 14,000 BV, taking place simultaneously with the tendency of lowering arsenic concentrations in the effluent from non-chlorinated water.

The expressions (6)-(21) were used to calculate concentrations of NOM and arsenic in the overall volume of groundwater in the experiments with chlorinated and non-chlorinated water, amounts of NOM and arsenic in the effluent during the experiment, as well as the amounts of NOM and arsenic adsorbed on the SBIX. Total amounts of NOM and As that passed through the resin as well as amounts that charged the resin are presented in Table 3.

Sorption characteristics	COD_{total} (gO_2)	COD_{load} (gO_2/g)	TOC_{total} (g)	TOC_{load} (g/g)	Ast_{otal} (g)	As_{load} (g/g)	BV
No chlorine	175.65	0.181	241.86	0.249	6.039	0.006	16,231
Chlorine present	179.90	0.185	247.71	0.255	6.186	0.006	16,623

Table 3. Experimentally obtained resins sorption characteristics

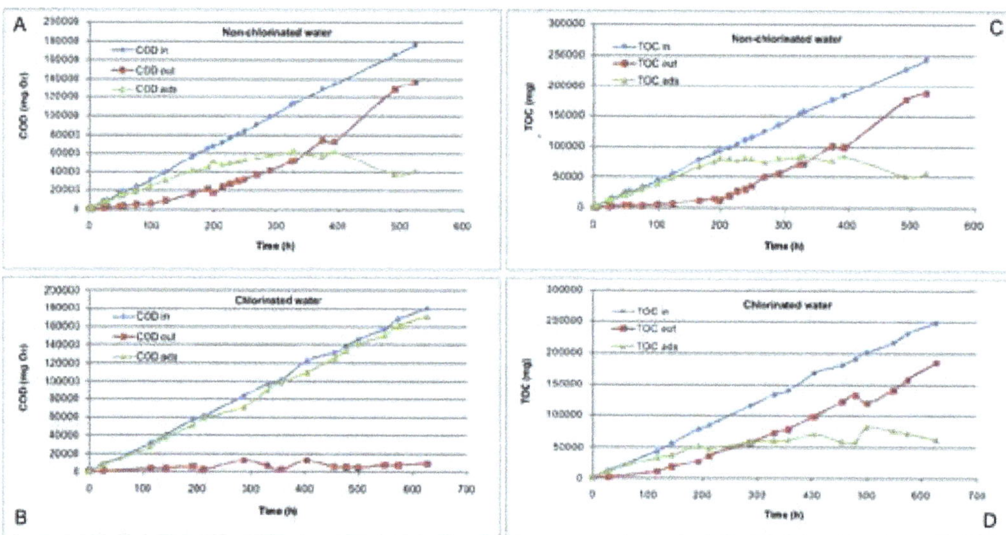

Figure 8. Dependency of NOM in non-chlorinated (A), (C) and chlorinated water (B), (D) adsorbed on the SBIX and one remained in the effluent during the course of the experiment

NOM fractions, which are more difficult to oxidize, remained in the effluent, and the more easily oxidized NOM compounds were adsorbed onto the resin. The assumption of NOM desorption is supported by the fact that at the longest experimental time, the COD_{ads} was 35% smaller than the adsorbed NOM at the intersection of the COD_{out} and COD_{ads} curves, as shown in Figure 8(A). Simultaneously, the COD in the effluent (COD_{out}) increased 2.2 times as large. The ratio between the maximum value for the COD_{ads} (before beginning NOM desorption) of

61,674 mg O_2 and the resin mass of 970 g describes the capacity of the resin for NOM sorption from the non-chlorinated water, with a value of 63.58 mg O_2/g. The NOM sorption process in chlorinated water (Figure 8(B)), without desorption, occurred throughout the experiment. The amount of the more easily oxidized fractions was very high; therefore, the COD_{ads} values were similar to the COD_{in} values. The total dissolved NOM sorption process in the non-chlorinated water, which was monitored by determining the TOC, was similar to the monitored COD (Figure 8(C)). After 10,900 BV, NOM in the effluent exceeded the NOM adsorbed onto the SBIX. The appearance of NOM desorption can be concluded from the fact that at the longest experimental time, the TOC_{ads} was 35% smaller than the TOC_{ads} at the intersection of the TOC_{out} and TOC_{ads} curves. At the same time, the TOC in the effluent (TOC_{out}) increased 2.2 times as large. As shown in Figure 8(D), the TOC_{out} increased 3.2 times as large, and the amount of NOM adsorbed on the resin remained nearly constant. The NOM desorption began at 7712 BV when oxidant was used. This means that the resin adsorption capacity for NOM in the non-chlorinated water was higher than in the chlorinated water.

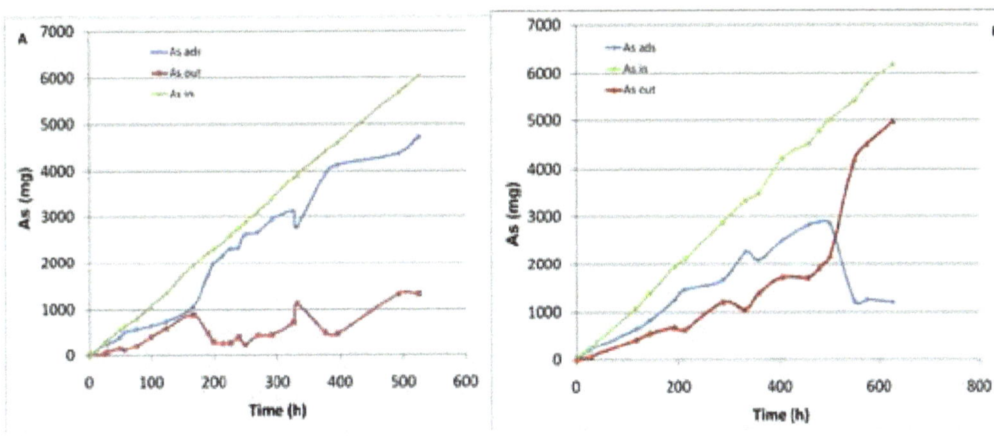

Figure 9. Dependence of the amounts of As contained in non-chlorinated (A) and chlorinated water (B) (As_{in}), adsorbed on the SBIX (As_{ads}) and remaining in the effluent (As_{out}) on the duration time of the experiment

Arsenic sorption process in the non-chlorinated water occurred throughout the entire experiment without the occurrence of desorption was shown in Figure 9(A). The only exception was at 166 h, when a short-lasting, dynamic equilibrium between the adsorbed and desorbed arsenic occurred. The amount of arsenic adsorbed during the experiment was very large, and the As_{ads} value was constantly approaching the value for As_{in}. The pH of the investigated groundwater was 8.06, at which the $HAsO_4^{2-}$ ions of As(V) and H_3AsO_3 molecules of As(III) dominate compared with the total arsenic content [20, 48]; therefore, As_{out} mostly represents the amount of As(III), and As_{ads} mostly represents the amount of As(V). The investigated resin was very effective at removing arsenates that were naturally present in the groundwater without the addition of the oxidant. The arsenates formed by the oxidation of arsenites with nascent oxygen simultaneously with competitive reactions of humin oxidation were adsorbed on the resin to a lesser extent, and after contact times exceeding 500 h, they desorbed and went into solution (Figure (9B)). Presence of arsenic desorption is supported by the fact that the As_{ads} was 2.13 times smaller than the As_{ads} at the intersection of the As_{out} and As_{ads} curves. It is at this point that desorption process for arsenic actually begins. Contemporaneously, the As in the effluent (As_{out}) increased 1.96 times as larger. The exchange capacities of arsenic sorbed were calculated for the highest values of As_{ads} for both parts of the investigation. Total exchange capacity of adsorbed arsenic As_{ads} for non-chlorinated and chlorinated water was 0.045 eq/L and 0.028 eq/L, respectively. Lower level of adsorption in the chlorinated water is consequence of competition of arsenic anions and smaller molecules originated from NOM oxidation for resin binding cites. Threshold limit value of arsenic [47] was met at 40.48 BV in experiment with chlorinated water. As can be concluded on the basis of Figure 9, the investigated resin is very effective in removing arsenates that are naturally present in groundwater, without addition of the oxidant. The arsenates formed by the oxidation of arsenite by nascent oxygen under the conditions of the simultaneous occurrence of the competitive reactions of humin oxidation are adsorbed on the resin to a lower extent, and at the contact times exceeding 500 h, they are desorbed and passed to the solution.

Equations 22-24 were used to calculate the sorption efficiency on the resin of the NOM (expressed via COD and TOC) and arsenic. Figure 10 shows the sorption efficiency of NOM expressed as E_{COD} and E_{TOC}.

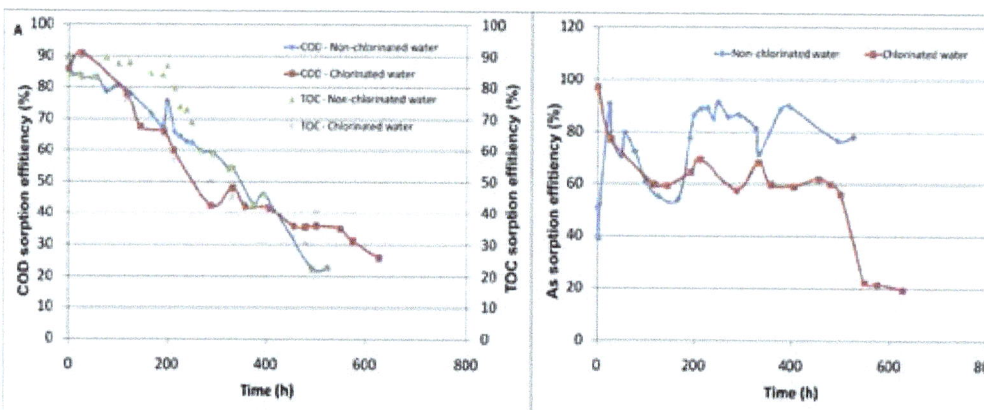

Figure 10. Changes of the efficiency of NOM sorption on SBIX expressed via COD and TOC (A) and arsenic (B) with the time of treatment of chlorinated and non-chlorinated water

As can be seen from Figure 10, the efficiency of NOM sorption expressed as E_{COD} decreases with time in the experiment with non-chlorinated water, and the changes are almost linear. On the other hand, the analogous dependence for chlorinated water is of a different pattern. In the initial part of the experiment, to 100 h, as well as after 400 h, the efficiency of NOM sorption from chlorinated water is more pronounced. This means that the conditions of the presence of the oxidant in the groundwater in the beginning of resin saturation and at the end of this process favor the sorption of NOM, expressed either as E_{COD} or E_{TOC}, compared to the desorption processes. The course of the changes of E_{TOC}, reflecting definitely the overall content of humic substances, shows that the preferential sorption of NOM is pronounced in one-third of the overall treatment cycle, that is between the 48th and 395th h of the experiment. The large humin molecules are more readily adsorbed than the smaller molecules formed by their degradation. Since the efficiency of NOM removal by the resin decreases with time, the outer sorption layers become more attractive to smaller molecules, so that they are more competitive, and at the longest time, the NOM sorption from chlorinated water is more efficient. As can be seen, the efficiency of arsenic removal from chlorinated water is high only in the beginning of the working cycle. Evidently, the presence of the oxidant, which converts As(III) to As(V), does not contribute to the efficiency of arsenic removal. If one considers the entire working cycle of the resin, the average efficiency of arsenic removal from chlorinated water is about 60%, and from non-chlorinated water it is about 80 %.

3.3. Results obtained by using a new approach for determination of equilibrium adsorption capacity of SBIX toward NOM

Obtained experimental results were applied in the equations 25-32 to calculate the important kinetic behavior of SBIX toward NOM. Calculated ADC values were decreased exponentially during the experiment as shown at Figure 11(A). Curve of exponential regression kinetically represents interface of passed (under the curve) and adsorbed (above the curve) quantity of NOM per mass unit of SBIX in time. The empirical coefficients β and δ (equation 29) were calculated at the base of equation 30, i.e. empirical linear equation (Figure 11(B)). Coefficients β and δ were calculated from linear equation (Figure 11(B)) lnA(DC) = −0.704 C + 1.471 and amounts 4.355 $Lg^{-1}h^{-1}$ and 0.70, respectively.

Empirical equation form of the obtained curve and regression coefficient were ADC = 4.355 $e^{-0.70C}$ and 0.961, respectively. By utilization of equations 31 and 32, respectively, the total specific adsorption capacity and adsorption capacity of the resin at pseudo equilibrium were calculated. The values of ADC_a and Q_{pe} were 14.41 mg/gh and 10.935,31 mg/g, respectively.

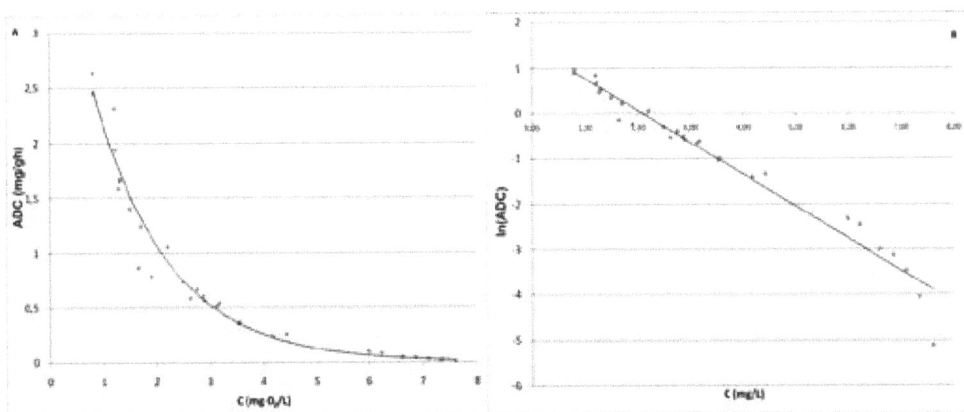

Figure 11. (A) Changes of specific adsorption capacity of SBIX to NOM during the experiment and (B) Linear form of exponential curve ADC vs. C for calculation of β and δ

4. Conclusion

A comparison of the efficiency of NOM removal from native groundwater and the same water with addition of sodium hypochlorite as oxidizing agent showed that at smaller SFR values up to 50 BV/h, the process was more effective in the presence of the oxidant. However, at the SFR values that were up to ten times higher compared to the maximum values given in the manufacturer's declaration, the NOM removal was more efficient in the absence of the oxidant. Namely, the NOM level in the effluent originated from the native groundwater was below MTC of 2 mg O_2/L at a SFR value that was 9.2 times higher, whereas under the oxidative conditions this was observed at the SFR that was by 6 times higher than the declared maximum SFR value. The arsenic removal to the level below the MTC of 0.01 mg/L was more efficient in the presence of the oxidant and shorter SFR compared to the minimum and at 10 times higher SFR values compared to the maximum value of SFR declared by the resin manufacturer. In the absence of the oxidant, only native As(V) was removed, and the process was essentially less effective because of the competition with sulfate ions. It was shown that at small BV values, of up to 700, the removal of NOM and arsenic is more efficient in the presence of the oxidant, whereas in the major part of the working cycle, the efficiency of their removal is higher from non-chlorinated water. The values obtained for the resin adsorption capacity for NOM from non-chlorinated water are by about 15 times higher than the corresponding value given in the specification of the resin manufacturer. New approach to the determination of resin sorption capacity, through definition of specific adsorption capacity, at pseudo-equilibrium showed that investigated resin posses very high affinity to NOM. Occurrence of NOM desorption, measured as COD, was observed in the treatment of non-chlorinated water, as well as by measuring the adsorbed TOC in the case of both chlorinated and non-chlorinated water. Desorption of arsenic from the resin surface was observed only in the case of treatment of chlorinated water. Namely, when the results for the optimum SFR are obtained for a pilot plant system for concrete groundwater of distinct physicochemical characteristics, it is simple to design a unique ion-exchange water treatment system of any capacity.

The obtained results make a sound basis for designing an appropriate plant for the removal of NOM from groundwater of the region of the town of Zrenjanin. The tested resin can also be used in the process of pretreatment of the same water, since it allows the removal of about 50% of the naturally occurring arsenic.

Author details

Miroslav Kukučka* and Nikoleta Kukučka

*Address all correspondence to: miroslav@envirotech.rs

Envirotech d.o.o., Kikinda, Serbia

References

[1] Sillanpää M., Matilainen A. NOM removal by coagulation. In: Natural Organic Matter in Water - Characterization and Treatment Methods, IWA Publishing; 2015. p55-80.

[2] Sillanpää M., Bhatnagar A. NOM removal by adsorption. In: Natural Organic Matter in Water - Characterization and Treatment Methods, IWA Publishing; 2015. p213-238.

[3] Świetlik J., Raczyk-Stanisławiak U., Biłozor S., Ilecki W., Nawrocki J. Adsorption of natural organic matter oxidized with ClO_2 on granular activated carbon. Water Research 2002;36(9) 2328-2336.

[4] Fettig J. Removal of humic substances by adsorption/ion exchange. Water Science and Technology 1999;40(9) 173-182.

[5] Sillanpää M., Levchuk I. Ion exchange. In: Natural Organic Matter in Water - Characterization and Treatment Methods, IWA Publishing; 2015. p239-273.

[6] Korth A., Bendinger B., Czekalla C., Wichmann K. Biodegradation of NOM in rapid sand filters for removing iron and manganese. Acta Hydrochimica et Hydrobiologica 2002;29(5) 289-295.

[7] Kukučka M., Habuda-Stanić M., Šiljeg M. Groundwater treatment by nanofiltration-pilot study in Kikinda, Northern Serbia. Proceedings of 11th Conference on Environmental Science and Technology, Chania, Crete, Greece, 2009.

[8] Koprivnjak J.F., Perdue E.M., Pfromm P.H. Coupling reverse osmosis with electrodialysis to isolate natural organic matter from fresh waters. Water Research 2006;40(18) 3385-3392.

[9] Krzysztof P., Kowalski. Advanced arsenic removal technologies review. In: Chemistry of Advanced Environmental Purification Processes of Water, Fundamentals and Applications. 2014. p285-337.

[10] Kukučka M., Kukučka N., Vojinović Miloradov M., Tomić Ž., Šiljeg M. A novel approach to determine a resin's sorption characteristics for the removal of natural organic matter and arsenic from groundwater. Water Science and Technology: Water Supply 2011;11(6) 726-739.

[11] Rohm & Haas, Product Data Sheet, (2008), PDS 0431A.

[12] Šmejkalová D., Piccolo A., Spiteller M., Oligomerization of humic phenolic monomers by oxidative coupling under biomimetic catalysis. Environmental Science and Technology 2006;40(22) 6955-6962.

[13] Stevenson, F.J. Humus Chemistry: Genesis, Composition, Reactions. 2nd Ed. John Wiley and Sons, New York; 1994.

[14] Piccolo A. The supramolecular structure of humic substances: A novel understanding of humus chemistry and implications in soil science. Advances in Agronomy 2002;75 57-134.

[15] Piccolo A., Stevenson F.J. Infrared spectra of Cu^{2+}, Pb^{2+} and Ca^{2+} complexes of soil humic substances. Geoderma 1982;27(3) 195-208.

[16] Stevenson F.J., Goh K.M. Infrared spectra of humic acids and related substances. Geochimica et Cosmochimica Acta 1971;35(5) 471-483.

[17] Blair G.J., Lefroy R.D.B., Lisle L. Soil carbon fractions based on their degree of oxidation, and the development of a carbon management index. Australian Journal of Agricultural Research 1995;46(7) 1459-1466.

[18] Anderson D.W., Schoenau, J.J. (1993) Soil humus fractions. In Carter M.R. (ed.), Soil Sampling and Methods of Analysis. CRC Press: Florida; 1993. p391-395.

[19] Smedley P.L. Kinniburgh D.G. A review of the source, behaviour and distribution of arsenic in natural waters. Applied Geochemistry 2002;17(5) 517-568.

[20] Brady P.V., Teter D.M., Khandaker N.R., Krumhansl J.L., Siegel M.D. Development of novel arsenic treatment approaches. SAND2004-0056, Sandia National Laboratories, Albuquerque, NM;2004.

[21] Choong T.S.Y., Chuah T.G., Robiah Y., Koay G.F.L., Azni I. Arsenic toxicity, health hazards and removal techniques from water: an overview. Desalination 2007;217(1-3)139-166.

[22] Ng J.C., Wang J., Shraim A. A global health problem caused by arsenic from natural sources. Chemosphere 2003;52(9) 1353-1359.

[23] Council Directive 98/83 EC, Official Journal of European Communities, L 330/32, 1998.

[24] US EPA, 2001. "National Primary Drinking Water Regulations; Arsenic and Clarifications to Compliance and New Source Contaminants Monitoring; Final Rule," Federal Register, 66:14-6976.

[25] Bolto B., Dixon D., Eldridge R., King S., Linge K. (2002) Removal of natural organic matter by ion exchange. Water Research 2002;36(20) 5057-5065.

[26] Naumczyk J., Szpyrkowicz L., Grandi F.Z. Organics isolation from fresh and drinking waters by macroporous anion-exchange resins. Water Research 1989;23(12) 1593-1597.

[27] Gottlieb M. The reversible removal of naturally occurring organics using ion exchange resins: Part 1. Industrial Water Treatment 1995;27 (3).

[28] Pandey A.K., Pandey SD., Misra V., Viswanathan P.N. Formation of soluble complexes of metals with humic acid and its environmental significance. Chemistry and Ecology 1999;16(4) 269-282.

[29] Warwick P., Inam E., Evans N. Arsenic's interaction with humic acid. Environmental Chemistry 2005;2(2) 119-124.

[30] Tan Y., Kilduff J.E. Factors affecting selectivity during dissolved organic matter removal by anion-exchange resins. Water Research 2007;41(18) 4211-4221.

[31] Fu P.L.-K., Symons, J.M. Removing aquatic organic substances by anion exchange resins. Journal of the American Water Works Association 1990;82(10) 70-77.

[32] Korngold E., Belayev N., Aronov L. Removal of arsenic from drinking water by anion exchangers. Desalination 2001;141(1) 81-84.

[33] Ficklin W.H. Separation of As(III) and As(V) in groundwaters by ionexchange. Talanta 1983;30 (5) 371-373.

[34] Edwards M., Patel S., McNeill L., Chen H.-W., Frey M., Eaton A.D., Antweiler R.C., Taylor H.E. Considerations in arsenic analysis and speciation. Journal of the American Water Works Association 1998;90(3) 103-113.

[35] Clifford D. Ion exchange and inorganic adsorption. In: Letterman A. (ed.) Water Quality and Treatment, American Water Works Association, McGraw-Hill, New York; 1999.

[36] Vasiljević M. Possibility of using groundwaters for the needs of water supply of the settlements and industry in the Pannonian Basin (Banat and Bačka), PhD Thesis, University of Belgrade (in Serbian); 1999.

[37] Lang B. The base metals-gold hydrothermal ore deposits of Baia Mare, Romania. Economic Geology 1979:74(6) 1336-1351.

[38] Perišić M. NOM and Arsenic Removal from Natural Water by Enhanced Coagulation, E-Water, Official Publication of the European Water Association (EWA) 2006.

[39] Greenberg A.E. (Ed.) Standard Methods for the Examination of Water and Wastewater, 19th ed., American Public Health Association Publications, Washington D.C. 1995.

[40] Kukučka M., Kukučka N., Vojinović Miloradov M., Tomić Ž., Šiljeg M. Effects of extremely high specific flow rates on the removal of NOM and arsenic from groundwater with ion-exchange resin: A pilot-scale study in northern Serbia. Journal of Environmental Science and Health, Part A; Toxic/Hazardous Substance & Environmental Engineering 2011;46(9) 952-959.

[41] Li M., Zhu X., Zhu F., Ren G, Cao G, Song L. Application of modified zeolite for ammonium removal from drinking water. Desalination 2011;271, 295-300.

[42] Alshameri A., Yan C, Al-Ani Y., Salman Dawood A., Abdullateef Ibrahim, Chunyu Zhou, Hongquan Wanga. An investigation into the adsorption removal of ammonium by salt activated Chinese (Hulaodu) natural zeolite: Kinetics, isotherms, and thermodynamics. Journal of the Taiwan Institute of Chemical Engineers 2014; 45 554-564

[43] Yan Y, Sun X., Ma F., Li J., Shen J., Han W., Liu X., Wang L. Removal of phosphate from etching wastewater by calcined alkaline residue: Batch and column studies. Journal of the Taiwan Institute of Chemical Engineers 2014;45(4) 1709-1716.

[44] Arslan A., Veli S. Zeolite 13X for adsorption of ammonium ions from aqueous solutions and hen slaughterhouse wastewaters. Journal of the Taiwan Institute of Chemical Engineers 2012;43(3) 393-398.

[45] Alshameri A., Ibrahim A., Assabri A., Lei X., Wang H., Yan C. The investigation into the ammonium removal performance of Yemeni natural zeolite: Modification, ion exchange mechanism, and thermodynamics. Powder Technology 2014;258 20-31.

[46] Balci S. Nature of ammonium ion adsorption by sepiolite: analysis of equilibrium data with several isotherms. Water Research 2004;38(5) 1129-1138.

[47] Serbian regulations, Pravilnik o higijenskoj ispravnosti vode za piće (1998) Službeni list SRJ, Vol. 42.

[48] Welch A.H., Oremland R.S., Davis J. A., Watkins S.A. Arsenic in ground water: a review of current knowledge and relation to the CALFED solution area with recommendations for needed research. San Francisco Estuary and Watershed Science 2006;4(2) 1-32.

7

Electric Field-Assisted Ion Exchange of Borosilicate Glass Tubes

Ali Talimian and Vincenzo M. Sglavo

Abstract

In this work, DC electric field-assisted ion exchange was carried out to enhance the sodium-potassium inter-diffusion and improve the mechanical performance of borosilicate glass. Electric fields with intensity varying between 100 V cm^{-1} and 3000 V cm^{-1} were applied in both direct and inverted polarizations. Four point bending test and the Vickers indentation method were used to characterize the mechanical properties. Energy dispersion x-ray spectroscopy was carried out to determine the potassium concentration within the surface layers of the samples.

The analysis of the potassium concentration profile near the surface shows that the external electric field governs the ion exchange process and it is possible to send potassium ions down to a depth of 45 μm in only 5 min. By plotting the electrical current versus time, it is revealed that the process stops after a certain saturation time. Vickers indentation measurements show that the compressive residual stress in the samples treated under electrical field is 3 times higher than that obtained by conventional chemical tempering. The bending strength of samples prepared by reversing the field direction is higher than that measured in specimens treated only on one side due to the symmetrical distribution of the stress on both sides.

Keywords: Field-assisted ion exchange, Borosilicate glass, Mechanical properties

1. Introduction

During the ion exchange process on glass, some ions in the materials are replaced by new ions from the liquid that can modify the physical and chemical properties locally [1-3]. The ion exchange process has been widely used to change the reflective index in a selected area of glass for producing waveguide devices [4, 5]. Starting about 50 years ago, ion exchange has been employed also to improve the mechanical properties of silicate glass [1].

Surface defects are responsible for the limited glass resistance and its large scatter [6]. The creation of a compressive stress layer in the surface of the material can limit the formation or propagation of flaws and improve the mechanical properties; thermal and chemical tempering of glass are two main methods for producing a compressive stress in the glass surface. The thermal method is widely used to make windows and other transparent flat structural components [2, 7].

The ion exchange or chemical strengthening of glass was almost abandoned for many years because of the high processing cost and long-duration process. In recent years, this method has been reconsidered because of the possibility of mechanical treatment after strengthening, the applicability to complicated shape and limited thickness components, and the absence of optical distortion [8].

The ion exchange process is typically carried out by immersing the components made of a glass containing lithium or sodium in molten potassium nitride salt. The process can be carried out at a temperature between the melting point of the salt and the transforming temperature of the glass and takes times in excess to 4 h, depending on the required depth for the compressive stress layer. After finishing the process, the samples are removed from the bath and the salt on the surface is simply washed out by water [2, 9].

The ion exchange process can be considered as an inter-diffusion reaction between the mobile ions in glass and the cations in the molten salt while the other glass components are considered as an immobile matrix of negative groups [10-13]. An external electric field can be the source of an extra driving force for the inter-diffusion of the mobile ions. This process is known as electric field-assisted ion exchange (EF-IOX) and it has been used especially for manufacturing waveguides [4, 5, 14, 15]. Three different procedures have been proposed. In the first one, a thin metal film as a source for ions is applied on the glass surface and the application of an electric field induces the oxidation/reduction at the interface with glass that generates cations that move into the glass on the anode side, creating the chemical concentration profile [5, 16, 17]. Alternatively, a molten salt can be used as the ion source at only the anode side of the sample, the cations at the anode side penetrates into the glass under the field [5]. In the last approach, the sample acts as a wall, separates two molten salt baths and each bath is connected to an electrode; by applying the electric field the cations in the salt start moving in the direction of the electric field [4, 5, 14].

By considering the ions flow and the electrical charge neutrality balance, different mathematical modelling have been proposed for the ion diffusion during the exchange process [18]. For a large enough field, the concentration of exchanged ions, at distance x from the surface and at time t, can be defined as:

$$C(x,t) = \frac{C_0}{2} erfc\left(\frac{x - \mu E t}{2\sqrt{Dt}}\right) \tag{1}$$

where C_0 is the surface concentration of the ions, E the applied electric field, μ the ion mobility, and D the inter-diffusion coefficient [8]. Thermal diffusion does not play an active role and the migration of ions is governed by the electric field, only. Consequently, a step-like profile is generated in a short period of time for the ions concentration [9]. The long range migration of cations in the glass allow to produce a deeper compressive layer [18, 19].

Field-assisted ion exchange has been limitedly studied with the aim of improving the mechanical properties during the 1970s [20]. The main problem was the unbalanced residual stress that caused the deformation of the samples.

In this work, we performed electric field-assisted ion exchange on commercial borosilicate glass tubes, thus avoiding the problems associated with the deformation of the samples. The aim was to analyze the possibility of using an external field for speeding up the ion exchange process and obtaining improved mechanical properties.

2. Experimental procedure

Borosilicate test tubes from commercial sources, Fiolax clear, Schott, were used in this study. The glass transition temperature was measured by the differential scanning calorimeter (DSC) (DSC2010, TA Instruments, USA) method [21]. The chemical composition and the physical properties of the glass are shown in Table 1 [22]. The tubes having nominal thickness and outer diameter of 0.5 mm and 11.8 mm, respectively, were cut in 100 mm long samples. The limited thickness of the tubes allows thermal equilibrium between the sample and the salt bath during the process.

Chemical composition (wt%)					Physical properties	
SiO_2	B_2O_3	Al_2O_3	Na_2O	CaO	Density (g/cm^3)	Glass transition temperature (°C)
75	10.5	5	7	1.5	2.34	582

Table 1. The chemical composition and the physical properties glass tubes used for ion exchange

The samples were ultrasonically cleaned in distilled water for 5 min, washed with acetone and air dried. They were treated in a modified lab furnace schematically reported in Figure 1. One sample at a time was treated but current and applied voltage were constantly monitored and controlled during the process to guarantee a similar procedure for all samples. The applied electric field varied between 100 to 3000 V cm^{-1} and the current density was limited to 4, 8, and 16 mA cm^{-2}. The samples were kept over the bath for 20 min before the process as a preheating step; then, they were filled with molten salt and immersed in the bath kept constant at 400±2°C. After applying the electric field. The tube was immediately emptied from the molten salt and held for 20 min over the bath. At the end of each cycle the samples were ultrasonically washed with distilled water. Ten samples were also treated by conventional ion exchange in a commercial lab furnace, Lema TC 20 A, Italy, for comparison, by holding 4 h at the identical temperature. Potassium nitrate salt (>99.5% pure) from commercial source was used.

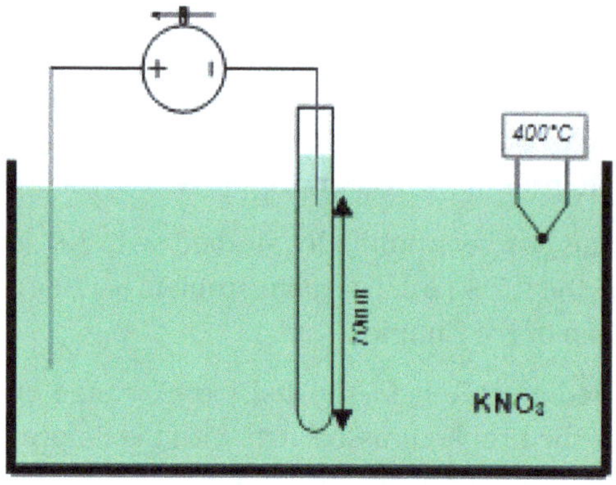

Figure 1. Schematic experimental setup used for field-assisted ion exchange

A four-point bending test was carried out to determine the mechanical strength, using inner (L1) and outer (L2) span equal to 10.3 and 40 mm, respectively. The test was carried out in lab air (temperature≈20°C, relative humidity ≈40%) with a constant loading rate of 1.1 MPa/s. The failure stress was calculated as:

$$S_b = \frac{4F_{max} r_2 a}{\pi\left(r_2^4 - r_1^4\right)} \tag{2}$$

where a is (L2-L1)/2, r_1 and r_2 the internal and the external radius of the tube, F_{max} the maximum load.

The presence of residual stresses on the surface layers was estimated by Vickers indentation method. The residual stress, σ_r, approximated as constant on the surface layers can be calculated as: [23]

$$\sigma_r = \frac{K_c}{\sqrt{\pi\Omega c}}\left(1 - \chi_r \frac{P^*}{K_c c^{3/2}}\right) \tag{3}$$

P^* being the indentation load, c the crack length, K_c the fracture toughness, χ_r and Ω two geometrical constants depending on the indenter and the crack shape, respectively.

The potassium penetration profile was measured on the fracture surface of test tube fragments. The fragments were fixed on an aluminum disk with an adhesive conducting tape and coated with Au-Pd alloy. The microscopic observation was carried out by a scanning electron microscope (SEM) (JSM 5500, Jeol, Japan). A clean path was chosen for analyzing the potassium $K\alpha$ on a certain length, around 70 μm long, by using energy dispersion x-ray spectroscopy

(EDXS) (EDS2000, IXRF System, USA) probe. The surface chemical composition of the glass before and after the treatment was also determined by EDXS.

3. Results and discussion

Figure 2 shows how the current density changes as a function of time during the EF-IOX tests. The current density limit was fixed at 20–25 A m^{-2} depending on the immersion depth of the test tube in the molten salt. For the samples subjected to fields with intensity lower than 500 V cm^{-1}, the current density shows similar trend, steadily decreasing with process time; for more intense electrical fields the current reaches a saturation limit, corresponding to the imposed current density limit, and rapidly decreases. During the process, it is thought that small sodium ions with high mobility are replaced by larger less mobile potassium ions, this determine an overall decrease in the ions movement and, consequently, a reduction of the current density.

The evolution of current density with time under different fields (500 to 3000 V cm^{-1}) in the condition of current limit = 16 mA cm^{-2} is plotted in Figure 3. All curves have a similar shape: the maximum current density is recorded at the initial application of the field and it is proportional to the applied field intensity. After 400 s, the current density decreases to 10% of the initial one.

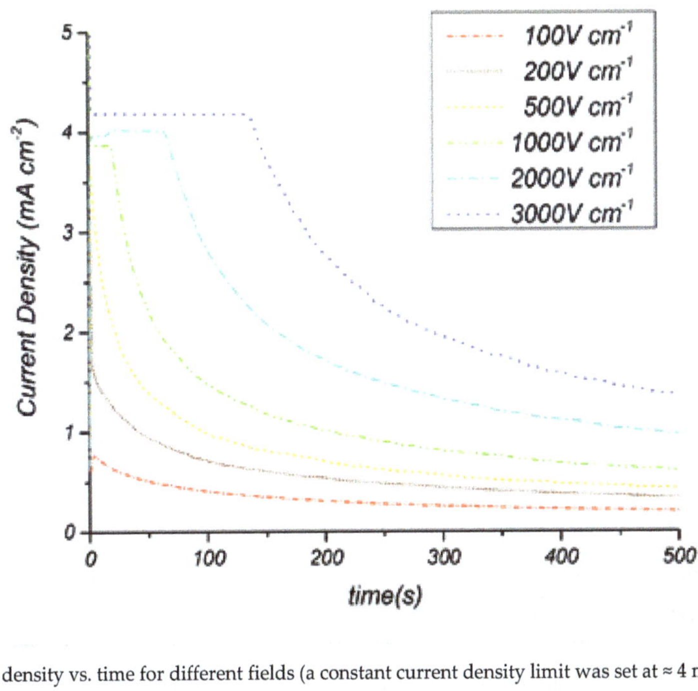

Figure 2. Current density vs. time for different fields (a constant current density limit was set at ≈ 4 mA cm^{-2})

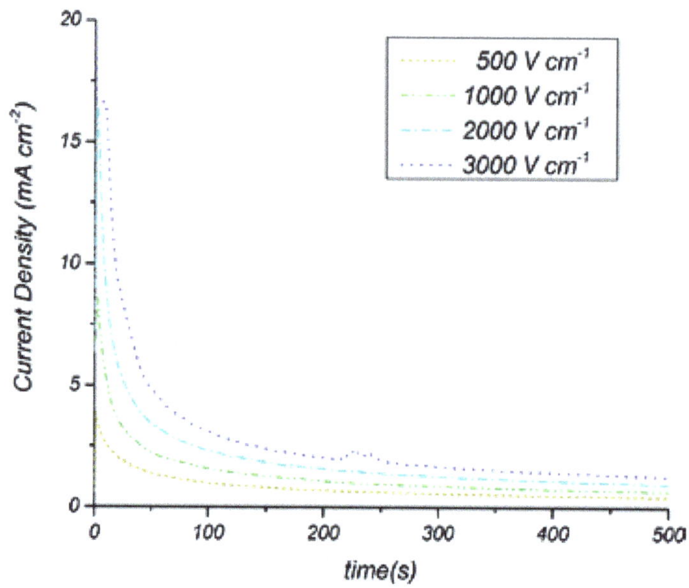

Figure 3. Current density vs. time for samples treated under different fields with current density limit of 16 mA cm^{-2}

The molar ratio between alkaline oxides ($K_2O/(Na_2O+ K_2O)$) as measured by EDXS is chosen as an indicative parameter of the surface chemical composition variation. The values for samples treated under fields of 500 and 2000 V cm^{-1} and current limit of 16 mA cm^{-2} are summarized in Table 2: It is clear that sodium ions are completely replaced by potassium in the outer surface, cathode side, while in the inner surface, anode side, sodium is detected.

By applying the electric field, potassium ions are moved into the glass in the direction of the field. Highly mobile sodium ions move faster in the glass than potassium ions and can reach the other side of the glass; as a result, one side of the glass is saturated with potassium and the other one has high amount of sodium. Some limited thermal ion exchange can occur during the process on the anode side and sodium ions are replaced by potassium.

Treatment conditions	Alkali Ratio
As-received	0
500 V cm^{-1} - Inner surface	0.7
500 V cm^{-1} - Outer surface	1
2000 V cm^{-1} - Inner surface	0.7
2000 V cm^{-1} - Outer surface	1

Table 2. $K_2O/ (Na_2O+ K_2O)$ molar ratio of test tubes after the field-assisted ion exchange

The area beneath the current density versus time curves represents the amount of ions moved through the unit area of the sample during the process. According to Eq. 1, the exchanged layer has a constant concentration of potassium ions and the following equation was proposed to calculated the thickness of the exchanged layer, Δ, in soda-lime silicate glass [24]:

$$\Delta(t_2,t_1)=\frac{Q(t_2,t_1)}{FC_0}=\frac{\int_{t_1}^{t_2} Jdt}{FC_0} \tag{4}$$

F being Faraday's constant, 96458.34 C mol^{-1}, C_0 the concentration of mobile ions in the base glass (i.e., the sodium concentration in the base glass = 2.64×10^{-3} mol cm^{-3}), and J the current density. By using Eq. 4, the depth of ion exchange varies between 8 to 120 μm depending on the intensity of the applied field. The depth of the exchanged layer is constant when the current density changes. Figure 4 shows how the current density varies with time for samples subjected to a field with intensity of 2000 V cm^{-1} and different current limits. By applying a current density limit, it assumes a constant value for a certain time and then decreases. After 200 s, all samples have similar behavior. The duration of the constant current step for the sample treated under a current limit of 8 mA cm^{-2} is longer; this makes the area underneath the curves and the amount of ions moved into the glass equal to the samples treated under current limit of 8 or 16 mA cm^{-2} and the amount of exchanged ions is the same for all samples after enough long time. Conversely, it is reported that the amount of exchanged ions depends on both the intensity of the electric field and the applied current limit for soda-lime silicate glass [24].

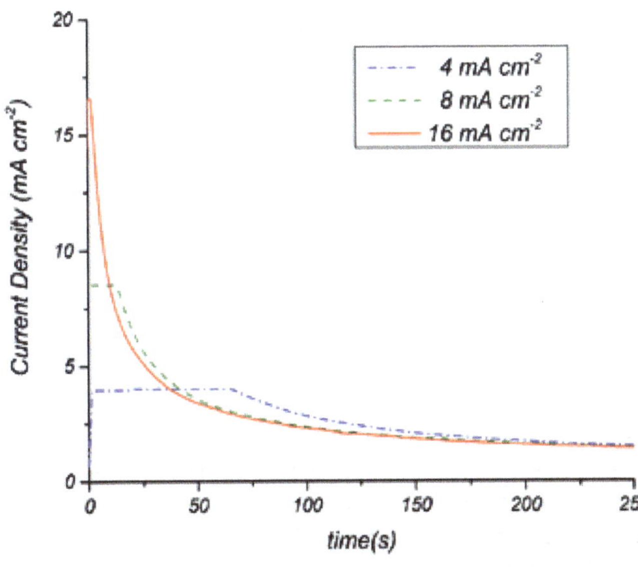

Figure 4. Current density vs. time for samples treated under field with intensity of 2000 V cm^{-1} and different current density

The potassium concentration profile near the surface of the samples subjected to the EF-IOX for 10 min are shown in Figure 5. In the outer surface a step in the concentration profile occurs for all conditions and the potassium penetration depth increases with the field intensity. The penetration depth is lower than that estimated by Eq. 4, according to mass balance and charge neutrality. During the process a local electric charge can be formed in the sample due to different mobility of sodium and potassium ions and, consequently, a local field is formed. Such local field can neutralize the external applied filed and prevent further exchange.

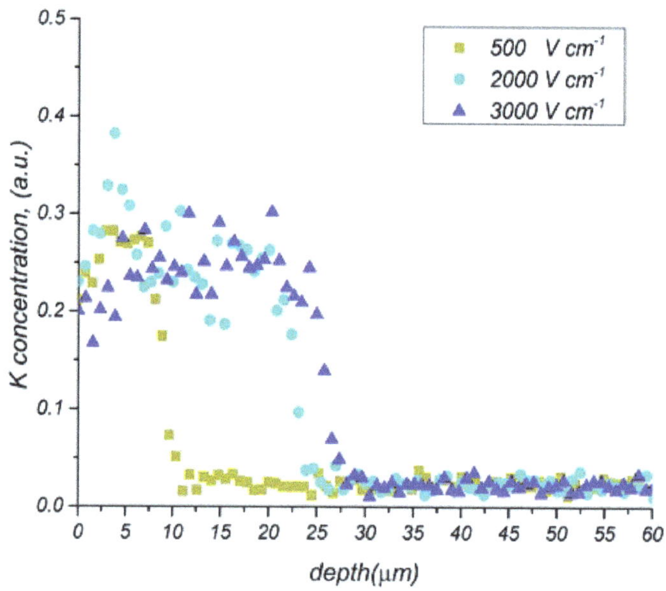

Figure 5. Potassium concentration profile in the external surface (cathode side) of tubes treated under different fields

Figure 6 shows the potassium concentration profile of glass tubes near the inner surface. Differently from Figure 5, similar profiles are recorded for different applied electrical fields and these resemble typical diffusion profiles obtained by conventional ion exchange process [7]. Probably, thermal ion exchange can be responsible for the observed diffusion concentration profile; nevertheless, more studies are required to reveal possible causes of such phenomenon.

All glass tubes subjected to the double (direct and inverse) field polarization—the initial one being out-to-in (Figure 7)—showed several surface cracks as in Figure 8; the cracks are on the inner surface and in some cases they reach the external one. Conversely, the samples treated first under a field with in-to-out polarization possess a structure similar to those subjected to conventional ion exchange. One possible reason for the formation of the observed cracks can be related to differential deformations between inner and outer surface upon the double ion exchange process. When potassium ions move into the outer surface of the tube first, the exchanged layer expands and compressive stress is formed due to the difference between the specific volume of the surface layer and the base glass. By exchanging the ions in the inner

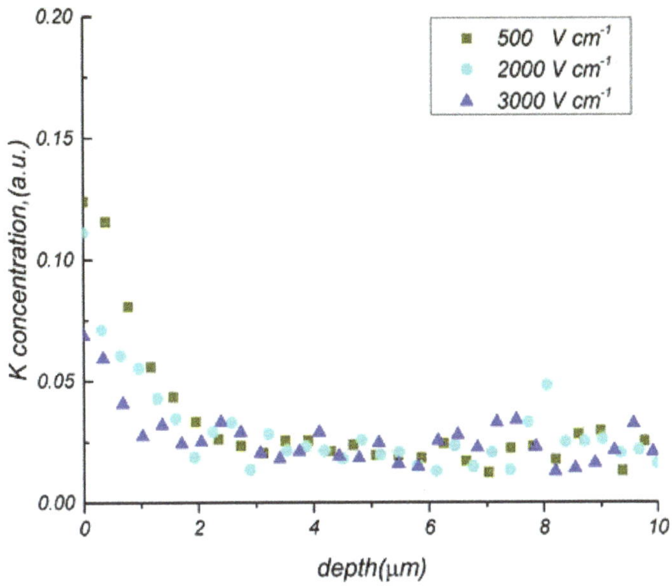

Figure 6. Potassium concentration profile in the internal surface (anode side) of tubes treated under different fields

surface during the second step, the expansion of the internal surface due to the geometry of the tube produces extra stresses in the interface of the exchanged layer and base glass. Conversely, by treating the inner surface first, the outer surface goes into tension because of the inner expansion and some cracks can be formed.

Figure 9 shows the bending strength of samples treated by different methods. The samples treated under the field on only one side are stronger than the as-received tubes but slightly weaker than the specimens subjected to conventional ion exchange processes. After the double-field polarization process, the strength increases further although the scatter becomes much larger.

Optical microscopy photographs of typical indentations on treated samples are reported in Figure 10. If the indentation load is 40 N no cracks are generated; conversely, at 50 N well-developed cracks are generated; therefore, a threshold load for crack formation can be pointed out for the strengthened samples, which is much higher than that for as-received glass, estimated equal to about 4 N. The effect of the compressive residual stress is absolutely clear.

The residual stress is built up by local replacement of small ions by larger ions during the ion exchange process; in the samples treated under the electric filed one should expect that the compressive stress is substantially built up only on one surface of the glass tube, it being balanced by tensile stresses in the remaining of the thickness [24]. The residual stress can be calculated by Eq. (3) from the measurement of the indentation crack length [25] and the results are shown in Table 3. The residual stress built up by the electric field-assisted process is significantly higher than that obtained in conventionally treated samples; the stress intensity increases further after the double-reversed electrical field treatment. Stress build up by

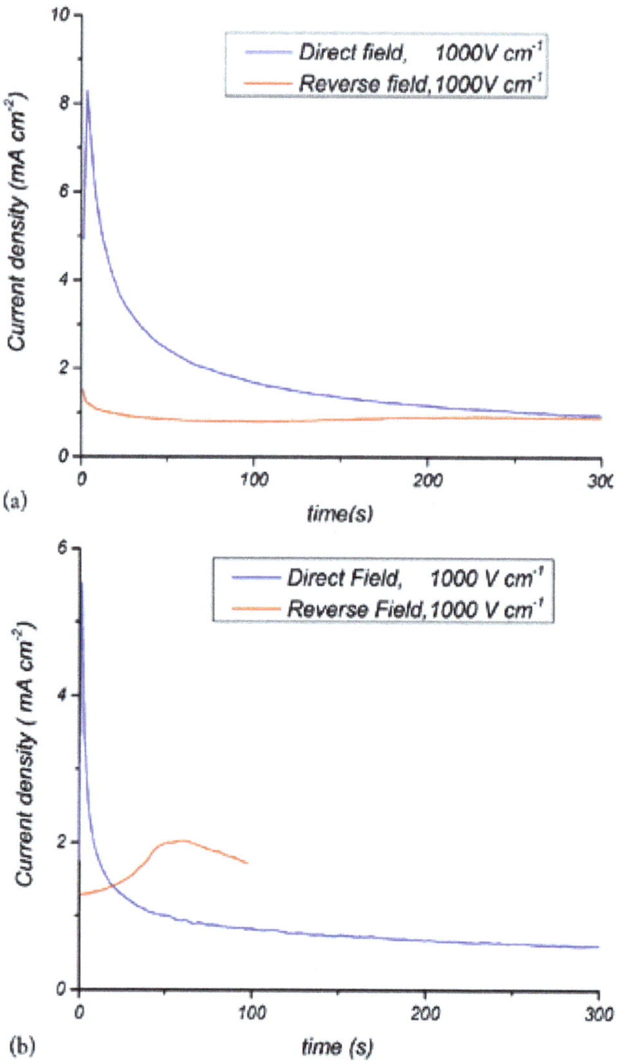

(a)

(b)

Figure 7. Current density vs. time for samples treated under direct and inverse electric field. The initial field polarization is (a) outside to inside, (b) inside to outside

exchanging ions is considered based on the volume change of glass during the process; the steep change of the potassium concentration profile produces higher residual stress compared to the gradual change in the thermally induced exchanged glass.

Treatment Condition	Conventionally-IOX	EF-IOX	Double & reversed EF-IOX
Residual stress (MPa)	26	49	91

Table 3. Calculated residual stress in the glass tubes treated by conventional ion exchange and EF-IOX

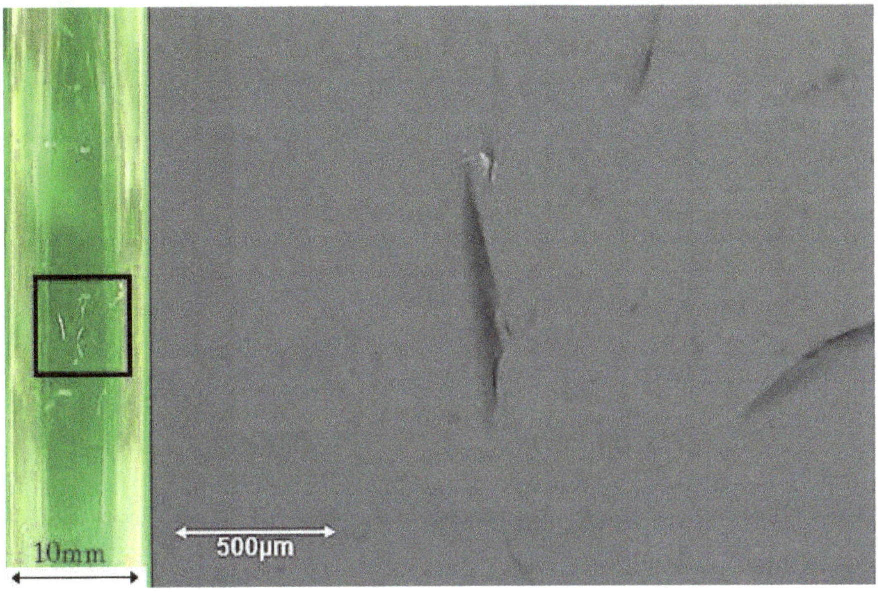

Figure 8. Crack formation in the glass tube during the EF-IOX with "out-to-in" electrical field polarization

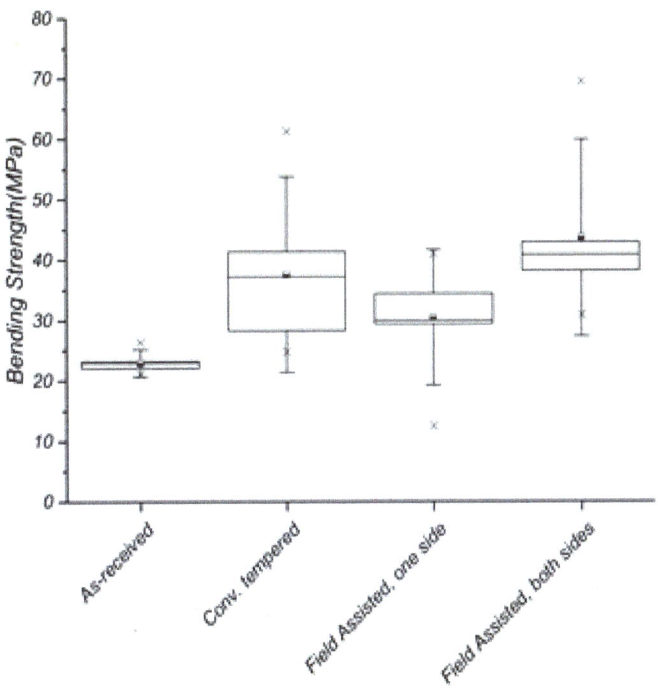

Figure 9. Four-point bending strength of glass tubes

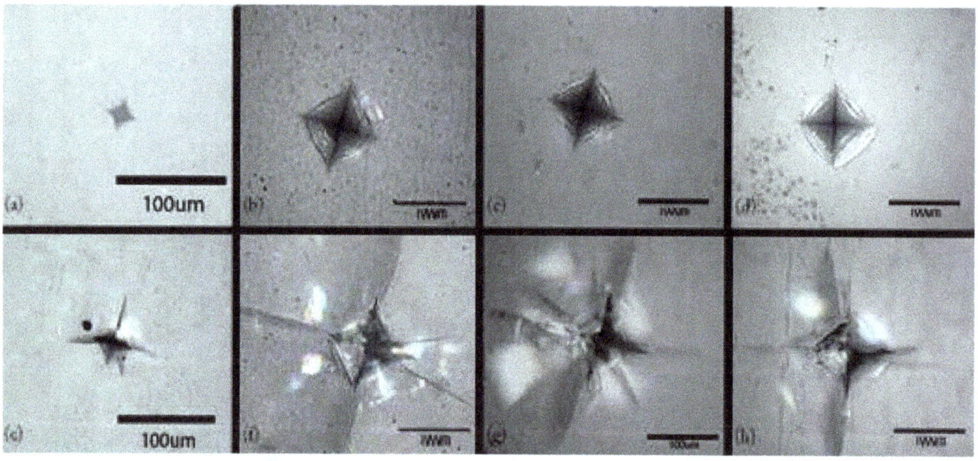

Figure 10. Vickers indentations on as-received samples subjected to (a) 3 N, (e) 5 N and samples subjected to 40 N indentation load, (b) ion exchanged (IOX), (c) EF-IOX, one side, (d) EF-IOX both sides, and 50 N indentation load, (f) ion exchange, (g) electric field-assisted ion exchange(EF-IOX), one side, and (h) EF-IOX both

4. Conclusions

Ion exchanged strengthening of borosilicate glass tubes was carried out under an electric field by using relatively short treatment times. The depth of the exchanged layer increases with the field intensity. Reversing the field polarization can modify the residual stress; moreover, the sequence of applying the fields with different polarizations is important to avoid the formation of cracks on the glass surface. Bending strength of samples treated by the double-reversed field is higher compared to those subjected to EF-IOX on one side or to conventional ion exchange.

Additional studies are required to find the effect of different parameters, such as salt and glass composition, treatment time and temperature, field intensity and other thermodynamic and electric factors on the assisted ion exchange process. The structural changes during the electric field should also be investigated. Nevertheless, the preliminary results obtained in the present work point out that electric field-assisted ion exchange can be employed for improving the mechanical properties by relatively short duration treatments.

Author details

Ali Talimian* and Vincenzo M. Sglavo

*Address all correspondence to: ali.talimian@unitn.it

Department of Industrial Engineering, University of Trento, Trento, Italy

References

[1] Varshneya AK. Chemical strengthening of glass: Lessons learned and yet to be learned. International Journal of Applied Glass Science. 2010;1(2):131-142. DOI: 0.1111/j.2041-1294.2010.00010.x.

[2] Karlsson S, Jonson B. The technology of chemical glass strengthening a review. Glass Technology.2010;51(2):41–54. ISBN: 0017-1050.

[3] Varshneya AK. The physics of chemical strengthening of glass: Room for a new view. Journal of Non-crystalline Solids. 2010;356(44-49): 2289–2294. DOI: 10.1016/j.jnoncrysol.2010.05.010

[4] Findakly T. Glass waveguides by ion exchange: A review. Optical Engineering. 1985; 24(2): 24-250.

[5] Tervonen A, West B, Honkanen S. Ion-exchanged glass waveguide technology: A review. Optical Engineering. 2011;50(7):711071-15.

[6] Varshneya AK. Fundamentals of Inorganic Glasses: Society of Glass Technology; 2006. 682 p. ISBN: 978-0-08-057150-8.

[7] René G. Ion exchange for glass strengthening. Materials Science and Engineering: B. 2008;149(2):159-165. DOI: 10.1016/j.mseb.2007.11.029.

[8] Mazzoldi P, Carturan S, Quaranta A, Sada... C. Ion exchange process: History, evolution and applications. Riv Nuovo Cimento. 2013;36(9):397-460. DOI: 10.1393/ncr/i2013-10092-1.

[9] Green DJ. Recent Developments in Chemically Strengthened Glasses. 64th Conference on Glass Problems: Ceramic Engineering and Science Proceedings, John Wiley & Sons, Inc.; 2008. p. 253–66. DOI: 10.1002/9780470294857.ch20v

[10] Liangbao J, Xintao G, Xiaoyu L, Lei L, Guanli Z, Yue Y. Different K+–Na+ inter-diffusion kinetics between the air side and tin side of an ion-exchanged float aluminosilicate glass. Applied Surface Science. 2013;265: 889–894. DOI: 10.1016/j.apsusc.2012.11.143.

[11] Morten MS, Qiuju Z, John CM, Marcel P, Steen M, Yuanzheng Y. Sodium diffusion in boroaluminosilicate glasses. Journal of Non-crystalline Solids. 2011;357(22-23): 3744–3750. DOI: 10.1016/j.jnoncrysol.2011.07.008

[12] Reiner K. On the mobility of alkaline earth ions in mixed alkali alkaline earth silicate glasses. Journal of Non-crystalline Solids. 2003;328(1-3): 157–163.DOI: 10.1016/S0022-3093(03)00474-5.

[13] Roling B, Ingram M. Mixed alkaline–earth effects in ion conducting glasses. Journal of Non-crystalline Solids. 2000;265(1–2):113–119. DOI: 10.1016/S0022-3093(99)00899-6.

[14] Ramaswamy RV, Srivastava R. Ion-exchanged glass waveguides: A review. Journal of Lightwave Technology. 1988;6(6):984–1002. DOI: 10.1109/50.4090.

[15] Hiroyuki Y, Takeshi K. Migration of two ions during electrolysis of glass waveguide. Journal of Applied Physics. 1985;58(5): 1739-1743.DOI: 10.1063/1.336022.

[16] Liu K, Pun EYB. K+-Na+ ion-exchanged waveguides in Er3+-Yb3+ codoped phosphate glasses using field-assisted annealing. Applied Optics. 2004;43(15):3179–3184. DOI: 10.1364/AO.43.003179.

[17] Tervonen A. A general model for fabrication processes of channel waveguides by ion exchange. Journal of Applied Physics. 1990;67(6):2746–2752. DOI: 10.1063/1.345440

[18] Abou-El-Leil M, Cooper AR. Analysis of field-assisted binary ion exchange. Journal of the American Ceramic Society. 1979;62(7–8):390–395. DOI: 10.1111/j.1151-2916.1979.tb19086.x.

[19] Sviridov SI, Eliseeva NP. Field-assisted diffusion of potassium ions in sodium silicate glass. Glass Physics and Chemistry. 2006;32(6):604–11. DOI: 10.1134/S1087659606060022.

[20] Urnes S. Na-K exchange in silicate glasses. Journal of the American Ceramic Society. 1973;56(10):514–517. DOI: 10.1111/j.1151-2916.1973.tb12400.x.

[21] International A. Standard Test Method for Assignment of the Glass Transition Temperatures by Differential Scanning Calorimetry. Calorimetry and Mass Loss. West Conshohocken, PA, 2014.

[22] Martienssen W, Warlimont H. Handbook of Condensed Matter and Materials Data. Springer; 2005. ISBN: 9783540304371.

[23] Marshall DB, Lawn BR. Residual stress effects in sharp contact cracking. Journal of Materials Science. 1979;14(8):2001–12. DOI: 10.1007/BF00551043.

[24] Abou-El-Leil M, Cooper AR. Fracture of soda-lime glass tubes by field-assisted ion exchange. Journal of the American Ceramic Society. 1978;61(3–4):131–136. DOI: 10.1111/j.1151-2916.1978.tb09255.x.

[25] Chuchai A, Kenneth JA, John JM, Jr. Residual stress in glass: Indentation crack and fractography approaches. Dental Materials. 2009;25(11):1453–8. DOI: 10.1016/j.dental.2009.07.001.

8

Ion Exchange Fundamentals and New Challenges

Maria Angélica Simões Dornellas de Barros, Marcelino Luiz Gimenes, Melissa Gurgel Adeodato Vieira and Meuris Gurgel Carlos da Silva

Abstract

Ion exchange is a stoichiometric phenomenon commonly used in water treatment as an end-of-pipe technique. Such process is highly influenced by mass transfer conditions and may be modeled by adsorption equations. Although widely applied in industries its theory has not been completely understood and depends on the exchanger characteristics. Moreover, competitive systems may add complexity and decrease removal efficiency and exchanger selectivity mainly in dynamic systems. In this chapter some general theory was presented and some detailed examples involving alginate biopolymer, bonechar and zeolite in single and competitive systems were discussed in batch and continuous state.

Keywords: ion exchange, kinetics, equilibrium data, fixed bed, multicomponent systems

1. Introduction

Ion exchange processes are the processes in which a solid phase (ion exchanger) reacts in a double exchange reaction when in contact with a liquid phase with electrolytes. Actually, this should not be considered as a true chemical reaction as it involves the redistribution of ions between two phases by diffusion. Chemical factors are almost negligible with small amount of heat, often less than 2 kcal/mol [1].

In an ion exchange process, the balancing ion (the one previously detected in the solid phase) is replaced by the counter ion (previously in the liquid phase) always when the exchange has a higher affinity to the counter ion. It is important to emphasize that the stoichiometric replacement involves charges. Nevertheless, normality is much more adequate to describe the phenomenon than molarity. Figure 1 presents examples of the monovalent and divalent exchange processes. In this diagram, it is easy to see the importance of charges in the stoichiometric process.

The first citation of an application of ion exchange can be found in Aristotle's, but the ion exchange processes became well known in the nineteenth century. In the 1930s, they were strengthened when organic cation exchangers were discovered. Nowadays, anion exchange resins are also commercially obtained [2].

Ion exchange is considered as an end-of-pipe technique used in wastewater and one of the best available techniques to remove heavy metal ions, which is of a great concern due to the toxic compounds constantly presented in bodies of water.

With such general aspects in mind, this chapter has the main goal to discuss the ion exchange phenomenon through already published results. Theory and fundamental aspects will be briefly revised.

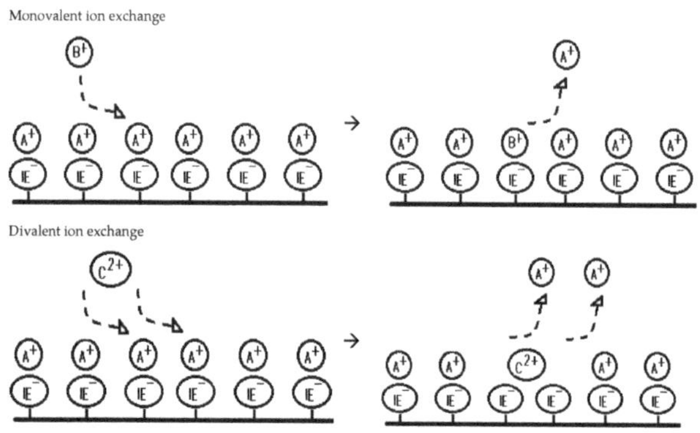

Figure 1. Cationic exchange examples. IE^- is the ionic exchanger charge. A is the balancing cation, B is the monovalent counter ion, and C is the divalent counter ion.

2. General theory

Ion exchange is generally controlled by diffusion, a consequence of the material structure. Ion exchange framework, size of the beads, and any other physical chemistry characteristics have important roles in this process. Nevertheless, in all cases, it is accomplished by transfer of ions to and from the interphase boundary, the exchange itself followed by the diffusion of the ion inside the solid phase and the diffusion in the surrounding solution.

Actually, the ion exchange process occurs in the following steps [2]:

1. Dissociation of the electrolytes in the bulk phase, originating the denominated counter ion
2. Diffusion of the counter ion from the bulk phase towards the interphase film
3. Diffusion of the counter ion through the interphase film
4. Diffusion of the counter ion inside the ion exchanger

5. Association between the counter ion and the functional group of the exchanger
6. Dissociation of the balancing ion and functional group of the exchanger
7. Diffusion of the balancing ion inside the exchanger towards the surface
8. Diffusion of the balancing ion through the interphase film
9. Diffusion and random distribution of the balancing ion in the bulk phase
10. Formation of the balancing ion complexes in the solution

Basically, the mechanism of ion exchange processes has some possible rate-controlling steps. The most important ones are related to steps 3 and 8, 4 and 7, 5 and 6. Steps 3 and 8 deal with film resistance and should be minimized through adequate agitation. The mass transfer in this case is defined by the diffusion coefficients. Steps 4 and 7 are related to intraparticle diffusion and depend on the physic-chemical properties of the system. Small particles of the ion exchanger may decrease such resistances. Steps 5 and 6 are the ion exchange processes properly. Almost always, the film and/or intraparticle resistances are considered the most important rate-controlling steps. Many kinetic models take into account these characteristics. More information regarded to kinetics can be seen elsewhere [3, 4]. The regeneration of the saturated ion exchanger may be also modeled [5]. It must be emphasized that many models are used to describe both adsorption and ion exchange mechanisms. Despite of the mathematical similarity, the significant differences related to these mechanisms should be in mind.

Besides kinetic data, ion exchange equilibrium data are also of great value. Isotherms may be classified in five different types [6], as shown in Figure 2.

Isotherm shapes indicate whether or not the ion is solution is preferably exchanged. However, they provide no information on the type of exchange sites or even whether they have similar energies. This is outstanding information as it is directly related to the ion exchange mechanism. The Kielland plot is an interesting thermodynamic tool to understand such process. Ion selectivity and the thermodynamic properties may be obtained from such data.

The Kielland plot may be obtained through $\log K \mid_B^A$ (equilibrium constant through the balancing ion A and the counter ion B) and $x_{A(Z)}$. Linear Kielland plots are a consequence of exchangers with only one kind of exchange site [7]. Nonlinear plots indicate different sites and different cavities where the exchanged cation occupies different positions in the framework [6].

Although batch operations where isotherms may be obtained are rarely used in industries, they are very common when investigating mechanisms such as the ones in equilibrium through isotherms. Continuous systems are almost often well suited for industrial purposes when scale-up process is needed. Continuous ion exchange uses packed beds where the mass transfer is of great importance. As the feed solution passes through the ion exchange packed bed, the outlet solution has different concentrations of the incoming ion as a function of time. Plots of the ratio outlet concentration of the incoming ion/concentration of the incoming ion in the feed solution versus time are well known as breakthrough curve. Mass balances in the column as well as the mass transfer parameters are reported elsewhere [8,9].

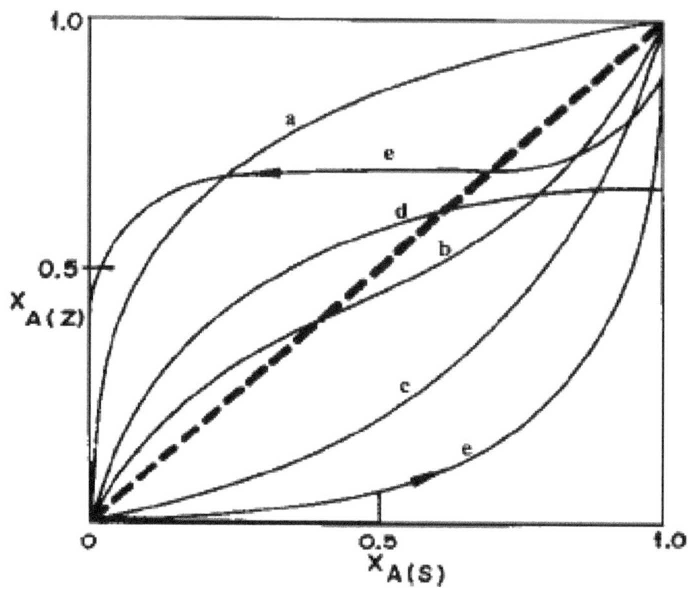

Figure 2. Ion exchange isotherms: $x_{a(z)}$: equivalent fraction of the counter ion in the exchanger; $x_{a(s)}$: equivalent fraction of the counter ion in solution [6]. (a) Favorable isotherm; (b) isotherm with reversal behavior, from favorable to unfavorable; (c) unfavorable isotherm; (d) incomplete favorable isotherm; (e) isotherm with hysteresis.

Besides the use of adsorption models, ion exchange systems may be more correctly represented by the mass action law (MAL). This is the most characteristic property of ion exchange and can be used as one of the possible modeling equations. Actually, it expresses the typical double exchange reaction where the balancing ion in the exchanger is replaced by the in-going ion according to the stoichiometry.

MAL is based on the definition of the chemical equilibrium of the chemical reactions first proposed by Cato Guldberg and Peter Waage in 1864. It was defined as the equilibrium constant K, which is the relationship of the activity coefficient of reagents and products in equilibrium at a given temperature.

If the chemical reaction presented in Eq. (1) is considered:

$$bB + cC \rightarrow dD + eE \tag{1}$$

The equilibrium constant may be written as follows:

$$K = \frac{\left[a_D\right]^d \cdot \left[a_E\right]^e}{\left[a_B\right]^b \cdot \left[a_C\right]^c} \tag{2}$$

where *a* stands for the ionic activity of each ion presented in Eq. (1).

It is obvious that monocomponent ion exchange only occurs when the solid in is contact with synthetic solutions. Other solutions, mainly wastewaters, always contain significant amount or other ions that may be also exchanged. That is why selectivity and affinity properties of the ion exchanger in relation to the specific incoming ion should be considered no matter if it is continuous or batch system. Of course, in such a case, modeling is more complex. Some examples of MAL can be seen in reference [10].

3. Ion exchangers

Ion exchangers are porous matrixes from different sources, with positive or negative excess charge, insoluble in aqueous solutions and in many organic solvents. The excess charge of the matrix should be compensated by the balancing ions, which may be replaced by the in-going ion depending on the selectivity and affinity of the exchanger to the ions involved.

Mechanical resistance as well as regeneration capacity is quite important when packed beds are considered. There are acid and basic exchangers being the anionic exchangers that have basic superficial groups and cationic exchangers those containing superficial acid groups. Exchangers may be also classified according to complete or incomplete dissociation based on the pH range where the exchange process is efficient.

Ion exchangers can be natural such as alginate, clay, algae, or even some zeolites. Alginate occurs in seaweeds as calcium alginate and is present in the cells of brown algae. Actually, the term alginate designates salts of alginic acid and its derivatives.

Clays are fine powders constituted by hydrated aluminosilicates that often tend to agglomerate. In clay materials alumina is presented in octahedral form whereas silica is found as tetrahedrons. Such materials are thermally stable and can be greatly improved by pillaring process. Zeolites are also aluminosilicates. Nevertheless, they have an open three-dimensional framework with interconnecting cavities. Both materials can be used as adsorbents, ion exchangers, catalysts, or catalysis support.

Natural exchangers have some disadvantages such as low exchange rate and rather poor mechanical properties and low abrasion resistance, which restrict their application, mainly in packed beds without any previous treatment.

Exchangers obtained specifically from synthetic materials are available commercially being. Zeolites and resins are the most famous representatives of such class. Many zeolites are related to cationic exchange process. Nevertheless, zeolites can act as anionic exchangers if tailored.

Resins are known since 1935. They can be used as ion exchangers or catalysts. Ion exchange resins may be found as in acids and bases, acting, therefore, as anionic or cationic exchangers. Cationic resins generally contain sulfonic acid groups, whereas anionic resins are commonly found in quaternary ammonium groups.

4. Experimental results of ion exchange

4.1. Case study: Metallic ion in calcium alginate biopolymer

4.1.1. Copper kinetics study of calcium alginate particles in a static system

Calcium alginate biopolymer was prepared by dropping sodium alginate into solution of calcium chloride (3% w/w) under continuous agitation. Calcium alginate particles formed (mean diameter, 1083 μm) were washed and dried to be used in adsorption/ion exchange experiments [11].

Kinetic studies were carried out using single nitrate solutions of Cu^{2+}, Cd^{2+}, Pb^{2+}, and Ni^{2+} of 3 mmol/L (Vetec, Brazil) and a bicomponent equimolar solution of $Cu^{2+}$and Cd^{2+} (1 mmol/L for each metal). Values of pH were corrected using NH_4OH and HNO_3.

All runs were conducted in finite baths at 25°C using 1 g of alginate immersed in 0.1 L of metallic solutions. Samples of these solutions were taken at different running times, and after filtration, the concentration of solutions was analyzed through atomic absorption spectrophotometry.

Mathematical modeling of the ion diffusion process in ion exchanger provided relevant information on mass transfer that is essential to the ion exchange/adsorption system design. The diffusion process in a solid matrix was described for the second Fick's law. In the spherical coordinate system, the concentration gradients are negligible in the angular direction, and the second Fick's law is represented by Eq. (3):

$$\frac{\partial q}{\partial t} = D_e \frac{1}{r} \frac{\partial}{\partial} \left(r \frac{\partial q}{\partial r} \right) \tag{3}$$

where q is the ion capacity in ion exchanger (mg/g), D_e is the ion diffusion coefficient in the adsorbent/ion exchanger (cm^2/s), t is time (s), and r is the radial direction (cm).

In this modeling, it was considered that the adsorbent was initially free of metal, the ion diffusivity was constant, the concentration in the fluid phase was homogeneous, and external resistance in the liquid film was negligible due to the agitation. The initial and boundary conditions used are described by Eqs. (4)–(6):

$$\text{at } t = 0 : q = q_0 \tag{4}$$

$$\text{at } r = R \text{ and } t > 0 : q = 0 \tag{5}$$

$$\text{at } r = 0 \frac{\partial q}{\partial r} = 0 \tag{6}$$

The average concentration of the metal ion exchanger is given by Eq. (7):

$$\frac{\bar{q}(t) - q_{\text{eq}}}{q_0 - q_{\text{eq}}} = 6 \sum_{j=1}^{\infty} \left(\frac{1}{\gamma_j^2} e^{-\gamma_j Fo_{\text{M}}} \right) \tag{7}$$

where

$$Fo_M = \frac{D_{\text{ef}} t}{R^2}; \gamma_j = j\pi$$

where γ is the j component of the activity coefficient, D_{ef} is the effective diffusivity (m^2/s), and R is the radius of the particle (m).

The metal concentration in the fluid phase is obtained from an overall mass transfer balance, represented by Eq. (8):

$$C(t) = \left(C_0 V - m_s \bar{q}(t)\right) / V \tag{8}$$

where C_0 is the initial concentration of metal in the fluid phase (mg/L), $\bar{q}$ is the metal incoming average capacity in the adsorbent ($mg_{\text{metal}}/g_{\text{alginate}}$), m_s is the mass of bioadsorbent/ion exchanger (g dry basis), and V is the volume of the solution (dm^3).

The method "golden search" was used to determine the effective diffusion coefficient of ions in bioadsorbent/ion exchanger to minimize the objective function given by Eq. (9):

$$F_{\text{obj}} = \sum_{j=1}^{n} \left(C_j^{\text{MOD}} - C_j^{\text{EXP}} \right)^2 \tag{9}$$

where n is the number of experimental data, C_j^{EXP} is the ion concentration in solution determined experimentally (mEq/L), and C_j^{MOD}: ion concentration in the solution calculated by the model (mg)

Figure 3 shows experimental results of the kinetics of the process and results obtained with the mathematical modeling for different metals in this study. The metal ions nickel, lead, and copper showed adsorption and desorption, indicating the competitiveness of metal ions (with calcium present in alginate) by the occupation of the active sites. The model used does not consider this competitiveness, but only the adsorption of the ions of interest, resulting in slightly different values when compared to the experimental data.

Table 1 shows the quantities of metal ions removed from the single solutions by adsorption/ion exchange process. It appears that nickel was less adsorbed (0.92 mmol $Ni/g_{calcium\ alginate}$) in equilibrium. Indeed, alginic acid has a greater affinity for calcium than for nickel [12]. In a study of removal of different metals using calcium alginate, it was obtained the following amounts adsorbed: 0.247 mmol Cu/g, 0.138 mmol Cd/g, and 0.247 mmol Pb/g [13]. Although the conditions used were different from those used in this work, it can be seen that the alginate showed the same order of adsorptive capacity, or $Cu^{2+} > Pb^{2+} > Cd^{2+}$; in the case of this study (Table 1), it was $Cu^{2+} > Pb^{2+} > Cd^{2+} > Ni^{2+}$.

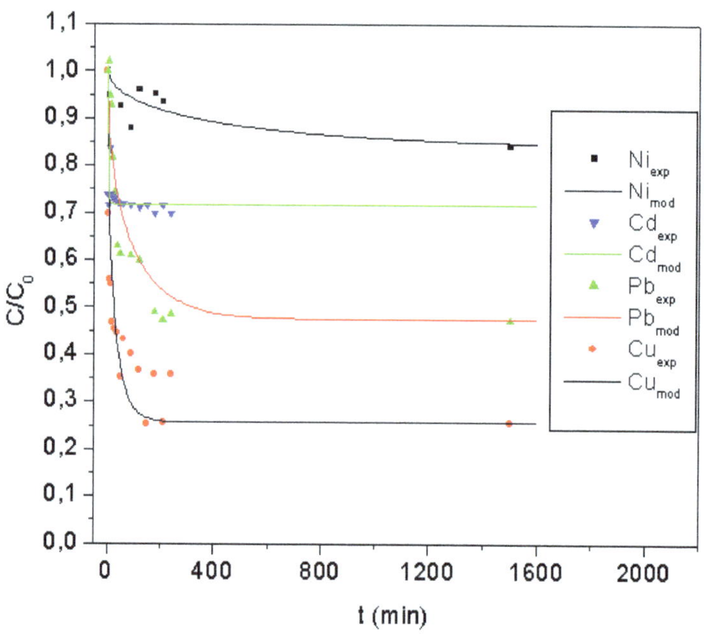

Figure 3. kinetic curves for different metal ions in monocomponent solutions for the adsorption/ion exchange into calcium alginate particles.

Copper	Cadmium	Nickel	Lead
4.658	1.792	0.987	3.106

Table 1. Adsorbed amount ($mmol_{metal}/g_{alginate}$) of different metals in the process of adsorption/ion exchange using single solutions.

The results obtained in finite bath for copper-cadmium binary mixture are shown in Figure 4. It is observed that copper was significantly more removed than cadmium. The adsorbed amount of copper and cadmium was 1.746 mmol $Cu/g_{calcium\ alginate}$ and 0.661 mmol $Cd/g_{calcium}$

alginate, respectively. The greater affinity of calcium alginate to copper may be due to the chemical parameters listed in Table 2. The higher the electronegativity and the reduction potential and the lower the ionic radius, the easier the ion exchange/adsorption [14]. In this case, copper is more susceptible to ionic interaction with the alginate than cadmium because it presents all the favorable parameters to the ion exchange.

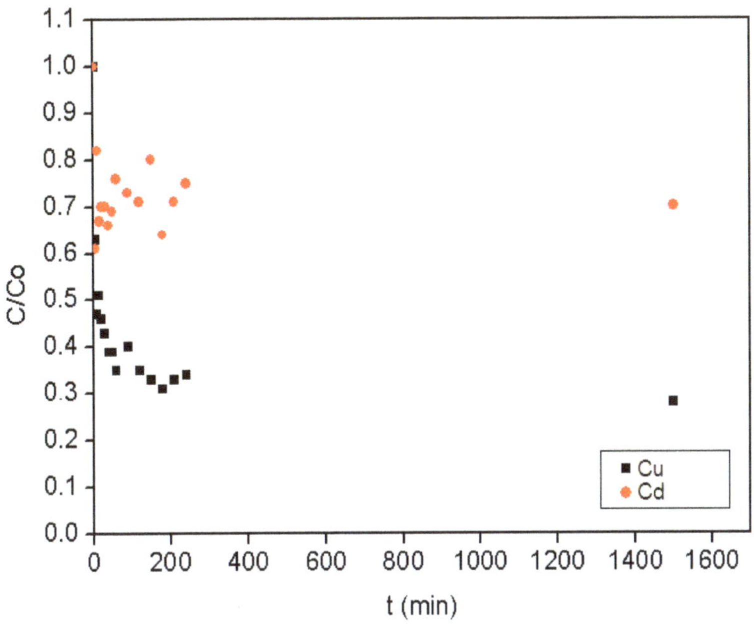

Figure 4. Kinetics of copper and cadmium in binary solution in the adsorption/ion exchange by calcium alginate particles.

	Copper	Cadmium	Lead	Nickel	Calcium
Atomic mass (g/mol)	63.546	112.411	207.200	58.710	40.080
Electronegativity	1.9	1.7	1.8	1.8	1.0
Ionic radius (Å)	0.73	0.97	0.132	0.69	0.99
Chemical reduction potential (V)	0.34	–0.40	–0.13	–0.23	–2.68

Source: reference [11].

Table 2. Chemical properties of divalent cations.

The diffusion capacity of the metal ions in the alginate particles analyzed was evaluated considering the Fick's law, and the respective values are shown in Table 3. It is observed that the cadmium showed higher diffusion capacity, and nickel is the metal resulting in an increased resistance to intraparticle diffusion; thus, the metal adsorption order was Cd^{2+} > Cu^{2+} > Pb^{2+} > Ni^{2+}. However, experimental results obtained with single and binary solutions were Cu^{2+} > Pb^{2+} > Cd^{2+} > Ni^{2+} and Cu^{2+} > Cd^{2+}, respectively, indicating that this parameter cannot solely describe the metal affinity of alginate. Therefore, it becomes necessary to evaluate other properties of the ion exchanger/adsorbent.

Metal	Diffusion capacity (cm^2/min)
Copper	7.44E–06
Cadmium	8.38E–05
Lead	2.19E–06
Nickel	4.57E–07

Table 3. Diffusion capacity of metal ions in calcium alginate.

4.1.2. Copper equilibrium study of calcium alginate particles in a static system

The isotherm of Langmuir model (L) is widely used due to its simplicity and theoretical basis and is expressed by Eq. (10). The parameter b is the ratio between the rate of adsorption and desorption and is directly related to Henry's constant. High values of b indicate high affinity of the ions by the active sites of the material. The parameter q_{max} indicates the total number of active sites available, and q^* and C^* represent the metal removal capacity at equilibrium in the solid and liquid phases, respectively. Adsorption is very favorable when values of $b.C^* >> 1$; however, if $b.C^* < 1$, the isotherm is almost linear.

$$q^* = \frac{q_{max} \cdot b \cdot C^*}{1 + b \cdot C^*} \tag{10}$$

The Freundlich isotherm is an empirical adjustment of a model, which considers that the energy of the active sites of the adsorbent material is heterogeneous and that the adsorption process is reversible. It corresponds to the exponential distribution heats of adsorption and is

d

the saturation of the adsorbent, allowing an infinite number of layers covering the ionic adsorbent [15]. When $n < 1$, it is typically a liquid adsorption [16].

$$q^* = k_d \cdot C^{*n} \tag{11}$$

Figure 5 shows the equilibrium data in finite bath by contacting 1 g of hydrated alginate with 100 mL of metal solution with different initial concentrations [11]. The Langmuir isotherm model (Equation 10) and the Freundlich model (Equation 11) were fitted to the experimental equilibrium data as shown in Figure 5, and the respective values of model parameters are presented in Table 4.

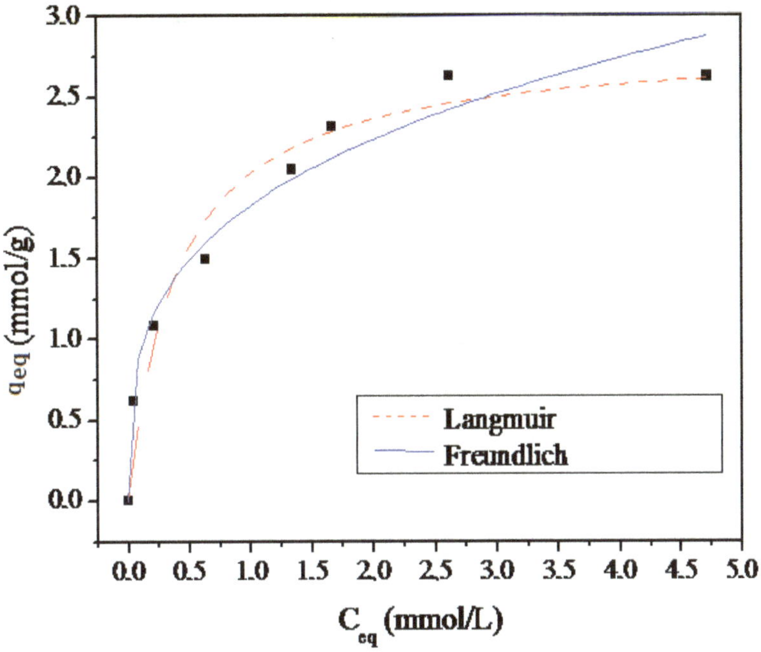

Figure 5. Cu adsorption isotherm in calcium alginate fitted by the Langmuir and Freundlich models.

Langmuir	q_{max} (mg/g)	b (mg/L)	R^2
	2.8323	0.3977	0.9623
Freundlich	k_d	n	R^2
	1.8269	0.2924	0.9739

Table 4. Parameters of Langmuir and Freundlich models.

According to Table 4, both models could satisfactorily adjust the equilibrium experimental data for copper ions. The isotherm obtained in Figure 5 can be classified as type I [17], which is characteristic of the Langmuir isotherm where adsorption occurs only in monolayer. Moreover, these models have also been adequately used to describe the process of removing metal ions using calcium alginate (Ca-Alginate) particles [18,19].

In the ion exchange process involving the binary system of copper and cadmium ions in calcium alginate (systems Cu^{2+}-Ca^{2+}, Cd^{2+}-Ca^{2+}, and Cu^{2+}-Cd^{2+}), it was considered only the presence of the higher affinity ion, or for the Cu^{2+}-Ca^{2+} system, only the presence of Cu^{2+} species, for Cd^{2+}-Ca^{2+}system, the presence of Cd^{2+} species, and for the Cu^{2+}-Cd^{2+} system, the presence of copper [11]. Many single adsorption isotherms were evaluated for the binary systems, as shown in Table 5 and in Figure 6.

Model	Parameters	Equation
Langmuir	q_{max}, b	$q^* = \frac{q_{max} \cdot b \cdot C^*}{1 + b \cdot C^*}$
Freundlich	q_{max}, n	$q^* = k_d \cdot C^{*n}$
Redlich–Peterson	q_{max}, b, n	$q^* = \frac{q_{max} b C^*}{1 + bC^{*n}}$
Toth	q_{max}, b, n	$q^* = \frac{q_{max} C^* b^{1/n}}{(1 + bC^{*n})^{1/n}}$
Radke–Praunitz	q_{max}, b, n	$q^* = \frac{q_{max} C^* b}{(1 + BC^*)^n}$
Sips	q_{max}, b, n	$q^* = \frac{q_{max} b C^{*n}}{1 + bC^{*n}}$

Table 5. Adsorption isotherm models used for the binary system by considering only the presence of the ion of higher affinity.

Model	Parameters	F_{obj}	R^2
Langmuir	q_{max} (mEq/g) = 14.540 b (L/mEq) = 0.139	0.069	0.990
Freundlich	K (mEq/g) = 2.020 n = 0.658	1.054	0.982
Redlich–Peterson	q_{max}(mEq/g) = 20.7841 b(L/mEq) = 0.091 n = 1.1331	0.56198	0.991
Toth	q_{max}(mEq/g) = 13.207 b(L/mEq) = 0.108 n = 1.1455	0.566	0.990
Radke–Prausnitz	q_{max}(mEq/g) = 40.555 b(L/mEq) = 4.677 n = 2.139	0.549	0.991
Sips	q_{max}(mEq/g) = 25.332 b(L/mEq) = 0.077 n = 0.847	0.592	0.988

Table 6. Parameters obtained for the single component isotherm models for the system Cu^{2+}–Ca^{2+}.

All models resulted in satisfactory determining coefficient values (R^2), indicating proper fit to the experimental data (Table 6).

The Sips model was successfully used to represent the removal of copper and cadmium ions in Ca-Alginate particles, mainly when compared to Langmuir and Freundlich models [19]. It happened due to the heterogeneity of the surface of the adsorbent, especially for metals of lower-affinity with the alginate.

Although in the system involving the Cd^{2+}-Ca^{2+} ions, experimental data have been acceptable adjusted, the interference of calcium on Cd^{2+}-Ca^{2+} system was higher when compared to Cu^{2+}-Ca^{2+}system, indicating that the alginate had a higher affinity for copper than for cadmium [11,20].

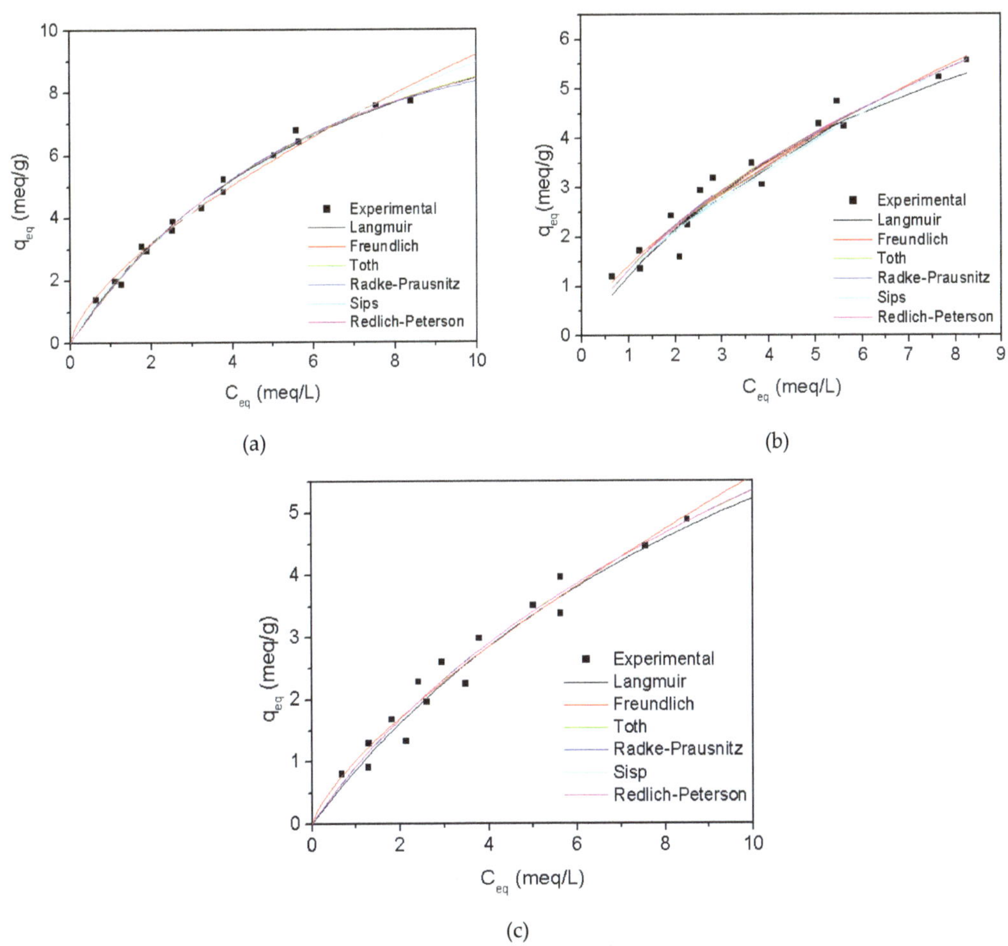

Figure 6. Cu^{2+} adsorption isotherms for Ca-Alginate system (a), Cd-Alginate system (b), and Cd^{2+} adsorption isotherms for Ca-Alginate system (c) in monocomponent system [11,20].

4.1.3. Copper removal in dynamic system: adsorption and desorption cycles

The study of removing copper ions in the calcium alginate particles porous bed column was assessed through adsorption and desorption cycles. The experiments were performed in a glass column (internal diameter, 1.4 cm) filled with 9.80 g of calcium alginate particles to reach a bed height of 13.3 cm. Flow tests were performed with 3 mL/min for adsorption and desorption cycles. The initial concentration of copper used in the adsorption step was 4.72 mmol/L. In the desorption step, calcium chloride solution with a concentration of 18 mmol/L was employed as eluent. The amount of copper removed in the column experiment was calculated by the Eq. (12). The experimental conditions were defined from fluid dynamic preliminary studies.

$$Q = \frac{C_0 F}{1000m} \cdot \int_0^t \left(\frac{C_0 - C}{C_0} \right) \cdot dt \tag{12}$$

where C_0 is the initial concentration of metal (mmol/L), $C(t)$ is the metal concentration at time t (mmol/L), F is the volumetric flow of the solution (mL/min), m is the mass of alginate (g), Q is the metal removal capacity at time t ($mg_{metal}/g_{bioadsorvente}$), and t is time (min).

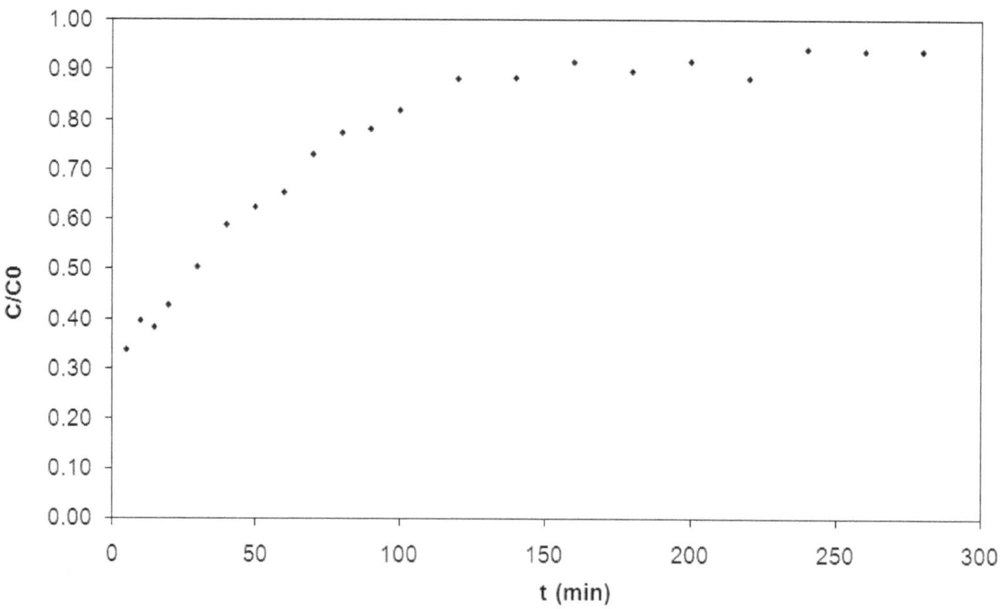

Figure 7. Kinetics of adsorption of copper ions in porous bed (Initial copper concentration 4.72 mmol/L).

Aliquots were withdrawn periodically, and the pH of feed metallic solution was monitored throughout the process, which was maintained between 4 and 4.5. The copper removal kinetics is shown in Figure 7. From the breakthrough curve, the complete saturation of the bed was reached at 140 min of process. The total amount of copper removed given by Eq. (12) was 2.83 mmol/g.

Copper was desorbed from alginate employing calcium chloride solution with a concentration of 18 mmol/L. Initially, it is known that alginate has a greater affinity for copper than the alginate. However, when alginate is saturated with copper ions and comes into contact with a solution containing a high concentration of calcium, copper alginate may desorb ions and adsorb calcium ions to achieve chemical equilibrium. However, in this case, which the ions being adsorbed has a lower affinity, the process occurs only when calcium is present in solution at high concentrations. The kinetics of this process step is in Figure 8. The flow rate used, as well as in adsorption cycle, was 3 mL/min. The calcium alginate recovery was 97%. The equilibrium time of the desorption system was close to 150 min.

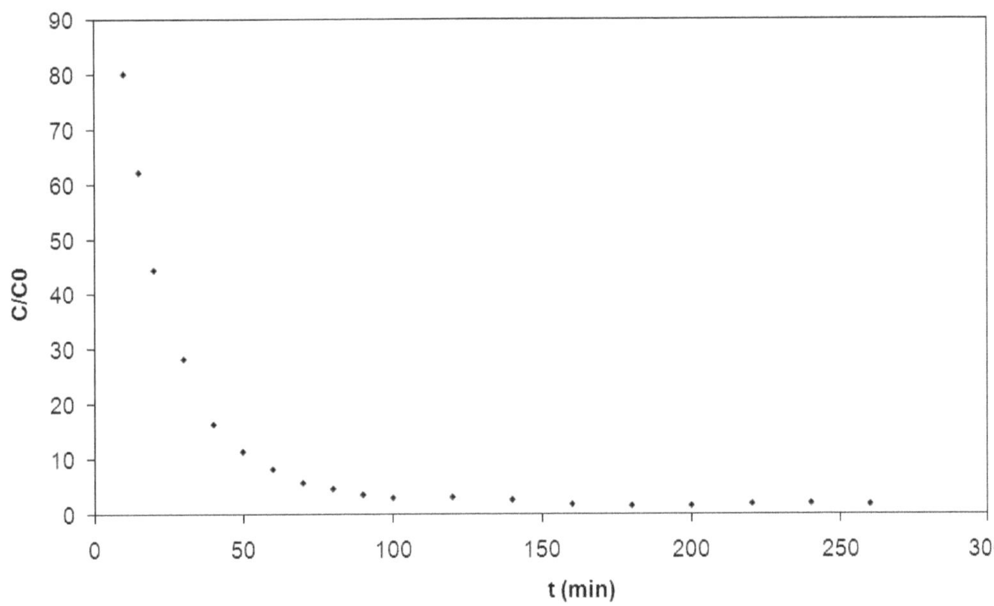

Figure 8. Kinetics of copper desorption in bed of porous calcium alginate particles.

4.2. Case study 2: Equilibrium and dynamic studies of Mn^{2+} and Cr^{3+} in bone char

It is already known that metal ions, like manganese and chromium, when present in wastewaters may contaminate the environment if not adequate treated. In particular, high concentrations of manganese ion in water promote corrosion of pipes, and as this metal is toxic to the brain, it may cause neurological disorders. The hexavalent chromium ion is another highly toxic metal present in wastewater, which is related to cancer diseases. The trivalent chromium specie is less toxic than the hexavalent one, and it can be easily oxidized in wastewater treatment through reduction of manganese ions. The removal of these metals from wastewater can be carried out by adsorption/ion exchange processes using bone char [21].

Bone char is an untypical kind of activated carbon due to its animal origin. It is composed by around 10% carbon and 90% calcium phosphate. Figure 9 illustrates the bone char structure. The calcium phosphate in bone char is present as hydroxyapatite—$Ca_{10}(PO_4)_6(OH)_2$ [22] with a calcium-to-phosphate ratio of 1.67, and unit cell dimensions of $a = b = 9.432$ Å and $c = 6.881$ Å [23].

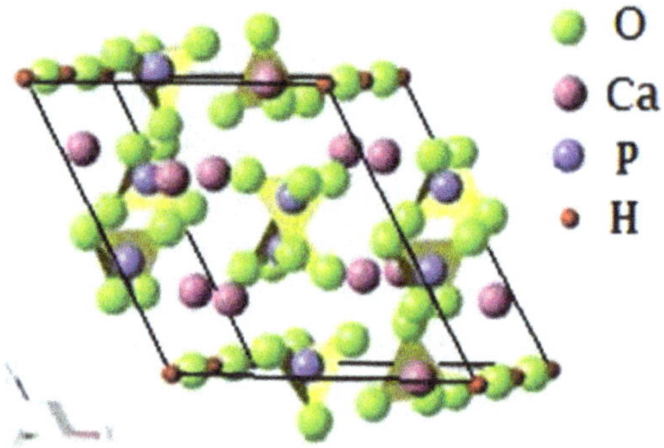

Figure 9. The hydroxyapatite structure viewed along the *c*-axis. The yellow polyhedrons represent the phosphate groups [23].

Cation exchange in bone char may occur preferentially with calcium ions, depending on radius and electronegativity of the incoming ion [24]. Under this consideration, bone char can be quite useful material to be used for removal of both Mn^{2+} and Cr^{3+} through calcium ion exchange.

4.2.1. *Materials*

The bovine bone char was crushed, sieved (20–28 mesh Tyler, average particle diameter of 0.725 mm), and elutriated with abundant water to remove fine particles and finally dried at 80°C for 24 h. The exchanger particulate material was characterized through N_2 adsorption, scanning electron microscopy (SEM), and infrared spectrophotometry (FTIR). Zero point charge (ZPC) was obtained based on references [25, 26].

Solutions of 15 mEq/L of $CrCl_3.6H_2O$ and $MnCl_2.4H_2O$ were used in single metal removal. Binary solutions containing 7.5 mEq/L of each cation were also used.

4.2.2. *Results and discussion*

N_2 isotherm showed that the bone char was predominantly mesoporous material with hysteresis and a BET area of 100 m²/ g, which is a typical for this kind of solid material (Figure 10).

Scanning electron microscopy (SEM) of the bone char sample (Figure 11) presented heterogeneity morphology of particles and channels similar to results already reported [27].

FTIR spectrum presented in Figure 12 depicts a characteristic band at 1380 cm^{-1} attributed to $^{v}NO_3$ group, bands at 3450 cm^{-1}, 603 cm^{-1} due to –OH group vibration, band at 1640 cm^{-1} due

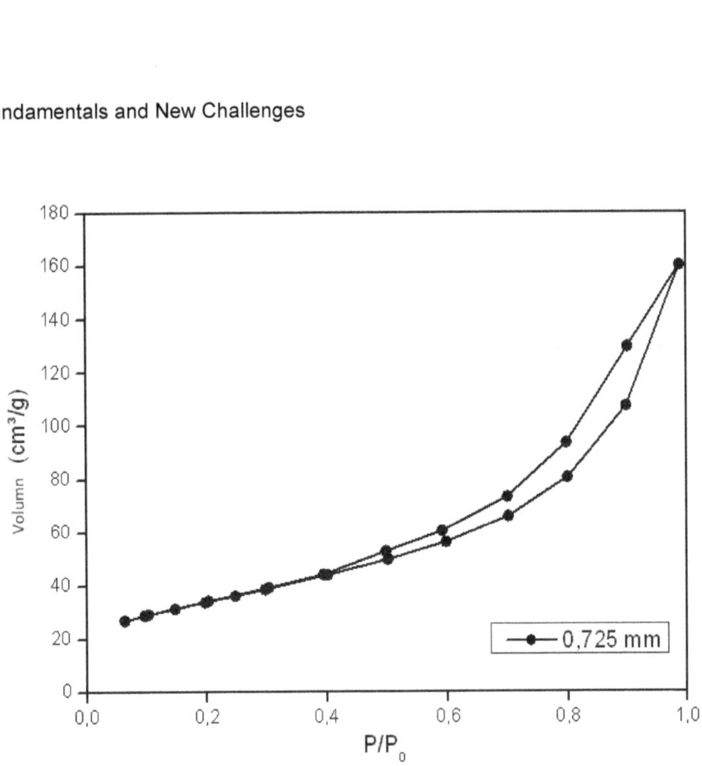

Figure 10. Nitrogen adsorption–desorption isotherm at 77K of bone char.

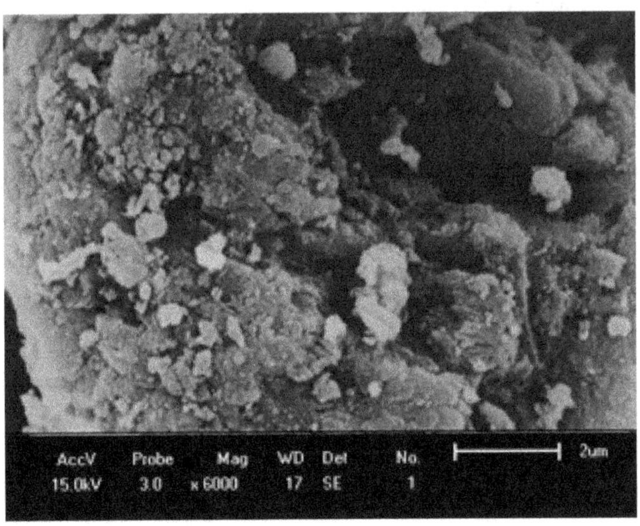

Figure 11. Micrograph obtained by SEM of the bone char.

to CO_3^{2-}, and a band at 1038 cm^{-1} attributed to the molecular vibration of PO_4^{3-} [22]. The balancing cations of such groups may be exchanged when in contact with electrolyte solutions.

The superficial zero point charge (ZPC) was pH 7.9. As the pH values solutions was 5–6, the surface charge was predominated by positively charged $\equiv CaOH_2^+$ and neutral $\equiv POH^0$ sites. The surface charge was then positive [28].

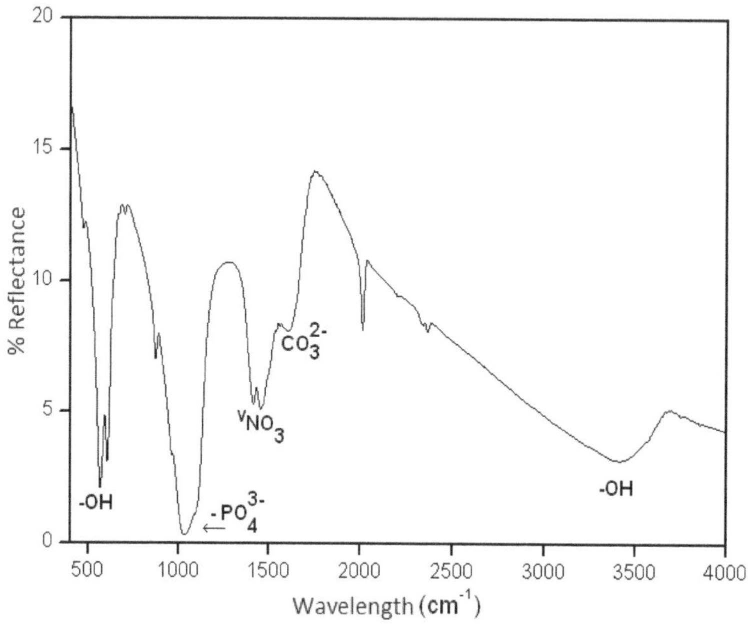

Figure 12. FTIR Spectrum of the bone char samples.

Single and bicomponent isotherms were obtained at 25°C, 30°C, and 40°C. Although bone char is a typical ion exchanger, ions may be also retained by adsorption mechanism. Actually, it is difficult to find out an equation that presents, mathematically, the contribution of all phenomena involved. Classical adsorption equations such as the Langmuir, Freundlich, and Langmuir–Freundlich models are commonly used [2].

Single isotherms are shown in Figure 13. Chromium isotherms are more favorable than the manganese ones. This can be seen through the steeper shape and higher amounts of ion retained, which is a consequence of higher ion charge and electronegativity [24].

Temperature seemed to have a higher influence in chromium removal when temperature was increased to 40°C. Probably, a reduction of the hydration radius occurred exposing the electronegativity and promoting the exchange process. According to the amount retained observed qualitatively in all temperatures, the selectivity order is $Cr^{3+} > Mn^{2+}$.

Cr^{3+} and Mn^{2+} ions may be located in site II, at the edge of the hydroxide channel of the hydroxyapatite. In site II, there would be a shift of the ion. In site I, Cr^{3+} and Mn^{2+} ions would be compressed within the local cluster.

The ion exchange process may be expressed as follows [24]:

$$\mathrm{CaHOAP} + \mathrm{Mn}^{2+} = \mathrm{MnHOAP} + \mathrm{Ca}^{2+}$$
$$3\mathrm{CaHOAP} + 2\mathrm{Cr}^{3+} = \mathrm{Cr}_2\left(\mathrm{HOAP}\right)_3 + 3\mathrm{Ca}^{2+}$$

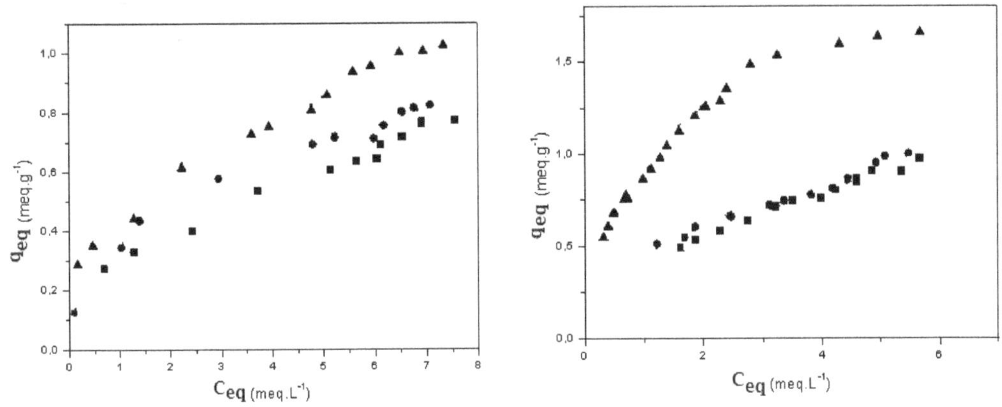

a) b)

Figure 13. (a) Mn^{2+} isotherms; (b) Cr^{3+} isotherms. (■)25°C, (▲) 30°C, (●) 40°C.

Due to pH < ZPC, replacements below may also happen:

$$\equiv POH^0 + Mn^{2+} = \equiv POMn^+ + H^+$$
$$\equiv POH^0 + Cr^{+3} = \equiv POCr^{+2} + H^+$$

Moreover, besides the ion exchange, the multilayer adsorption may occur as the typical plateau of ion exchange monolayer is not seen in the isotherms. In agreement with such phenomena, experimental data may be better represented by the Langmuir–Freundlich isotherms [29]. Tables 7 and 8 show such results.

Parameter	**25°C**	**30°C**	**40°C**
		Langmuir	
q_{max}	0.885 ± 0.073	0.986 ± 0.036	0.962 ± 0.076
B	0.495 ± 0.136	0.539 ± 0.066	1.258 ± 0.523
R^2	0.9136	0.9787	0.8199
	Freundlich		
k_f	0.301 ± 0.017	0.342 ± 0.013	0.472 ± 0.023
N	0.446 ± 0.033	0.446 ± 0.023	0.367 ± 0.029
R^2	0.9736	0.986	0.9698
	Langmuir–Freundlich		
q_{max}	1.341 ± 0.573	1.435 ± 0.395	1.796 ± 0.687
B	0.279 ± 0.144	0.321 ± 0.112	0.281 ± 0.138
$1/n$	0.643 ± 0.159	0.714 ± 0.141	0.700 ± 0.134
R^2	0.996	0.993	0.995

Table 7. Equilibrium parameters for manganese ions.

Parameter	25°C	30°C	40°C
	Langmuir		
q_{max}	1.507 ± 0.082	1.305 ± 0.097	1.836 ± 0.066
B	0.287 ± 0.033	0.450 ± 0.087	1.090 ± 0.121
R^2	0.9772	0.9217	0.9612
	Freundlich		
k_f	0.381 ± 0.011	0.444 ± 0.019	0.897 ± 0.020
N	0.531 ± 0.020	0.452 ± 0.031	0.408 ± 0.019
R^2	0.9886	0.9627	0.9678
	Langmuir–Freundlich		
q_{max}	1.792 ± 0.216	1.263 ± 0.589	3.789 ± 0.013
B	0.261 ± 0.036	0.503 ± 0.029	0.318 ± 0.145
$1/n$	0.779 ± 0.052	0.871 ± 0.049	0.571 ± 0.091
R^2	0.987	0.989	0.983

Table 8. Equilibrium parameters for chromium ions.

The binary isotherms are shown in Figure 14. As expected, chromium is again more retained, although the shape of the isotherms is completely different. This is a consequence of competition of both ions for site II.

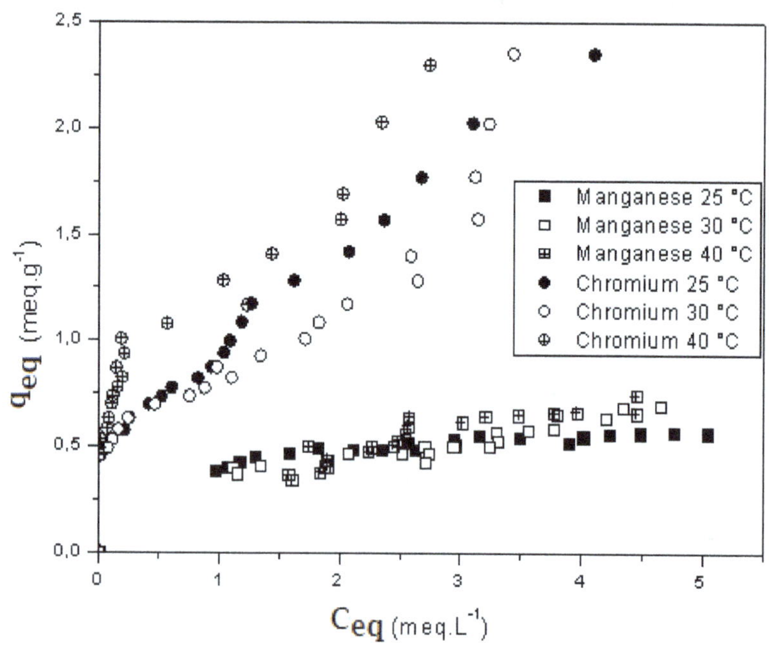

Figure 14. Isotherms for Mn^{2+}/Cr^{3+} binary system in bone char.

Again, the Langmuir–Freundlich model was the one that best represents bicomponent exchange in bone char sample, as seen in Table 9 with the lowest objective function. The parameter b_i for the bicomponent adjustment for the Langmuir–Freundlich model is lower than the ones estimated for the single isotherms due to the competition to site II. Nevertheless, q_{max} are much higher than the single values, indicating a physisorption process where more ions are retained with weaker energy.

Parameters	25°C	30°C	40°C
		Extended Langmuir [30]	
b_1 (L/mEq)	2.597	0.993	3.000
b_2 (L/mEq)	0.925	0.102	0.007
q_{m1} (mEq/g)	2.914	2.097	2.184
q_{m2} (mEq/g)	0.967	4.121	39.644
Objective function	1.880	1.550	4.287
	Jain and Snowyink [31]		
b_1 (L/mEq)	130.1	11.542	16.16
b_2 (L/mEq)	71.19	0.211	0.047
q_{m1} (mEq/g)	1.222	1.015	1.303
q_{m2} (mEq/g)	0.606	2.304	5.565
Objective function	1.183	0.317	1.658
	Langmuir–Freundlich		
b_1 (L/mEq)	0.296	0.178	0.251
b_2 (L/mEq)	0.394	0.053	0.026
n_1	0.260	0.297	0.203
n_2	0.370	0.550	0.828
q_{m1} (mEq/g)	5.775	6.324	6.045
q_{m2} (mEq/g)	1.592	7.192	10.931
Objective function	0.241	0.148	0.374

Cr^{3+} = subscript 1; Mn^{2+} = subscript 2.

Table 9. Model parameters for Mn^{2+}/Cr^{3+} binary system.

4.3. Case study 3: Dynamic ion exchange in multicomponent solution

Zeolites NaY and NaX are well-known ion exchangers. In this study, these zeolites were used for chromium uptake from bicomponent solutions. As it will be seen, multicomponent ion exchange in dynamic systems may provide some overshooting ($C/C_o > 1$) related to a sequential ion exchange, where a more preferable cation with lower diffusion will be exchanged not by the balancing ion but by the competing ion previously retained [32].

4.3.1. *Materials and methods*

NaY has the unit cell composition of $Na_{51}(AlO_2)_{51}(SiO_2)_{141}$ on a dry basis, and a cation exchange capacity (CEC) of 3.90 mEq/g NaX zeolite has a unit cell composition of $Na_{81}(AlO_2)_{81}(SiO_2)_{111}$, which corresponds to a higher cation exchange capacity of 5.96 mEq/g.

First, the zeolite samples were added to 1 mol/L solution of NaCl four times at 60°C under stirring. Samples were then washed at the end of each addition with 2 L of hot deionized water and finally oven-dried at 100°C. Such procedure aimed to originate a homoionic sodium zeolite. Samples were pelletized (average diameter size of 0.180 mm).

Reagent-grade $CrCl_3$ $9H_2O$, $MgCl_2$ $6H_2O$, $CaCl_2$ $2H_2O$, and KCl were used to obtain binary solutions, always containing chromium and another cation in an equivalent ratio of 1:1. The concentration of chromium solution was 18 mg/L·.

The dynamic runs were conducted in a packed bed of a clear glass tube 0.9 cm ID and 30 cm long. The zeolite pellets were located in the middle of the column that operated isothermally at 30°C. The packed bed was composed of 1.60 g of NaY or 1.04 g of NaX in order to provide the same cation exchange capacity, whereas the system was fed at 9 mL min^{-1} of ion solution. Although the packed bed heights were different for NaY and NaX zeolites, experiments conducted with the same cation exchange capacity will generate results to compare the performance of the zeolites, mainly when uptake efficiency is aimed.

4.3.2. *Results of the dynamic binary runs*

Breakthrough curves of metals ions in packed beds of NaY and NaX zeolites are presented in Figures 15 to 17. In all cases, except for the Cr/Mg system in NaX zeolite, some overshooting could be seen as C/C_o reached values higher than 1. In such cases, the incoming ions were first uptaken. As their hydrated radii are smaller than the one for chromium, their diffusion into the zeolitic channels were improved. Rapidly, the incoming ions were exchanged. Chromium ions are much more preferred due to the trivalent charge, and after reaching the exchanging sites, they could replace the ion previously retained.

Besides the sequential ion exchange, it is also important to emphasize the influence of the ion exchanger. NaY and NaX zeolites are isomorph materials, and the only difference between them is the charge density in the packed bed. Denser sites of NaX may promote higher attraction with electrostatic instability due to repulsion of ions closely attracted. As a consequence, the ratio of chromium uptake up to its breakpoint time (tb)/cation exchange capacity (CEC) is less retained than in NaY system, as it can be seen in Table 10.

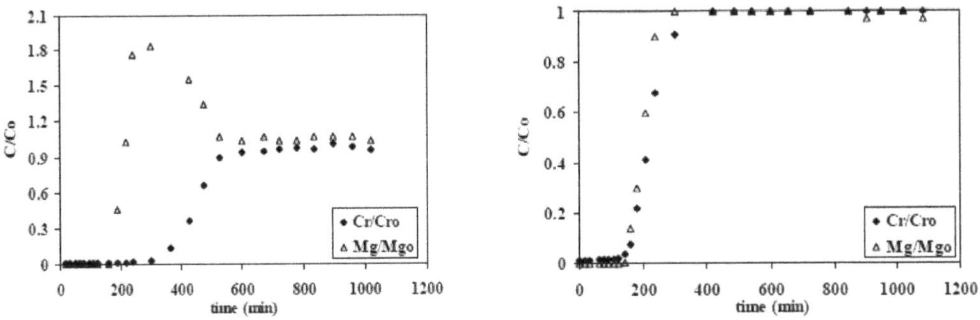

a) b)

Figure 15. Breakthrough curves for the Cr-Mg competitive system: (a) NaY and (b) NaX.

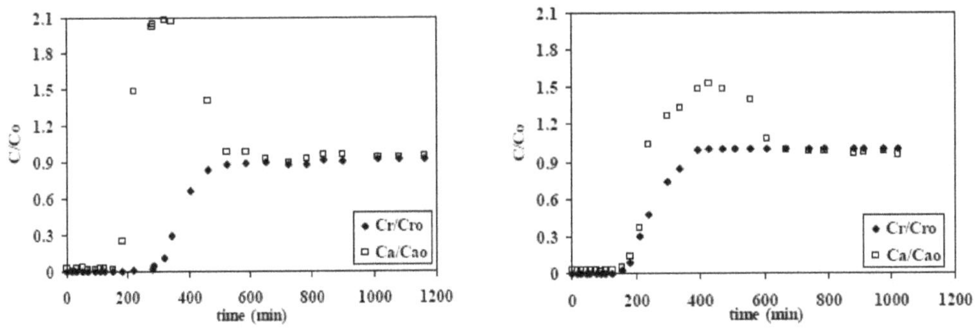

a) b)

Figure 16. Breakthrough curves for the Cr-Ca competitive system: (a) NaY and (b) NaX.

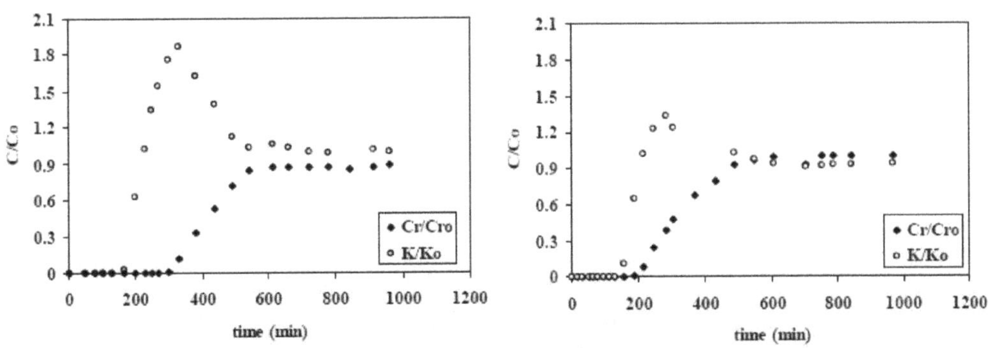

a) b)

Figure 17. Breakthrough curves for the Cr-K competitive system: (a) NaY and (b) NaX.

System	u_{tb}^{Cr}/CEC	System	u_{tb}^{Cr}/CEC
Cr-NaY	0.69	Cr-NaX	0.71
Cr/Mg-NaY	0.54	Cr/Mg-NaX	0.27
Cr/Ca-NaY	0.51	Cr/Ca-NaX	0.31
Cr/K-NaY	0.66	Cr/K-NaX	0.41

u_{tb}^{Cr}/CEC= ratio of chromium uptake up to chromium breakpoint time (tb) and the cation exchange capacity of the column.

Table 10. Amount of chromium ions retained up to the breakpoint of 5% of the feed concentration.

5. General conclusions

Ion exchange is a much more complex phenomenon than adsorption as many ion exchangers may also act as adsorbents, increasing the difficulty in understanding the sorption removal.

Moreover, electrolytes add complexity in fluid phase and also in the solid phase. Ion charges and hydration energy of the incoming ions as well as charge density of the ion exchanger are undoubtedly factors that deserve detailed investigation, mainly for scaling-up proposals.

Author details

Maria Angélica Simões Dornellas de Barros[1], Marcelino Luiz Gimenes[1], Melissa Gurgel Adeodato Vieira[2] and Meuris Gurgel Carlos da Silva[2]

1 Chemical Engineering Department, State University of Maringá, Maringá, PR, Brazil

2 School of Chemical Engineering, University of Campinas, Campinas, SP, Brazil

References

[1] Helferich, F., Ion Exchange, EUA, Dover Publications Inc., 1995.

[2] Zagorodni, A.A., Ion Exchange Materials Properties and Applications, Great Britain, Elsevier, 2007.

[3] Febrianto, J., Kosasih, A., N., Sunarco, J., Ju, Y-H., Indraswati, N., Ismadji, S., J. Hazard. Mater., 2009, 162, 2–3, 616.

[4] Qui, H., Lu, L., Pan, B., Zhang, Q., Zhang, W., Zhang, Q., J. Zhejiang Univ. Sci. A, 2009,10, 715.

[5] Tseng, J., Chang, C., Chang, C., Chenc, Y., Chang, C., Ji, D., Chiud, C., Chiang, P., J. Hazard. Mat., 2009, 171, 370.

[6] Breck, D. W., Zeolite Molecular Sieves—Structure, Chemistry and Uses, Robert E. Kriger Publishing Co., USA, 1974.

[7] Barrer, R.M., Bull. Soc. Fr. Minéral. Cristallogr., 1974, 97, 89.

[8] Barros, M. Arroyo, P.A., Silva, E.A., Chap. 14. General aspects of aqueous sorption process in fixed beds, In: Mass Transfer—Advances in Sustainable Energy and Environment Oriented Numerical Modeling, InTech, Croatia, 2013, 361–386.

[9] Ostroski, I. C., Dantas, J. H., Silva, E. A., Arroyo, P. A., Barros, M. A. S. D., Adsoption Sci. Tech., 2012, 30, 4, 275.

[10] Ostroski, I.C., Silva, E.A., Arroyo, P.A., Barros, M.A.S.D., Fluid Phase Equilib., 2014, 372, 76.

[11] Welter, R. A., Estudo do equilíbrio dos sistemas binários e ternário de troca iônica dos íons cobre, cádmio e cálcio pelo biopolímero alginato. Campinas: Universidade Estadual de Campinas, UNICAMP, 2009. 197p. D.Sc. Thesis (in Portuguese).

[12] Davis, T. A., Volesky, B. Mucci, A. Wat. Res., 2003, 37, 4311.

[13] Huang, C., Cheng, Y.C., Liou, M. R., J. Hazard. Mater.,1996, 45, 265.

[14] Sag, Y., Kaya, A., Kutsal, T., Hydrometallurgy, 1998, 50, 297.

[15] Reed, B. E., Matsumoto, M. R., Sep. Sci. Technol., 1993, 28, 13, 2179.

[16] Ciola, R., Fundamentos da Catálise, Ed. Moderna, São Paulo, 1981. Davis, T. A., Volesky, B. Mucci, A., Water Res., 2003, 37, 4311.

[17] Brunauer, S., Deming, L.S., Deming, W. E., Teller, E., Bureau of Agricultural Chemistry and Engineering and George Washington University, 1940, 1723–1732.

[18] Lai, Y.L., Annaduarai, G., Huang, F.C., Lee, J.F., Bioresource Tech., 2008, 99, 6480.

[19] Papageorgiou, S. K., Kouvelos, F. K., Katsaros, F. K., Desalination, 2008, 224, 293.

[20] Vieira, M. G. A., Silva, M. G. C. Adsorção/Bioadsorção de Metais Pesados, Compostos Orgânicos e Corantes em Solução Aquosa In: Aplicações tecnológicas em sistemas particulados. 1 ed.São Carlos: ANIMERIS, 2011, 171–212 (in Portuguese).

[21] Oliveira, C. M., Remoção de Metais Pesados Utilizando Carvão Ativado. Maringá: Universidade Estadual de Maringá, UEM, 2012. 120p. MSc thesis (in Portuguese).

[22] Hassan, S. S. M., Awwad, N. S., Aboterika, A. H. A., J. Hazard. Mater. 2008, 154, 992.

[23] Snyders, R., Music, D., Sigumonrong, D., Schelnberger, B., Jensen, J., Schneider, J. M., Appl. Phys. Letters, 2007, 90, 193902.

[24] Takeuchi, Y., Arai, H. J., Chem. Eng. Jpn., 1990, 23, 75.

[25] Park, J., Regalbuto, J. R., J. Colloid. Interface Sci., 1995, 175, 239.

[26] Babić, B. M. S. K., Milonjic', B. M. B., Polovina, M. J., Kaludierovi, B. V., Carbon 1999, 37, 477.

[27] Smičklas, I., Dimović, S., Šljivić, M., Plećaš, I., Lončar, B., Mitrić, M., J. Nucl. Mater., 2010, 400, 15.

[28] Pan, X., Wang, J., Zhang, D., Desalination, 2009, 249, 609.

[29] Tóth, J. Adsorption Theory, Modeling, and Analysis, Marcel Dekker, Inc., USA, 2001.

[30] Markhan, E. C., Benton, A. F., J. Am. Chem. Soc., 1931, 53, 497.

[31] Jain, J. S., Snowyink, V. L., J. Water Pollut. Control Fed., 1973, 45, 2463.

[32] Barros, M.A.S.D., Zola A.S., Arroyo, P.A., Sousa-Aguiar, E.F., Tavares, C.R.G., Braz. J. Chem. Eng., 2003, 20, 04, 413.

9

Chitosan-Derived Synthetic Ion Exchangers: Characteristics and Applications

Rajendra Sukhadeorao Dongre

Abstract

Today growing science and technological needs explored various biopolymers to procure novel utilities in its modern developments. Consequently, polysaccharides embraced huge prospective and vastly caters such desired growing needs. Amid, chitin the second most ubiquitous after cellulose comprise of β-[1,4]-2-acetamido-2-deoxy-d-glucose flexible skeleton undergo alteration for requisite physico-chemical features and its highly sophisticated utility superseded counterpart cellulose. Chitosan have unique parameters namely bio-compatibility, non-toxicity, hemeostaticity, anti-microbials which offer competent solutions of many challenging problems. Thus, many products namely biomarkers, biosensors, quantum dots are fabricated via adoptable productive chitosan matrixes. Advancement in chitosan chemistry proffers unambiguous industrial utility in cosmetics, pharmaceuticals, nanobiotechnology, water purifications etc. Chitosan composites own enhanced muco-adhesivity that aids pharmacological safe and successful DNA/SiRNA/tissue releases with bioavailability at target specific carriers. ZnO, ZnS, TiO2 filled/imposed in chitosan and resultant hybrids, quantum dots, surface active microcapsules and nanoparticles are used as biosensors, bio-markers, adsorbents that proffers revolutionary medical usage. Nanointegrated chitosan own complementary strengths and possess assorted utility namely nano-electronic high-resolution devices, for in-vivo imaging, diseases diagnosis, generating new therapeutic and smart tissue engineering scaffolds. Novel modalities with innovative formulations are skillfully designed via chitosan matrix for myriad benefit in biology, chemistry, polymer, and pharmaceutics are displayed in this chapter.

Keywords: chitosan, chitin, biopolymers, biomarker, biosensor, drug delivery, water, membrane, nanotechnology, pharmaceuticals

1. Introduction

Bioactive molecules and natural polymers obtained from animals, plants, fungi and bacteria have ever fascinated the global scientists due to their survival profile for our environment and life [1]. Amid these bio-polymers, polysaccharides namely cellulose, chitin and starch pertaining functions in genetics as cell wall architecture, in biology as energy storage, in material science for bio-composite designing/discovery and for physic-chemical adsorption in chemistry respectively. Today S&T indispensably practices biopolymers for inclusive development in perspective of connectivity, communicative transactions and economic progression in medical sciences besides to cope up advanced biotechnology demands. Twenty-first century scientific innovations and technological throughput ingeniously boosts myriad fields like pharmaceutics, environment and nanotechnology besides uplift our living by enabling reachable services and numerous reliable products in health care. Nevertheless, the requisite new innovative materials are being fabricated from both, chitin and chitosan matrixes due to inherent characteristics uniqueness as futuristic, multi-functional; novel, versatile and peculiar diversity which is devoid in conventional/counterparts.

Chitin is the second most abundant biopolymers after cellulose as explored for R&D besides numerous remarkable papers/patents catering futuristic demands in S&T besides utility in biochemistry, organic/polymer chemistry, pharmacology and medicines. In fact chitin polysaccharide comprise of β-[1,4]-2-acetamido-2-deoxy-D-glucose repetitive units i.e., *N*-acetylglucosamine linked via glucosidic bonds that are facile for physico-chemical adaption to yield into new bio-composites, hybrids and blends owing superior capabilities widespread over cellulose. Deacetylated chitin called chitosan can be categorized by means of purity/quality found to possess exclusive features namely bio-compatibility, bio-degradable, non-toxicity, antimicrobials and hemeostatic. Advance nano-technology exploited superior, productive and widely derived matrixes from flexible chitin/chitosan, thus signifies their chemistry in science and technology [1]. About 1500 tons/year of chitin is usually produced throughout the world. In past decades, progressive bendable chitin/chitosan makes it an ideal model matrix for desired physic-chemical and/or enzymatic alterations/modulations as advantageous in modern S&T [1, 2]. Incredibly practical impact of chitosan in modern S&T is attributed to its liberal super-active $-NH_2$/$-OH$ groups with subsequently performed diverse chemical modifications namely *N*-acylation/alkylation, *N*-quaternization and C-6 carboxylation. Literature reported that such substantial adaptation onto chitin/chitosan matrix persuades inclusive cationic charge via amine protonation to ammonium ion/NH_3^+ further improve its pH dependency (acid to alkaline) and resultant solubility [2]. Thus certain chitosan based materials owing wide putative applications are portrayed as below.

Chitosan matrixes extensively provides generous opportunities in biotechnology, theranostic and pharmaceuticals for drug/gene release, tissue engineering and wound healing respectively due to exceptional features like biodegradability, biocompatibility, antimicrobial profile, besides low toxicity, and immunogenicity [2–8]. This mini review signifies chitosan for advanced formulations in nano-science/biotechnology with reference to quantum dots, and nanoparticles, carbon dots, biosensors and biomarkers.

2. Advances in chitosan chemistry: unambiguous myriad applications in modern S&T

Chitosan own adaptable promising applications de-acetylated chitin derivative, despite its limited mechanical strength and solubility. The proactive functionalities of native chitosan like primary amino and hydroxy groups are facile to undergo many possible physical/chemical modifications/alternations through grafting and ionic interactions so as to yield assorted derivatives which can offer specific and requisite commercially utilities. The following sections comprise contemporary research in chitin/chitosan matrix towards applications in numerous industrial and clinical fields.

2.1. Chitin/chitosan processing

Chitin/chitosan are commercial natural resources, as well 150,000 metric tons as a sea-food industry wastes generate via processing from crab, shrimp, shell-fish, krill, clam, oyster and squid processing beside extracted from certain fungi. Compared to synthetics/counterparts, chitin/chitosan own exclusive nitrogen content and high concentrations of proteins along with calcium carbonate as procured via physic-chemical/biological extractions/treatments. Further chitin undergoes de-acetylation by means of alkali refluxing yields 85% product called chitosan. Chronological step-wise treatments like deproteinization, demineralization, decolouration and de-acetylations performed to produce chitosan from sea-food shown below:

Crustacean shell-size reduction—protein separation by NaOH/alkali reflux—washing—demineralization by HCl/mineral acid—dewatering—chitin decolouration—deacetylation by NaOH—washing/dewatering yield 85% chitosan.

2.2. Solubility of chitin/chitosan

Chitin/chitosan owes anhydro-glucoside link akin to cellulose, yet characteristics are vastly different as being strong hydrophobic insoluble in water/aqueous state. Many organic solvents, too cannot dissolve chitin/chitosan instead solvent mixtures are employed for its dissolution namely hexa-fluoroisopropanol, hexa-fluoroacetone, chloro-alcohol, 5% LiCl added dimethylacetamide, 30% aqueous acetic acid, aqueous *N*-methyl morpholine-*N*-oxide and aqueous mineral acids. Chitin and chitosan being highly basic (pH > 7) are soluble in various aqueous-organic/mineral-acid mixtures to form invariably poly-electrolytic jelly that's easily transformed into myriad compositions including salt, film, hybrid, chelate/complex and gels [2].

2.3. Chitin/chitosan salient treatments

Concentrated acid hydrolysis and drastic treatment onto chitin/chitosan produces serviceable β-ᴅ-glucosamine: amino sugar functional unit. Chitin/chitosan, based on deacetylation degree owns 6–8% w/v nitrogen/free $-NH_2$ being facile for varied chemical transformations/modifications namely *N*-acylation and Schiff-base reactions. Chitosan on treatment with keto-acids followed by sodium borohydride reduction yields proteic/non-proteic amino acid (1,3/1,6)-β-D-glucan and melanin linkage pertaining effective immuno-stimulant behaviour to boost immunity in human/animals. Chitosan undergoes many chemical reactions like forms *N*-acylation with acid anhydrides/acyl halides aldimine and ketimine with aldehyde and ketone respectively at NTP. Hydrogenation and *N*-alkylation with more bulky substituent weakens hydrogen bonding in chitosan skeleton; so *gets* swelled in water and retains film forming tendency in spite of hydrophobic alkyl chains. Further chitosan is quite versatile than chitin due to the presence of amino groups at the C-2 skeleton.

2.4. Chemical and biological properties of chitosan

Chitosan is linear polyamine skeleton owing free-active functionalities like primary amino and hydroxyl both can chelate transitional metals to form complexes. Biological chitosan is biocompatible natural polymer which is biodegradable to normal body constituents, hence safe and non-toxic to environment upon disposal. Chitosan binds to mammalian and microbial cells assertively own connective gum tissue regeneration and accelerate osteoblast formation as vital for bone regeneration, Moreover, chitosan possess board microbiological profile namely as hemostatic, fungistatic, spermicidal, antitumor, anticholesteremic besides acts as central nervous system depressant and immune adjuvant.

2.5. Chitosan derivatives

Chitin/chitosan is readily derivatized at primary amine, primary and secondary hydroxyl functionalities [3]. These derivatives own potential significance via further bi/poly-functional chemical reaction yields polymeric composites, blends, gels and industrially applied polyampholytes effective in remediation of pollution. Chitosan chelated complexes with transition metals acts as matrix for enzyme immobilizations [12]. Reactions with pure chitin have been carried out mostly in the solid state owing to the lack of solubility in ordinary solvents. Chitin/chitosan of about 50% degree of de-acetylation found to be water soluble [17] and used as feedstock for smooth modifications, through various solution phase transformation as enumerated below:

A. *N-phthaloylation of chitosan*: phthalic anhydride-DMF reacts with chitosan to yield *N*-phthaloylated chitosan which found to enhance solubility and affixed bulkiness due to breaking hydrogen of primary amine and averts hydrogen bonding to firm backbone/skeleton.

B. *Chitosan-sialic acid dendrond hybrid*: the water-solubility of chitosan matrix gets improved effectively via gallic acid and tri-ethylene glycol dendronized chitosan-sialic acid hybrid synthesis. Residual amine functionality can be further *N*-succinylated to impart great water solubility of such novel chitosan derivatives.

C. *Alkyl/aryl-thiocarbamoyl chitosan*: methyl/phenyl-thiocarbamoyl chitosan derivatives are prepared for selective metal ionic sorptions from aqueous solution/contaminated waters.

D. *Chitosan hydrogels*: chitosan hydrogels can be straight grafted by treatment with D,L-lactic and/or glycolic acid owing great interfacial water-chitosan interaction due to grafting with such acids. Chitosan side chains can physically cross-linked/aggregated to yield pH-sensitive hydrogels that owe potential biomedical functions namely wound dressing/healing, gene and drug delivery.

E. *Chitosan composite base quantum dots*: nano-cadmium sulphide doped chitosan improved thermal and mechanical besides aqueous solubility in parent bio-polymeric matrix. Efficient chitosan based CdS-QD/film/composite can be easily obtained by mixing chitosan, $Cd(Ac)_2$ and CdS in 1% aqueous acetic acid solution.

F. *Chitosan based nano-particals/composites*: gadolinium neutran capture therapy (Gd-NCT) induced chitosan-gadopentetic acid nanoparticles used for cancer therapy appropriate for intra-tumoral treatment. Flexible chitosan skeleton endure facile blends with synthetic polymers via mechano-chemical routes yields novel composites/nano-particles. Thermal performance and molecular signaling of synthetic polymeric matrix/composites are exclusively dissimilar from each/both components. These blends impart strong integrated chitosan cascades owing talented compatibility under mild conditions as utilized in smart and advanced drug-gene delivery.

2.6. Biological profile of chitin/chitosan

Human body fluids enzyme, lysozyme can dissolve some bacteria via cell-wall material cleaving, also facile to assemble in chitin/chitosan matrixes and impose myriad therapeutic usages in past decade. Certain vital medical usages of such chitosan matrixes including fibroplasias inhibition resulted wound healing/dressing, as absorbable sutures and supports in tissue/cell growth or differentiating tissues. Such chitosan based sutures found to resist attack in urine, bile and pancreatic juice while hasten to enhance wound healings dressings/textures that's hardly achieved by absorbable counterparts.

2.7. Industrial utility of chitin/chitosan matrixes

Apart from this chitin/chitosan apparel industrial usages and impending functions are exploited in wastewater/water purifications especially heavy metal and organic pollutant removal via chelation with wide scope and possibility. Chitosan base derivatives impart range of pharmaceutical and cosmetic products way from water treatment to plant/food protections.

2.7.1. Cosmetics

Unlike to other polyanionic hydrocolloids, this natural amino-polysaccharide chitosan encompassed hydrocolloids are exclusively cationic that gets viscous/semi-solid upon neutralization with aqueous organic acid solutions. Such viscous/semisolid chitosan facilitates intervene common integuments like skin covers and hairs. Chitin/chitosan both are fungicidal and fungi-static and thus compatible with other highly integrated biological counterparts used in such beauty products. Chitosan-alginate (1–10 μ size) microcapsules embodied various hydrophobic materials are utilized in range of cosmetic/beauty products. Substances absorbing harmful UV light and assorted dyes get linked to amino functionality of chitosan via covalent bonding. Consequently, chitosan enclosed compositions were developed by *Sonat* Company USA that can insert anti-oxidants, anti-allergic, and anti-inflammatory agents as novel depilatory designed in many areas of cosmetics like skin care, hair care, and curling hairs besides oral care. Chitin/chitosan and human hairs can harmonize mutually owing to opposite electrical charges as chitosan is cationic and hairs are anionic. Chitosan containing solution forms clear, elastic film or foam and create emulsifying action on hairs so, boots softness, smoothness, and mechanical strength. Chitosan based compositions also forms hydro-gel in aqueous alcohol thus used in shampoo, rinse, permanent wave agent and styling lotion besides hair spray/colorant and tonic. Several chitosan based compounds namely glyceryl chitosan, *n*-hydroxypropyl chitosan, quaternary hydroxypropyl chitosan, polyoxyalkylene chitosan, chitin sulphate and carboxymethyl chitosan own potential in hair care products. Chitosan derivatives are good candidate for skin care as they impart positive electrical charge, and due to high molecular weights it cannot infiltrate human skin, so act as a moisturizer and might compete with hyaluronic acid in this perspective. Chitin/chitosan both own wide utilities in formulation like creams, pack material, lotions, foundation, eye shadow, lipstick, cleansing, bath agents and nail enamel/lacquers. Rather diacid anhydride treated chitosan derivatives are best employed in many skin care products besides usage in paste, mouthwash and chew gum. Chitosan based salts are supplemented in toothpaste for mask silicon oxide unpleasant taste besides acts as powder binder to uphold granular shapes. Chitin materials are used as dental fillers that vitally absorb candida/thican, teeth sticking fungi, so recover cleaning false teeth.

2.7.2. Chitosan based scaffolds in medical/clinical/pharmaceuticals

Chitosan based skeletons are well-known in biomedical utility namely drug/gene release, wound dressing and advanced nanotechnology [1]. Chitosan based 2D/3D scaffolds like sponge, foam, gels fibers/film are more developed for modern tissue and bone engineering [2]. Notable *N*-trimethyl-*N*-octyl-chitosan scaffolds have established its potential utility as controlled drug delivery of hydroxyl-camptothecin anticancer agent [3]. Antitumor drug delivery is also reported with specially fabricated magnetic nano-iron doped maltosyl chitosan hybrids [2, 3]. Several clinical reports endorse impact chitosan matrixes in cancer chemoprevention due to precise nutrients encapsulation which imparts enhanced drug delivery in blood with fewer side effects on healthy cells [1, 2].

2.7.3. Chitosan based matrixes in science and technology

Synthetic plastic are produced @ 300 million tons/year with only 3% recyclability, rather 97% plastic wastes break down in oceans or in landfills and harms the green environment. But, polysaccharide derived 3D bio-plastics offered predictable hardiness e.g., Wyss Institute, USA, has derived chitin based bio-degradable plastic "*Shrilk*" silk protein substitute. Shrilk's hardness and peculiarity offered myriad utilities namely implantable medical devices, bone-tissue gallows, laminated silk fibroin, bio-composts/fertilizer (release nitrogen nutrients), implantable foams, films, surgical closure scaffolds, wound healing, and FDA-approved devices.

2.7.4. Fabricated chitosan matrix via nano-biotechnology

Biotechnology can interfaces between science, engineering and technology. Amid, nanobiotechnology is complementary yet untapped science field that can exploits/improves biotechnology and aids to fabricate natural/bio-mimetic nano-structures. Nano-science technology covers and merges biological R&D with various fields as it alters material's parameters including biological, physicochemical and cellular electrochemical responses along with molecular motions. Advanced nano-biotechnology fantastically utilizes natural polysaccharides like chitin/chitosan in such perspectives [1, 2]. Thus, varied matrixes are formulated through chitin/chitosan skeleton owing clinical, biomedical and industrial applicable bio-polymeric prominence like quantum/carbon dots, nano-particle/composite and biomarkers usages in cancer detection. Nano-biotechnology exploited unique and admirable features of chitosan namely biodegradability, biocompatibility, low/no toxicity, antimicrobial activity and low immunogenicity in drug delivery, siRNA/DNA delivery, tissue engineering, and wound healing, biosensors besides theranostics utilities **Figure 1**. Chitosan based many novel nano-materials/devices owns vast beneficial applications to mankind. Strategic chitosan nano-matrixes carriers vital impact on global pharmaceutical use to control drug release due to enhance drugs solubility, superior protein bioavailability, and better uptake of hydrophilic substances across epithelial layers besides great intracellular drug delivery [2–5].

2.7.4.1. Chitosan-carbon dots

Quantum dot are 'nanometer scale' i.e., zero-dimensional particles own semiconducting, optical and electronic characteristics emitted at specific frequency of light that can be adjusted via matter utilized and size, shape and arrangements of dot/particle. Nano-chitonous based carbon quantum dot empowered fluorescent benefits for bio-sensing or imaging due to prominent features namely extremely tune-ability, brilliant water solubility, biocompatibility and better photo-stability [1–4]. Certain chitosan-carbon dots [2–3] yield as smooth, soft films which are robust to UV-visible blockage exploited in biomedical usages imparting low cytotoxicity and excellent biocompatibility with enhanced swelling, better thermal/mechanical properties over pure chitosan film. Amino functionalized fluorescent carbon nitride dots own improved water

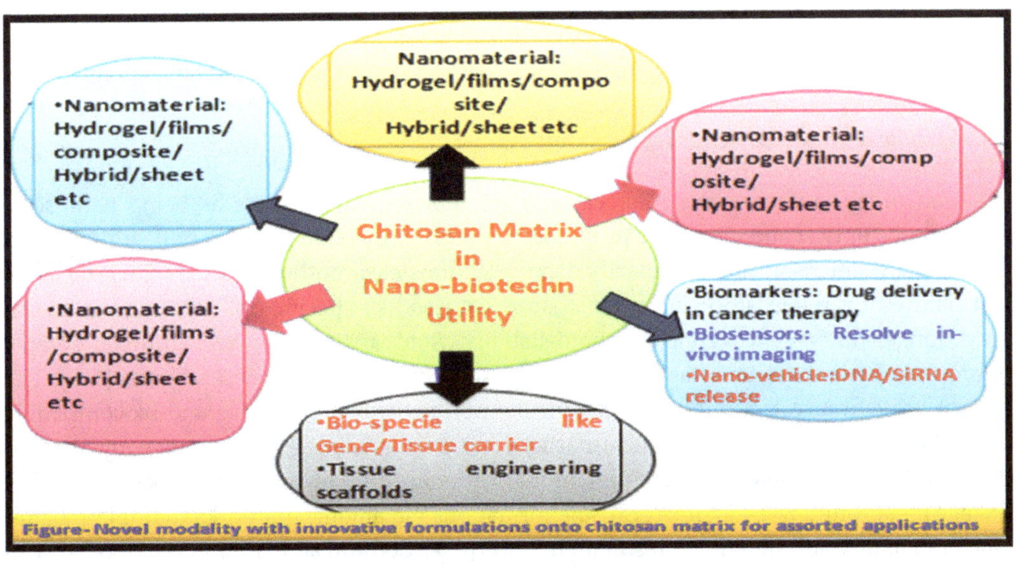

Figure 1. Chitosan matrix in nano-biotechnology.

solubility and strong fluorescent effect as advantageous in medical diagnosis and cancer treatments. Multi-color chitosan-carbon dots exhibited bio-labelling potential with varied bacterial model methods for biomedical usages [2–4]. Solid nano-solar cells are designed via chitin, chitosan and glucose based carbon quantum dot, hybrids as competence of mixed utmost with layer-by-layer sensitizer nanozinc-oxide coating. Such fluorescent nano-crystals/quantum dots can acts as an imaging agent for diseases detection with significant utility. Thus, novel chitosan-zinc sulphide based quantum dots obtained with pH dependent/tuneable optical/electrical properties as probed in pharmaceutical usages [2]. Luminescent chitosan-L-cysteine impregnated cadmium-tellurium films/dots showed antibacterial profile for broad range of biomedical utility. Chitosan-cadmium-telluride quantum dots generated onto indium-tin-oxide coated glass acts as electrochemical biosensor for culprit DNA in chronic myelogenous leukemia treatment or cancer detection. CdS- and/or nano-gold doped chitosan quantum dots are used to formulate antibody immobilization that own brilliant control and bioactive profile compared to other immune-sensors for protein detection studies [2–4]. Such nano-chitosan formations are frequently developed for delivery of safe, effectual plasmid DNA, siRNA, and oligo-nucleotides as gets rapid noticed by genetic materials so as to treat silencing unwanted gene's expression, defects and substituted missing in diseases curing therapeutics [1–5].

Nano-particles can be easily deposited on porous chitosan surfaces to offer homogeneous QDs for multiple active sites adsorption and desorption of hydrogen gas. Certain nano-metallic particles gets integrated into flexible chitosan to exploit for many purpose like catalytic activity, manipulated electronic and chemical characteristics in modern S&T [5]. Polyaniline-chitosan-platinum films are designed on pencil graphite electrode showed brilliant synergic performance in water electrolysis for electro-catalytic hydrogen generation. Thus, chitosan based cheap well-designed gels, films and composites yields via immobilizing desired quantum dots for facilitated hydrogen evolution reaction (HER) and practically skillful hydrogen production with long-term durable electro-catalyst. Flexible organic matrixes of chitosan can be altered for its superior fluorescent, semiconductor behaviour by hyper-branched ligand stabilizer own gelated quantum dots. These fabricated QDs offers solution to multi-responsive novel applications in nano-science for imaging, bio-sensing and drug delivery actions.

2.7.4.2. Chitosan based surface active microcapsules

Chitosan based surface active microcapsule resembles quantum dots as being nano-scale size pertains optical and electronic properties rather intermediate between bulk and discrete. Surface active microcapsules compactly detain electrons/electron-holes varied with size/shape and nature of raw/feedstock to own tuneable features like specific opticity, huge quantum yield, longer fluorescence, enhance photo-stability as advantageous over traditional organic fluorophores in recognition, tagging and imaging in biology and clinical sciences [1–5]. The surface active integrated chitosan skeletons are non-toxic, biodegradable and biocompatible owing liquid-core microcapsule, micro-particle, and macromolecule matrixes own myriad utilities in advanced nanotechnology [6]. Laminar cationic chitosan skeleton owns $-NH_2-OH$ linkages that found to assist microcapsule formations via anionic interlocks to implant better biocompatibility and stability in resultant hybrids [6, 7]. Sodium alginate and/or nano-ZnS can micro-capsule chitosan to yield nano-hybrid/nano-gel responsible for many pharmaceutical applications including bio-imaging, bio-labelling and gene/drug release purpose. Flexible chitosan skeleton can catatonically interacts with anionic surface-active surfactants via apparent loss of positive charge to yield insoluble composites or complexes. Such, rational design chitosan based homogenized microcapsule acquired via co-acervation, emulsification, solvent-evaporation, gas-liquid micro-fluidic and layer-by-layer assembly techniques impart improved bioavailability, reproducibility and repeatable drug release/delivery in today's advanced nano-biotechnology [7]. Host-guest interactive and responsive external stimuli sensibility imparts hydrophobic tails/cavities in resultant microcapsules that crucially control its inherent functions as induced via intrusion or doping with surfactants like cyclodextrins and sodium dodecyl sulphate. Strategically controlled/uniform size nano-cadmium sulphide entrapped chitosan quantum dots or microcapsules with stimuli-accountable α-hydrophobic cavities are developed for detection of toxic/hazardous chemicals as beneficial in remediation of ever increasing environmental pollution. Requisite chemical's stimuli-response gets formulated as in rationally controlled chitosan-CdS and/or ZnS quantum dots as liquid-core microcapsules encapsulated with poly(DL-lactide-co-glycolide) showed good fluorescent stability in aqueous condition. Rather usage of surfactants like α-cyclodextrin induces changes/slight influence on shape and fluorescent color in rationally designed chitosan-bends with CdS/ZnS in resulted mono-disperse micro-capsulation. These surfactant induced chitosan stimuli-responsive micro-capsules are cost effective micro-detectors for assorted chemicals than traditional counterparts.

2.7.4.3. Chitosan based nanoparticle

Chitosan based nano-particle performs efficient exogenous gene release into primary chondrocytes and imparts immense potential for requisite delivery of therapeutic drugs used in treatment of various diseases. Chitosan skeletal alteration found to enhance transfect ion efficiency by virtue of enhanced self-branching achieved in resultant composites/matrixes/hybrids. Tri-saccharide substitution linear chitosan counterparts compel better gene transfer with intact biocompatibility for cellular uptake with superior stability [3–7]. *N,N,N*-trimethylated chitosan scaffolds encourages *in-vivo* intracellular si-RNA delivery with improved extracellular competence and fine silencing profile along with effective DNA-drug release [1, 8]. Glycol-chitosan scaffolds that can entrap chemo-theraptic like doxorubicin and DOX drugs attenuated utmost si-RNA delivery via surmounts resistance observed in adorn dose-dependent *in-vivo* analysis [8, 9]. Chitosan-poly-D,L-lactide-co-glycolide matrix acts as precise non-viral devices for lot of uses like pulmonary si-RNA release, *in-vitro* H1299 gene silencing and fluorescent protein cell expression [9].

2.7.4.4. Biosensors & bio-markers

Sensor receives and responds to signals by converting into magnetic/electrical fields that further detected by an electronic device. Biosensor for biological entities comprises bio-polymeric coalesce and utilizes for physicochemical detection of assorted sensitive natural species like; tissues, cells microorganisms, organelles, enzymes, antibodies and nucleic acids [9]. Biosensor interacts with these analytes and performs recognition/diagnosis based on corresponding interactive evaluation as employed for DNA/RNA, enzymes, antibodies and signal transduced/immobilized tissues [9]. Bio-engineered analytical gizmo aids to offer myriad chitosan based bio-sensores which are beneficial due to uniqueness namely cheap, bio-compatible, eco-friendly, adaptable, portable, high sensitive, intrinsic selective and benign to use in moderately complex environments by virtue of quick responses compared to traditional sensors [10]. Flexible chitosan skeleton recognizes the sample species via immobilization in fabricated matrixes to be exploited for complicated enzyme sensing [11]. Tyrosinase-Fe_3O_4 dope chitosan used to bio-sense/detects certain organic pollutants like catechol onto specifically designed porous nano-iron oxide proactive surfaces [12]. Chitosan diffusion into graphene skeleton induces huge surface area and electrical conductivity in resultant matrix which is used for effectual bio-sensing or immobilization of enzymes and glucose estimation with splendid sensitivity besides durable stability [13]. Nanocarbon dope chitosan yields amperometric matrix to be used for assorted purpose like biosensors, biomarkers, to encapsulate lactase and bio-fuel cells besides as bio-electrochemical devices [14]. Polyaniline-nano-chitosan entrapped creatinine amidino-hydrolase shown good immobilization of CAH enzyme with better stability and durability.

Biomarker is biological indicator use to perform characteristic objective measurements, detections and/or indications for validity of certain phenomenon namely biological state, living organism existence, pathogenic processes and therapeutic intervening pharmacologic responses besides manage cancers and other diseases [15]. Biomarker established doubt in advanced pharmaceutical for facile screening and risk assessment before its diagnosis besides detect diseases with staging, grading and preliminary treatment options for monitoring supplementary therapy [16]. Gold coated chitosan-xanthan biosensor is used for bio-imaging in numerous diseases diagnosis [17] and signal improvement for melanoma. Chitosan-graphene nano sphere marked horseradish peroxidase in α-fetoprotein induce cancer diagnosis with enhanced signal augmentation than electrochemical immuno-sensors [18]. Nano-chitosan biomarkers are preferred detect alpha-fetoprotein and carcino-embryonic antigen with more precision and accuracy over ELISA test. Thus, chitosan integrated biomarkers own stupendous including haemostatic, fungi-static, bacteriostatic, spermicidal, anti-cholestermic and

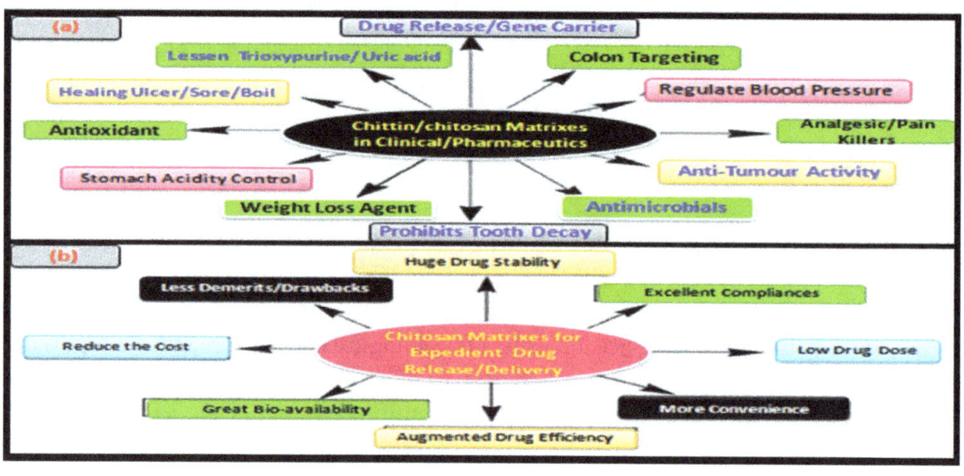

Figure 2. Chitosan base matrixes for pharmaceutical/clinical and expedient drug delivery.

anti-carcinogenic features fascinated for skillful healings, disease prevention/diagnosis, drug delivery and tissue engineering. Primary —NH_2 link of chitosan smartly offers surface active cationic charge variations (below pH 5.5) achieved via H-bonding imparts adhesion for tissues and subsequent fabricated matrixes offers novel clinically applied products. Some pharmaceutical/clinical expedient utilities of chitosan matrix includes dentistry, orthopedic, ophthalmology, surgical measures, optical and wave guiding utilities as shown in **Figure 2**.

2.7.4.5. Chitosan-doped $Ca(OH)_2$ microcapsules

Liberal —NH_2 groups of chitosan gets protonated in acidic conditions, thus chitosan based microcapsules, micelles and hydrogels can be synthesized to favor varied biomedical/clinical applications. Certain pH-triggered calcium hydroxide coated chitosan hybrids are used for endodontic treatment with long-lasting antibacterial activity against *Enterococcus faecalis* refractory strains. Ethylcellulose-calcium hydroxide coated onto chitosan yields [CS-EC@$Ca(OH)_2$] microcapsule found to have unique features namely non-toxic, biocompatible, possess superior host immunity, tender controlled drug/gene release besides usage for apical periodontitis in confronted customary root canal therapy. Such CS-EC@$Ca(OH)_2$ microcapsule recuperates innate calcium hydroxide properties which approved osteogenesis effect along with lessening inflammation as beneficial for apical periodontitis in bone defects healing.

Vitapex, calcium hydroxide coated chitosan paste/gels is clinically used to accomplish prominent apical periodontitis effect via stagnant release $Ca(OH)_2$ than other aqueous counterparts. Thus, amid gel vehicles chitosan based microcapsules like Ca (OH)2 loaded propylene glycol-chitosan/agar formulation releases calcium hydroxide via altering gel bases in chronic inflammatory lesion around apex of tooth treatment, besides offered controlled drug delivery with decrease cytotoxicity. Chitosan EC@$Ca(OH)_2$ scaffolds/pH-triggered microcapsule release $Ca(OH)_2$ rapidly in acidic conditions own many significances like inhibits bacterial infections, endorses AP repairs/periapical periodontitis, wide antibacterial activity against *Enterococcus faecalis* refractory strains, great anti-inflammatory profiles and evade bone resorptions. Such, facile and sustainable pH-triggered calcium EC@calcium hydroxide-CS-matrixes own significantly extended antibacterial profile and showed amazingly diminish inflammation to persuade osteogenesis capable in endodontic diseases therapy.

3. Drawbacks and remedies

Amid, all above meritorious usages of chitosan chemistry, it also own certain drawbacks namely as weak basicity at low physiological pH (pka = 6.2) besides, only soluble in aqueous organic acid solution due to protonated —$NH_3^+X^-$ salt formation. Chitosan shows elevated swelling in aqueous medium and carry out unavoidable quick/instant drug delivery, needs to modify/alter chitosan inherent framework. Certain above mentioned demerits in chitosan utility can be trounced by —NH_2 and/or —OH glucosamine derivative formation which solve few complications to cater wide utility. The imperative and characteristics applications of chitosan based bio-polymeric adsorbents established in the environmental pollution remediation is [19] summarized in **Table 1**.

SN	Material types	Practical characteristics		Applications
		Useful property	Adverse property	
1	Nano-filtrations (RO/FO)	Reliable, automated, charge based repulsion, high selectivity, low pressure, costly	High energy, costly, membrane chock, intense polarized, nano-pore dimension	Lessen hardness, color, odor, remove heavy metals, sea water desalination and wastewater treatments
2	Nano-composite membranes	Viable hydrophilic/hydrophobic, huge porosity, more water flux/permeability, thermal stable, mechanically robust, foul resistant, stay at high pressure	Nano-particle leakage, bulk nano-materials needed for oxidation, and composite dependent	Reverse osmosis; eliminate pollutants, composites in ultra-filtration, cartridges nano-fiber composites
3	Self-assemble membrane	Uniform, nano-porous, tuneable size/shape, facile design	Applied on lab/small scales	Ultra-filtrations and process scale up
4	Dendrimer/dendrons (arborol cascade)	Inner hydrophobicity, outer hydrophilicity, water soluble, bio-mimics, easy encapsulation, handy, non-toxic and reusable	Dendrimer/dendron formations are complex multistage processes	Organic/heavy metal removals, biodegradable, biocompatible dendrite
5	Metals magnetic nano-particles	Biocompatible, tuneable colloidal nano-particle, super-paramagnetic, highly recyclable, facile residual separation and huge surface: volume ratio	Stabilization (surface modification needed to enhance its potential	Remediation of hazardous/toxic pollutants, cation/magnetic sensors, nano-beads used for sorption

Table 1. Characteristics of chitosan base membranes in water treatments/purification.

4. Futuristic prospective applications of chitosan matrixes in water purifications

Certain rationally engineered chitosan based adsorbents are vulnerable and flexibly adjustable for facile elimination of contaminants from water/wastewater [14, 20] than compatibly integrated traditional treatments. Bio-materials are beneficially benign to integrate into diversified multifunctional membranes to facilitate both particle retention and contaminant mitigations than conventional materials based water treatments [20]. Further, such chitosan based materials usage own superior process efficiency and high adsorption profile besides adaptable mass scale utility in point-of-use devices [12, 14, 15].

Membrane-based water purifications or treatment processes can address the global challenges of water scarcity and aquatic environmental pollutions [14, 18]. Conservative water purification membranes owe constrained due to inherent limits of conventional materials used in their fabrications [18]. Advanced nano-technology has developed sophisticated methods to control the structural and chemical functionality in chitosan based film, sheet, hydrogel, microcapsule and dots to new classes of materials for water purification [12]. Technologically manufactured chitosan based materials caters the need for futuristic advanced water purification/treatments. Such molecularly designed well selective and focus materials are highlighted by surface modification to minimize interfacial interactions, inherent limitations and enhance efficiency over customarily used materials as in **Table 2**.

Utility	Nature of work		Features
Energy	Sunlight conversion (like DSSCs, PBs)		• Efficient light-harvesting, especially in biomaterials replica or biocomposites examples; • fast charge separation, high current density; • high gas permeability; • high storage density; • fast electron and ion transport; • small resistance
Life sciences	Engineered/designed Biomaterials/films/composites		• Biocompatibility; • promoting cell adhesion; • good mechanical property; • controlled shapes/sizes
Chemical sciences	Pre-concentration device For separation and adsorption	Bioreactors	• High permeability; • homogeneous flow-through pore structure; • control pore structures and surface properties; • used monolithic column

Table 2. Applications of chitosan matrixes.

5. Conclusion

Chitosan biopolymer is preferred for biomedical and pharmacological promising purposes as its safe/harmless besides impart successful drug delivery/releases. Muco-adhesive chitosan improved abode span and consequently provide drug's bioavailability in target specific carriers. Certain fillers like ZnO, ZnS and TiO_2 chosen to intrude into chitosan skeleton to yield resultant matrixes/hybrids/composites with widen utilities in clinical/pharmaceuticals. Systematically performed R&D in chitosan based scaffolds/matrixes owing interactive filler/dopant added to improve its applicability which opens subsistence revolutionary and advanced medical usages.

Nanobiotechnology integrated technology and complementary strengths of bio-molecule chitosan with the nano-electronics to own assorted applications/outcomes as biosensor and biomarker proliferation as high-resolution devices (for *in-vivo* imaging) and drug delivery treatment in cancer respectively. Further advance biotechnology exploited excellent biological characters of chitosan so as to introduce novel modality with innovative formulations namely nanovehicles for gene/DNA/SiRNA delivery and quantum dots/nanobio-composites for various diseases diagnosis, generating new therapeutic techniques besides development of tissue engineering scaffolds. Thus, nano-biotechnology skillfully explored chitosan matrix to design and device never-ending clinical and scientific applications for the benefit of human besides environment/nature.

Acknowledgements

The author is thankful to the Head, Department of Chemistry, R.T.M. Nagpur University, Nagpur, for laboratory facilities and to the Vice Chancellor, Nagpur University, Nagpur, for the sanction of a research project under University Research Project Scheme, No. Dev/RTMNURP/AH/1672 (9) dated 24 September 2016.

Author details

Rajendra Sukhadeorao Dongre

Address all correspondence to: rsdongre@hotmail.com

Department of Chemistry, RTM Nagpur University, Nagpur, Maharashtra, India

References

[1] Dongre RS. Biological Activities & Application of Marine Polysaccharides. Vol. 1. Croatia: In-Tech Open; 2017. pp. 181-206. DOI: 10.5772/65786

[2] Jain1 T, Kumar S, Dutta PK. Chitosan in the light of nano-biotechnology: A mini review. Journal of Biomedical Technology and Research. 2015;**1**(1):101-107

[3] Bănică F-G. Chemical Sensors & Biosensors: Fundamentals & Applications. Vol. 176. UK: John Wiley Sons; 2012. ISBN: 978-0-470-71067-8

[4] Jianc H, Su W, Caracci S, Bunninc TJ, et al. Optical wave guiding and morphology of chitosan thin films. Journal of Applied Polymer Science. 1996;**61**:1163-1171

[5] Burkatovskaya M, Tegos GP, Swietlik E, Demidova TN, P Castano A. Use of chitosan bandage to prevent fatal infections developing from highly contaminated wounds in mice. Biomaterials. 2006;**27**:4157-4164

[6] Kurita K. Chitin & chitosan: Functional biopolymers from marine crustaceans. Marine Biotechnology. 2006;**8**(3):203-226

[7] Mourya VK, Inamdar NN. Chitosan-modifications and applications: Opportunities galore. Reactive and Functional Polymers. 2008;**68**(6):1013-1051

[8] Sashiwa H, Shigemasa Y. Chemical modification of chitin & chitosan 2: Preparation & water soluble property of N-acylated or N-alkylated partially deacetylated chitins. Carbohydrate Polymers. 1999;**39**(2):127-138

[9] Dutta PK, Dutta J, Tripathi VS. Chitin and chitosan: Chemistry, properties & applications. Journal of Scientific and Industrial Research. 2004;**63**:20-31

[10] Peter MG, Domard A, Muzzarelli RAA. Advances in Chitin Science. Vol. IV. Potsdam, Germany: Universität Potsdam; 2000. p. 395. ISBN: 3-9806494-5-8

[11] Hudson SM, Jenkins DW. Chitin & Chitosan. Encyclopedia of Polymer Science and Technology-1. 3rd ed. New York: Wiley Interscience. DOI: 10.1002/0471440264.pst052

[12] Dutta PK, Ravikumar MNV, Dutta J. Chitin and chitosan for versatile applications. Journal of Macromolecular Science-C, Polymer Review. 2002;**42**(3):307-354

[13] Noipa T, Ngamdee K, Tuntulani T, Ngeontae W. Cysteamine CdS quantum dots decorated with Fe^{3+} as a fluorescence sensor for the detection of PPi. Spectrochimica Acta—Part A: Molecular and Biomolecular Spectroscopy. 2014;**118**:17-23

[14] Yan J-J, Wang H, Zhou Q-H, You Y-Z. Reversible and multisensitive quantum dot gels. Macromolecules. 2011;**44**(11):4306-4312

[15] Hardison D, Deepthike HU, Pathirathne T, Wells MJ. Temperature-sensitive microcapsules with variable optical signatures based on incorporation of quantum dots into highly biocompatible hydrogel. Materials Chemistry. 2008;**18**(44):5368-5375

[16] Sá-Lima H, Caridade SG, Mano JF, Reis RL. Stimuli-responsive chitosan-starch injectable hydrogels combined with encapsulated adipose-derived stromal cells for articular cartilage regeneration. Soft Matter. 2010;**6**(20):5184-5195

[17] Thongngam M, McClements DJ. Influence of pH, ionic strength & temperature on self-association & interactions of sodium dodecyl sulfate in the absence & presence of chitosan. Langmuir. 2005;**21**(1):79-86

[18] Chen Y, Yao R, Wang Y, Chen M, Qiu T, Zhang C. CdS QDs-chitosan microcapsules with stimuli-responsive property generated by gas–liquid microfluidic technique. Colloids and Surfaces. B, Biointerfaces. 2015;**125**:21-27

[19] Li X, Yang Z. Chitosan-decorated calcium hydroxide microcapsules with pH-trigger release for endodontic application. Journal of Materials Chemistry B. 2015;**3**:8884-8891

[20] Kayan DB, Koçak D, İlhan M. The activity of PAni-Chitosan composite film decorated with Pt nanoparticles for electrocatalytic hydrogen generation. International Journal of Hydrogen Energy. 2016;**41**(25):10522-10529

10

Ion Exchange in Geopolymers

José Ramón Gasca-Tirado, Alejandro Manzano-Ramírez, Eric M. RiveraMuñoz, Rodrigo Velázquez-Castillo, Miguel Apátiga-Castro, Rufino Nava and Aarón Rodríguez-López

Abstract

Geopolymers have been widely used for construction and building materials. Nevertheless, some other applications have been found from their ability to be ion exchanged. An example is the encapsulation of heavy metals, but some others involve the ion exchange of the aluminosilicate structure to form photoactive particles or to link copper ions. In this chapter, we summarize some of the properties which make aluminosilicate inorganic polymer (geopolymers) ion exchangeable: the synthesized temperature, its effect over their porosity and their stoichiometric nature. Also, the effects of ion exchanging a geopolymer with an NH_4^+Cl, $(NH_4)_2TiO_2(C_2O_4)_2$ and $(CH_3)_4N^+Br$ are presented. The geopolymer was characterized by FT-IR, XRD, BET and MAS NMR, showing how a 100% of replacement was achieved for NH_4^+Cl. On the contrary, the efficiency was reduced in $(NH_4)_2TiO_2(C_2O_4)_2$ and $(CH_3)_4N^+Br$, effect ascribed to the fact of the molecular size that did not allow the counterions to reach the aluminum atoms in the geopolymer. Finally, the procedure followed to ion exchange a metakaolinite-based geopolymer is described, and the potential applications related are presented.

Keywords: geopolymer, ion exchange

1. Introduction

Since the origin of the term geopolymer to define those materials synthesized by the alkali activation of aluminosilicate monomers at ambient temperature, studies about geopolymers have been widely spread [1].

Examples include environmental alternatives for Portland cement [2], refractories [3] and ceramic precursor, for instance [4, 5].

Nevertheless, while these studies continue, others have explored new applications as immobilizations of waste, radioactive and toxic materials [6, 7], medicinal applications, catalytic materials [8, 9], catalytic materials for volatile organic degradation [9–11] and support material for optical applications like color holder, color pH indicator and photoluminescent material.

These new applications take advantage of the nonordered AlO_4-SiO_4 structure of geopolymers that results in a negative charged framework where alkali or alkaline earth cations (Na^+, K^+, Ca^{2+}, Li^+ and rarely Cs^+) need to be present to balance the charge of the geopolymer.

The chemical similarity of geopolymers to zeolites makes them an important former material to synthesize a wide variety of new ceramics through the ion exchange of the charge-balancing cation [8, 12].

To improve the studies and increase the amount of applications of geopolymers, it is necessary to understand clearly the core variables that determine a good ion exchange. Therefore, in this chapter, we review some of these core variables which make geopolymers ion exchangeable: the synthesis temperature was varied from 40 to 90°C to define a specific pore size distribution (PSD) within the samples, the removal of soluble species from the samples and the use of the right chemical compound according to the desired ion to be exchanged to avoid any change in the geopolymer structure after ion exchange. Finally, we describe the procedure followed to ion exchange a metakaolinite-based geopolymer and their potential applications.

2. Synthesis of geopolymers

Geopolymers are synthesized by the dissolution and polycondensation of an aluminosilicate powder in an alkaline silicate solution under hydrothermal conditions [13]. Even though there are different sources for solid aluminosilicate like slag, fly ash and mine tailing [2], the metakaolin is the preferred starting material chosen to synthesize geopolymers. This is mainly due to its high purity that results in a suitable geopolymer to be ion exchanged.

On the other hand, the available alkaline silicate solutions needed to synthesize a geopolymer are limited to solution in which cations are alkaline enough for synthesizing geopolymers (e.g. Na^+ or K^+) [14]. Therefore, the variety of geopolymers that can be synthesized is constrained to this issue.

The samples synthesized for this study were prepared by mechanically mixing stoichiometric amounts of metakaolin, sodium hydroxide, distilled water and sodium silicate (Na_2O/SiO_2 wt. ratio: 0.31) to follow the recommended molar ratios: $SiO_2:Al_2O_3$ = 2.89, $NaO_2:SiO_2$ = 0.29 and $H_2O:Na_2O$ = 10.01 which gave geopolymers good mechanical properties (3.16, 0.34 and16.2 for the SiO_2/Al_2O_3, K_2O/SiO_2 and H_2O/K_2O, respectively, in the case of potassium geopolymers) [15, 16].

To have a homogeneous slurry, a mechanical mixing followed by 15 min of vibration was done. Final samples were casted into plastic containers, and they were left under constant temperature at 40 and 90°C for 24 h [16–19]. After cooling, geopolymers were cut with a diamond disc to have samples of around 1 g weight.

3. Geopolymer structure

As all geopolymers, the sample synthesis takes place in local coordination environment like zeolites. The silicate and aluminate units that form the geopolymer are completely disordered on length scales longer than 10 Å and are connected to form a disordered network

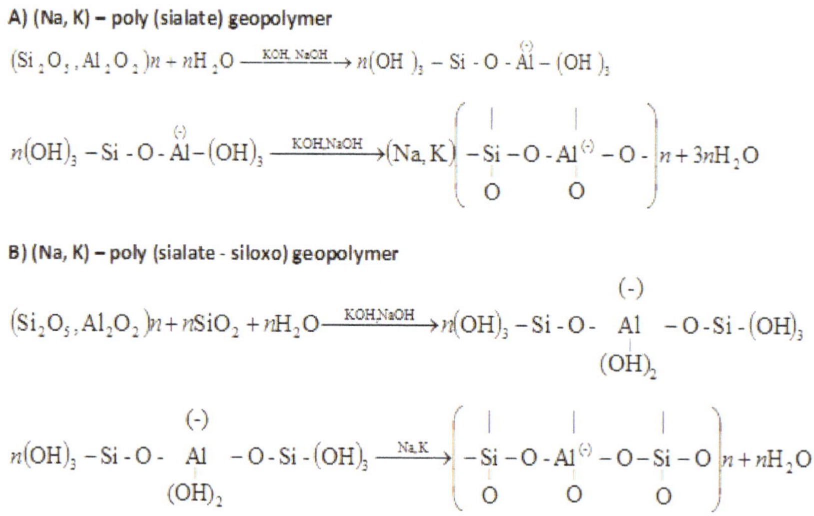

Figure 1. Formation of geopolymers: (A) poly(sialate) and (B) poly(sialate-siloxo).

Si:Al ratio	Potential applications	Si:Al ratio	Potential applications
Si-O-Si-O-Si-O-Si-O-Si-O / O / Al / O / Si-O-Si-O-Si-O-Si-O-Si-O 20:1 < Si:Al < 35:1	Fire resistant materials	(-Si-O-Al-O-Si-O) Si:Al 2:1	Cementitious material. Toxic waste encapsulation.
(Si-O-Al-O-Si-O / O / Si / O / (F, SiO₂)) Si:Al >3:1	Thermal insulation material (200 to 600 °C). Tooling for SPF aluminium.	(-Si-O-Al-O) Si:Al 1:1	Fire protection construction Materials. Building bricks.
(Si-O-Al-O-Si-O / O / Si / O) Si:Al 3:1	Thermal insulation material (200 a 1000 °C). Tooling for titanium processing.		

Figure 2. Potential applications of geopolymers according to the Si:Al ratio.

with the alkali cations in irregular sites [14]. These cations are present inside the geopolymer framework, and they are needed to balance the negative charge of Al_3^+ in $[AlO_4]^{-1}$ [20].

As it can be seen in **Figure 1** where geopolymers are classified according to the ratio of aluminates and silicates, the geopolymer synthesized was poly(sialate) geopolymers which are mainly used as cementitious and toxic waste encapsulation material (**Figure 2**).

According to their structure, the applications of geopolymer may be summarized in **Figure 2**.

4. Background of ion exchange

Ion exchange is a chemical reaction where an ion carrier material (ion exchanger) exchanges their own ions with others disposed in solution (counterions). Even though the ion exchanger is not dissolved, the ion exchanger must have an open network structure to let ions be dissolved in the aqueous solution [21].

The ion exchangers can be cationic, anionic or amphoteric in accordance with the charge of the balancing ion, and the number of ions exchanged should be equal to the proportion of free counterions in solution in agreement with the stoichiometric ratio of exchange.

In the case of geopolymers, which are X-ray amorphous materials formed by non-localized but well-distributed negative tetrahedral silicate and aluminate units [1, 15], they can be considered as cationic exchangers. Their negatively charged structure, balanced by alkali metals such as Na^+ or K^+ (**Figure 3**), can be completely hydrated and mobilized [3, 22]. This gives a lower bonding strength in comparison with zeolites [20] and an ease to be ion exchanged when they are in contact with solutions of chloride or nitrate of a desired cation.

The geopolymer structure is made up by silicon and aluminium atoms linked by oxygen (**Figure 3**), and it is not modified when the sodium atom, the exchangeable ion linked to the aluminum atom, is replaced with another cation. The maximum number of possible exchangeable cations, known as theoretical ion exchange capacity (TEC), is equal to the negative charge of the geopolymer network and can be measured by elemental analysis of aluminum. Nevertheless, it is not always possible to ion exchange the total amount of available exchangeable cations (TEC value), and a real exchange capacity (REC value) should be determined by an ion-exchange method [23, 24].

The general procedure of ion exchange can be formulated by Eq. (1). When the geopolymer, represented as Na^+G^- (ion exchanger), is ionized in an aqueous solution, the released cation (Na^+ or K^+) diffuses inside the material, transfers through the interphase boundary and finally arrives to the solution. If a salt MY is also dissolved in the solution (M^+ and Y^- ions), the Na^+ in geopolymer is replaced by an equivalent amount of counterion M^+ due to the electroneutrality requirement:

$$Na^+G^- + M^+Y^- \rightarrow M^+G^- + Na^+ + Y^- \qquad (1)$$

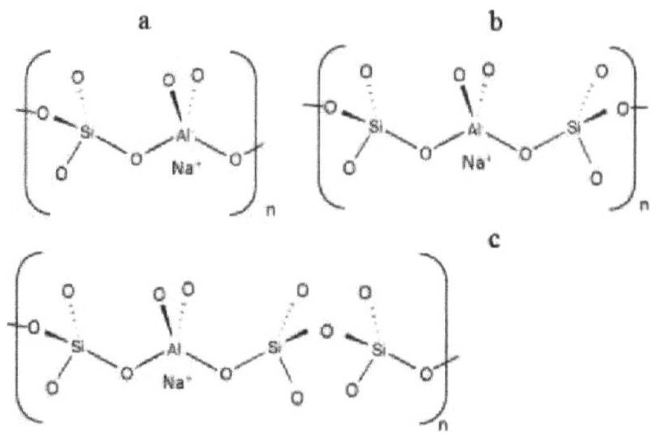

Figure 3. Position of the cation in the (a) sialate geopolymer, (b) sialate-siloxo geopolymer and (c) sialate-disiloxo geopolymer.

5. Procedure of ion exchange in geopolymers

Before an ion exchange is done, some soluble species from the samples must be removed. In this case, we followed a procedure described by Skorina [18]. The samples were rinsed several times with deionized water and a sequence of wash cycles, followed by a final treatment with a solution of KNO_3 until neutral "pH" was reached.

According to the desired ion to be exchanged with, the geopolymer was kept in contact for 12 h with solutions of NH_4Cl, $(NH_4)_2TiO_2(C_2O_4)_2$ and $(CH_3)_4NBr$ in concentration of 0.1 M with the aim of having a larger proportion of free counterions in solution and to guarantee a complete ion exchange. Finally, the samples were removed from the solution, rinsed several times with deionized water and dried for 1 week at ambient temperature.

6. Test performed over geopolymers

The pore diameter and surface area of samples were done under water desorpted samples (120° C for 24 h) by nitrogen adsorption on a NOVA 2000e Quantachrome instruments employing the Barrett–Joyner–Halenda method of analysis.

The geopolymer external surface was observed with a Philips XL-30 ESEM scanning electron microscope coupled with an EDAX energy-dispersive X-ray microanalyzer. The extent of ion exchange was determined by calculating the ratio of sodium with aluminum in the geopolymer as it was described by O'Connor [14].

The modification to the geopolymer structure was followed with a RIGAKU Ultima IV X-ray diffractometer using Cu irradiation (λ = 1.5406 Å), and the chemical composition was analyzed with a Perkin Elmer Spectrum Two using the IR-KBr technique.

Finally, ^{27}Al and ^{29}Si MAS NMR was performed to distinguish the atoms surrounding the aluminum atom with a Bruker Avance III HD 400 MHZ MAS NMR. This equipment was operated at 15 KHz for ^{27}Al and 8 KHz for ^{29}Si.

7. Effect of ion exchange over geopolymers

The total surface area and pore diameter of geopolymers synthesized at 40 and 90°C are summarized in **Table 1**. As it can be observed, the main factor that modified these parameters was the temperature synthesis. In addition, a reduction in pore width and the increase in specific surface were observed in all samples when the geopolymer was ion exchanged.

Modifications on the geopolymer surface were also observed in **Figure 4**. The ion exchanged samples were settled of small particles that occupied the void sites of the former geopolymer but leaving unchanged the geopolymer structure.

Counterion	Geopolymers at 40°C				Geopolymers at 90°C			
	Before ion exchange		After ion exchange		Before ion exchange		After ion exchange	
	S ($m^2\ g^{-1}$)	D (nm)	S ($m^2\ g^{-1}$)	D (nm)	S ($m^2\ g^{-1}$)	D (nm)	S ($m^2\ g^{-1}$)	D (nm)
NH_4Cl	45.88	19.56	123.84	2.19	27.21	30.52	491.33	2.45
$(NH_4)_2TiO_2(C_2O_4)_2$	45.88	19.56	526.54	2.19	27.21	30.52	195.99	2.19
$(CH_3)_4NBr$	45.88	19.56	150.14	2.19	27.21	30.52	450.17	2.7

Table 1. Surface area and pore diameter of geopolymers at 40 and 90°C, before and after ion exchange.

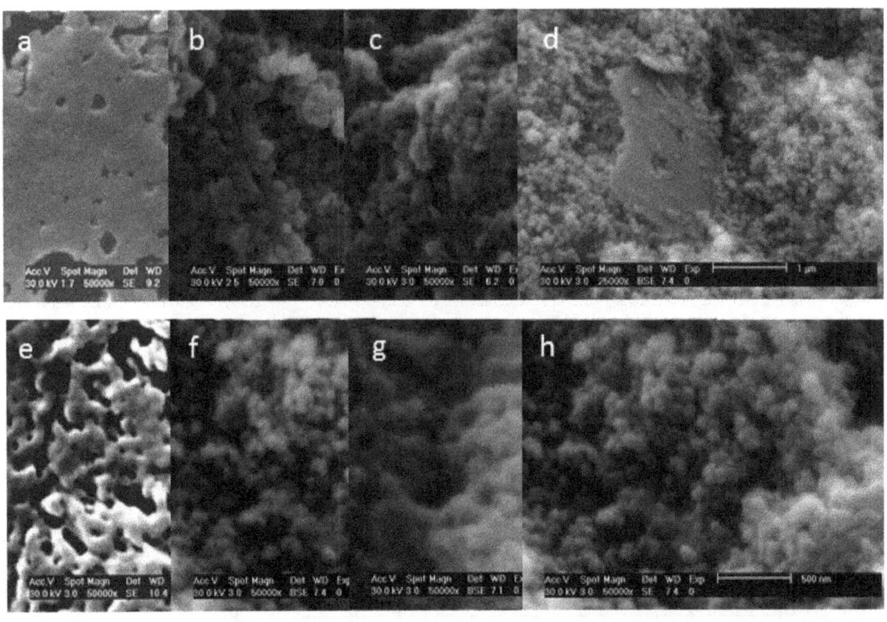

Figure 4. Effect of the ion exchange over the geopolymer surface. (a) Geopolymer at 40°C, (b) geopolymer at 40°C ion exchanged with NH_4Cl, (c) geopolymer at 40°C ion exchanged with $(NH_4)_2TiO_2(C_2O_4)_2$, (d) geopolymer at 40°C ion exchanged with $(CH_3)_4NBr$, (e) geopolymer at 90°C, (f) geopolymer at 90°C ion exchanged with NH_4Cl, (g) geopolymer at 90°C ion exchanged with $(NH_4)_2TiO_2(C_2O_4)_2$ and (h) geopolymer at 90°C ion exchanged with $(CH_3)_4NBr$.

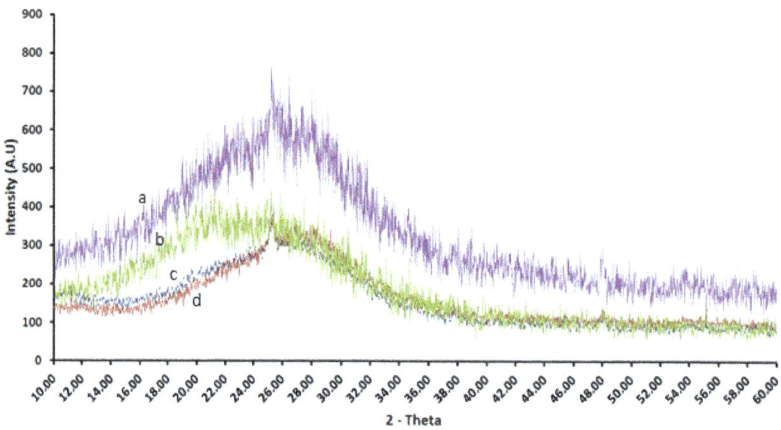

Figure 5. XRD pattern of (a) former geopolymer, (b) geopolymer ion exchanged with $(NH_4)_2TiO_2(C_2O_4)_2$, (c) geopolymer ion exchanged with NH_4Cl and (d) geopolymer ion exchanged with $(CH_3)_4NBr$.

This observation was supported by results obtained by XRD (**Figure 5**), where the amorphous structure of the geopolymer, represented by a hump at $2\Theta = 28°$, remained after ion exchange.

On the other hand, the FT-IR results of geopolymers at 40°C and 90°C before and after being ion exchanged are shown in **Figures 6** and 7. As it can be observed in all spectra, the vibration modes corresponding to amorphous sodium aluminosilicate (1645.44, 878.56 cm^{-1}), symmetric stretching of Si-O-Al, Si-O-Si (706.9 cm^{-1}) and flexion of Si-O-Si and O-Si-O (571.73 cm^{-1})

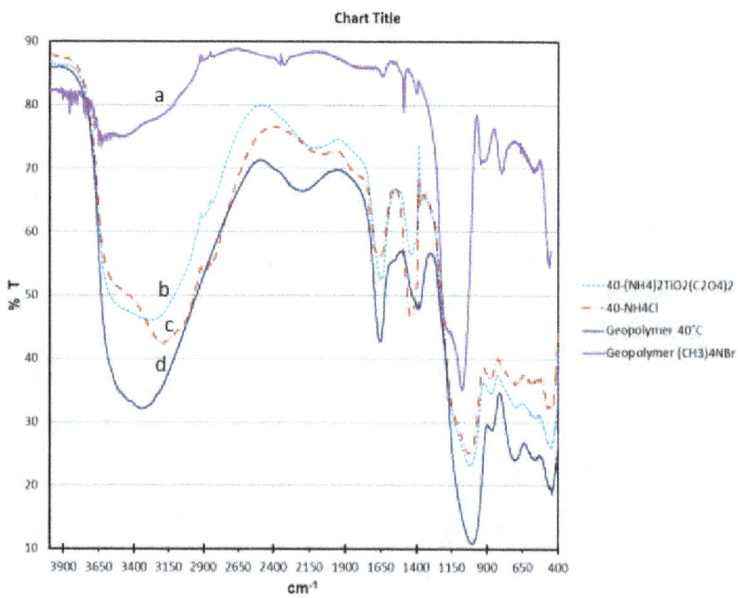

Figure 6. FT-IR of (a) geopolymer at 40°C ion exchanged with $(CH_3)_4NBr$, (b) geopolymer at 40°C ion exchanged with $(NH_4)_2TiO_2(C_2O_4)_2$, (c) geopolymer at 40°C ion exchanged with NH_4Cl and (d) geopolymer at 40°C.

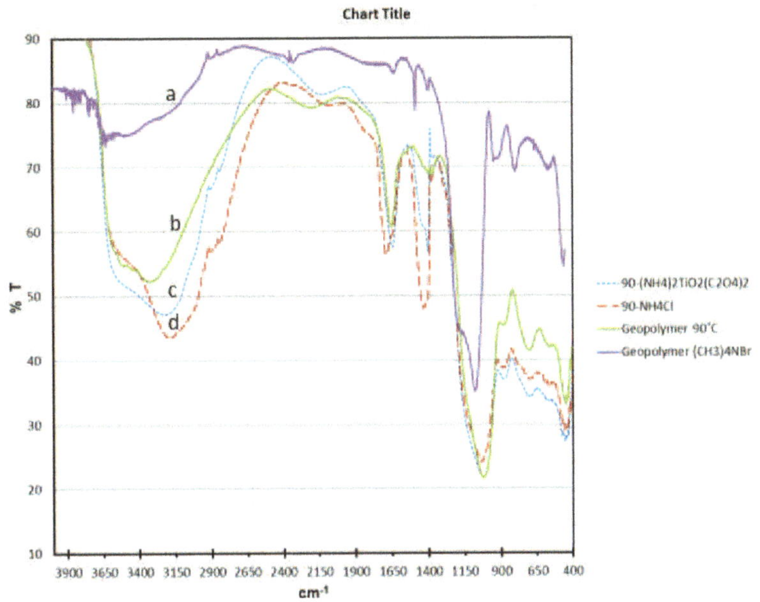

Figure 7. FT-IR of (a) geopolymer at 90°C ion exchanged with $(CH_3)_4NBr$, (b) geopolymer at 90°C, (c) geopolymer at 90°C ion exchanged with $(NH_4)_2TiO_2(C_2O_4)_2$ and (d) geopolymer at 90°C ion exchanged with NH_4Cl.

[25, 26] remained as a former geopolymer. The only modifications observed between spectra were those ascribed to vibration modes representative of their counterions.

Whenever a geopolymer is synthesized, the three-coordinated aluminum in metakaolin, Al (IV), Al (V) and Al (VI) (chemical shift: 49–80, 35–40 and −5–15 ppm, respectively), are transformed into Al (IV) to form the geopolymer structure [27]. As it can be observed in **Figure 8**, this structure remained in all ion exchanged samples, and in some occasions, Al (V) and AI (VI) arose at 30 ppm and 2 ppm as in the case of $(CH_3)_4NBr$ because of the size of the counterion mainly.

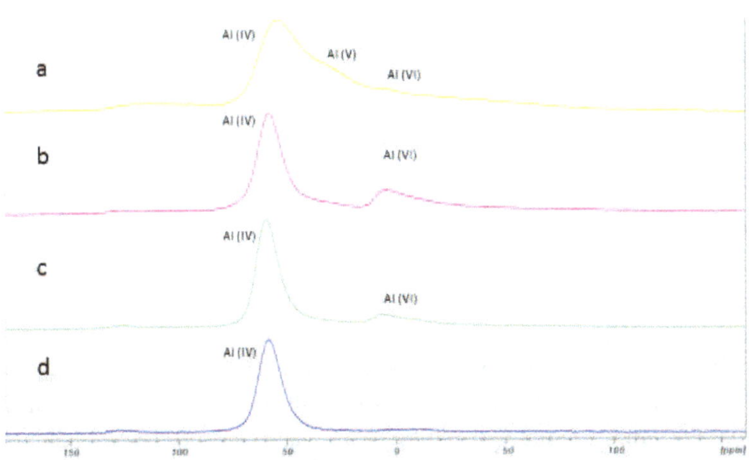

Figure 8. ^{27}Al MAS NMR of (a) geopolymer ion exchanged with $(CH_3)_4NBr$, (b) geopolymer ion exchanged with NH_4Cl,

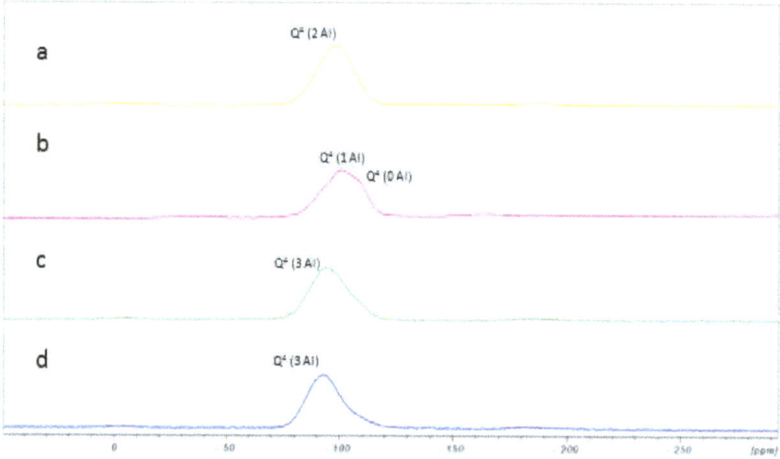

Figure 9. ^{29}Si MAS NMR of (a) geopolymer ion exchanged with $(CH_3)_4NBr$, (b) geopolymer ion exchanged with NH_4Cl, (c) geopolymer ion exchanged with $(NH_4)_2TiO_2(C_2O_4)_2$ and (d) former geopolymer.

Counterion	Real exchange capacity (%)
NH_4^+	100
$(NH_4)_2TiO_2(C_2O_4)_2$	80
$(CH_3)_4N^+$	85
Li^+	82
Cd^+	78
Mg^{2+}	57

Table 2. Real exchange capacity of geopolymers.

In the case of the ^{29}Si MAS NMR, all the geopolymer spectra presented a resonance peak between −90 and −100 ppm which were ascribed to tetrahedral silicon surrounded by two aluminum atoms [28] (**Figure 9**).

Finally, the real exchange capacity of samples is summarized in **Table 2**. A total replacement of sodium cation by NH_4^+ and a reduction on samples ion exchanged with $(NH_4)_2TiO_2(C_2O_4)_2$ and $(CH_3)_4N^+$ were determined. Results reported by O'Connor for Li^+, Cd^+ or Mg^{2+} are summarized, as a comparative, in **Table 2** [14].

8. Conclusions

In this chapter, the effects of ion exchanging a geopolymer with NH_4^+Cl, $(NH_4)_2TiO_2(C_2O_4)_2$ and $(CH_3)_4N^+Br$ were analyzed. Results were compared with the ion exchange with Li^+, Cd^+ and Mg^{2+} reported by O'Connor. A 100% of replacement was achieved for NH_4^+Cl. Nevertheless, the efficiency was reduced in $(NH_4)_2TiO_2(C_2O_4)_2$ and $(CH_3)_4N^+Br$ probably due to their molecular size that inhibited the counterions to reach the aluminum atom.

As it was observed by FT-IR, the ion exchange procedure did not modify the amorphous structure of geopolymers and the vibration modes ascribed to the symmetric stretching of Si-O-Al and Si-O-S, and the flexion energy of Si-O-Si and O-Si-O remained unmodified after the process. This can be explained if it is considered that the exchanged atom, Na^+, does not take part of the rigid structure of the geopolymer.

A similar conclusion was obtained from the XRD pattern, where the characteristic hump of amorphous geopolymer remained in all samples. Same results were observed in geopolymers synthesized at 40 and 90°C. The only observable change in this case was the superficial morphology which was measured by BET. Results gave a more porous material for samples synthesized at 40°C, but the extent of ion exchange was not affected by this parameter as it was summarized in **Table 2**.

In addition, the ^{27}Al and ^{29}Si MAS NMR spectra exposed the presence of Al (IV) and tetrahedral silicon, respectively, as an indicative of a structural order. In some cases, Al (V) and Al (VI) arose as it was reported by O'Connor for Mg^{2+}, Pb^{2+} and Cd^{2+} ions.

Finally, because the variety of geopolymers that can be produced by direct synthesis is limited to the selection of Na^+ or K^+ cations, the aim to find new application for geopolymer should lead research to find more exchange counterions.

Author details

José Ramón Gasca-Tirado[1*], Alejandro Manzano-Ramírez[2], Eric M. RiveraMuñoz[3], Rodrigo Velázquez-Castillo[4], Miguel Apátiga-Castro[3], Rufino Nava[4] and Aarón Rodríguez-López[5]

*Address all correspondence to: ragatsi99@yahoo.com

1 Universidad de Guanajuato, Campus Celaya-Salvatierra, Celaya, Guanajuato, México

2 Centro de Investigaciones y Estudios Avanzados del I.P.N., Unidad Querétaro, Querétaro, México

3 Centro de Física Aplicada y Tecnología Avanzada, Universidad Nacional Autónoma de México, Querétaro, México

4 División de Investigación y Posgrado, Facultad de Ingeniería, Universidad Autónoma de Querétaro, Querétaro, México

5 Universidad Politécnica de Santa Rosa Jáuregui, Santa Rosa Jáuregui, Querétaro, México

References

[1] Davidovits J. Geopolymers. Journal of Thermal Analysis. 1991;**37**:1633-1656. DOI: 10.1007/BF01912193

[2] Duxson P, Provis JL, Lukey GC, van Deventer JSJ. The role of inorganic polymer technology in the development of 'green concrete'. Cement and Concrete Research. 2007;**37**:1590-1597. DOI: 10.1016/j.cemconres.2007.08.018

[3] Barbosa VFF, MacKenzie KJD. Synthesis and thermal behaviour of potassium sialate geopolymers. Materials Letters. 2003;**57**:1477-1482. DOI: 10.1016/S0167-577X(02)01009-1

[4] Kyoichi O, Yoshinori N, Swaroop V. Optical spectra of nickel-bearing silicate gels prepared by the geopolymer technique, with special reference to the low-temperature formation of liebenbergite (Ni_2SiO_4). Journal of the American Ceramic Society. 2004;**84**: 1717-1720. DOI: 10.1111/j.1151-2916.2001.tb00904.x

[5] Bell JL, Driemeyer PE, Kriven WM. Formation of ceramics from metakaolin-based geopolymers. Part II: K-based geopolymer. Journal of the American Ceramic Society. 2009;**92**:607-615. DOI: 10.1111/j.1551-2916.2008.02922.x

[6] Van Jaarsveld JGS, Van Deventer JSJ, Schwartzman A. The potential use of geopolymeric materials to immobilise toxic metals: Part II. Material and leaching characteristics. Minerals Engineering. 1999;**12**:75-91. DOI: 10.1016/S0892-6875(98)00121-6

[7] Khalil MY, Merz E. Immobilization of intermediate-level wastes in geopolymers. Journal of Nuclear Materials. 1994;**211**:141-148. DOI: 10.1016/0022-3115(94)90364-6

[8] O'Connor SJ, MacKenzie KJD. Synthesis, characterisation and thermal behaviour of lithium aluminosilicate inorganic polymers. Journal of Materials Science. 2010;**45**: 3707-3713. DOI: 10.1007/s10853-010-4383-x

[9] Sazama P, Bortnovsky O, Dědeček J, Tvarůžková Z, Sobalík Z. Geopolymer based catalysts—New group of catalytic materials. Catalysis Today. 2011;**164**:92-99. DOI: 10.1016/j.cattod.2010.09.008

[10] Gasca-Tirado JR, Manzano-Ramírez A, Reyes-Araiza JL. The Potential Use of Geopolymer for Cleaning Air. 2017. DOI: 10.1016/B978-0-08-102001-2.00009-7

[11] Gasca-Tirado JR, Manzano-Ramírez A, Villaseñor-Mora C, Muñiz-Villarreal MS, Zaldivar-Cadena AA, Rubio-Ávalos JC, et al. Incorporation of photoactive TiO_2 in an aluminosilicate inorganic polymer by ion exchange. Microporous and Mesoporous Materials. 2012;**153**. DOI: 10.1016/j.micromeso.2011.11.026

[12] Oleg B, Jiri D, Zdenka T, Zdeněk S, Jan Š. Metal ions as probes for characterization of geopolymer materials. Journal of the American Ceramic Society. 2008;**91**:3052-3057. DOI: 10.1111/j.1551-2916.2008.02577.x

[13] Cengiz B, Gregory PK, Kevin CS, Waltraud MK. Synthesis and characterization of silicon carbide powders converted from metakaolin-based geopolymer. Journal of the American Ceramic Society. 2016;**99**:2521-2530. DOI: 10.1111/jace.14254

[14] O'Connor SJ, MacKenzie KJD, Smith ME, Hanna JV. Ion exchange in the charge-balancing sites of aluminosilicate inorganic polymers. Journal of Materials Chemistry. 2010;**20**:10234-10240. DOI: 10.1039/c0jm01254h

[15] Barbosa VFF, MacKenzie KJD, Thaumaturgo C. Synthesis and characterisation of materials based on inorganic polymers of alumina and silica: Sodium polysialate polymers. International Journal of Inorganic Materials. 2000;**2**:309-317. DOI: 10.1016/S1466-6049(00)00041-6

[16] Rogers JJ, MacKenzie KJD, Trompetter WJ. New phosphors synthesised by ion exchange of a metakaolin-based geopolymer. Applied Clay Science. 2018;**157**:1-7. DOI: 10.1016/j.clay.2018.02.025

[17] Gasca-Tirado JR, Manzano-Ramírez A, Vazquez-Landaverde PA, Herrera-Díaz EI, Rodríguez-Ugarte ME, Rubio-Ávalos JC, et al. Ion-exchanged geopolymer for photocatalytic degradation of a volatile organic compound. Materials Letters. 2014. DOI: 10.1016/j.matlet.2014.07.090

[18] Skorina T. Ion exchange in amorphous alkali-activated aluminosilicates: Potassium based geopolymers. Applied Clay Science. 2014;**87**:205-211. DOI: 10.1016/j.clay.2013.11.003

[19] Singhal A, Gangwar BP, Gayathry JM. CTAB modified large surface area nanoporous geopolymer with high adsorption capacity for copper ion removal. Applied Clay Science. 2017;**150**:106-114. DOI: 10.1016/j.clay.2017.09.013

[20] He P, Fu S, Yuan J, Rao J, Xu J, Wang P, et al. Celsian formation from barium-exchanged geopolymer precursor: Thermal evolution. Journal of the European Ceramic Society. 2017;**37**:4179-4185. DOI: 10.1016/j.jeurceramsoc.2017.04.074

[21] Kumar S, Jain S. History, introduction, and kinetics of ion exchange materials. Journal of Chemistry. 2013;**2013**:13. DOI: 10.1155/2013/957647

[22] van Deventer JSJ, Provis JL, Duxson P, Lukey GC. Reaction mechanisms in the geopolymeric conversion of inorganic waste to useful products. Journal of Hazardous Materials. 2007;**139**:506-513. DOI: 10.1016/j.jhazmat.2006.02.044

[23] Inglezakis VJ. The concept of "capacity" in zeolite ion-exchange systems. Journal of Colloid and Interface Science. 2005;**281**:68-79. DOI: 10.1016/j.jcis.2004.08.082

[24] Lehto J, Harjula R. Experimentation in ion exchange studies—The problem of getting reliable and comparable results. Reactive and Functional Polymers. 1995;**27**:121-146. DOI: 10.1016/1381-5148(95)00038-H

[25] Rees CA, Provis JL, Lukey GC, van Deventer JSJ. Attenuated total reflectance Fourier transform infrared analysis of fly ash geopolymer gel aging. Langmuir. 2007;**23**:8170-8179. DOI: 10.1021/la700713g

[26] Serra J, González P, Liste S, Chiussi S, León B, Pérez-Amor M, et al. Influence of the non-bridging oxygen groups on the bioactivity of silicate glasses. Journal of Materials Science. Materials in Medicine. 2002;**13**:1221-1225. DOI: 10.1023/A:1021174912802

[27] Škvára F, Kopecký L, Šmilauer V, Bittnar Z. Material and structural characterization of alkali activated low-calcium brown coal fly ash. Journal of Hazardous Materials. 2009;**168**:711-720. DOI: 10.1016/j.jhazmat.2009.02.089

Permissions

All chapters in this book were first published by InTech Open; hereby published with permission under the Creative Commons Attribution License or equivalent. Every chapter published in this book has been scrutinized by our experts. Their significance has been extensively debated. The topics covered herein carry significant findings which will fuel the growth of the discipline. They may even be implemented as practical applications or may be referred to as a beginning point for another development.

The contributors of this book come from diverse backgrounds, making this book a truly international effort. This book will bring forth new frontiers with its revolutionizing research information and detailed analysis of the nascent developments around the world.

We would like to thank all the contributing authors for lending their expertise to make the book truly unique. They have played a crucial role in the development of this book. Without their invaluable contributions this book wouldn't have been possible. They have made vital efforts to compile up to date information on the varied aspects of this subject to make this book a valuable addition to the collection of many professionals and students.

This book was conceptualized with the vision of imparting up-to-date information and advanced data in this field. To ensure the same, a matchless editorial board was set up. Every individual on the board went through rigorous rounds of assessment to prove their worth. After which they invested a large part of their time researching and compiling the most relevant data for our readers.

The editorial board has been involved in producing this book since its inception. They have spent rigorous hours researching and exploring the diverse topics which have resulted in the successful publishing of this book. They have passed on their knowledge of decades through this book. To expedite this challenging task, the publisher supported the team at every step. A small team of assistant editors was also appointed to further simplify the editing procedure and attain best results for the readers.

Apart from the editorial board, the designing team has also invested a significant amount of their time in understanding the subject and creating the most relevant covers. They scrutinized every image to scout for the most suitable representation of the subject and create an appropriate cover for the book.

The publishing team has been an ardent support to the editorial, designing and production team. Their endless efforts to recruit the best for this project, has resulted in the accomplishment of this book. They are a veteran in the field of academics and their pool of knowledge is as vast as their experience in printing. Their expertise and guidance has proved useful at every step. Their uncompromising quality standards have made this book an exceptional effort. Their encouragement from time to time has been an inspiration for everyone.

The publisher and the editorial board hope that this book will prove to be a valuable piece of knowledge for researchers, students, practitioners and scholars across the globe.

List of Contributors

Roman Rogoziński
Optoelectronics Department, Silesian University of Technology, Gliwice, Poland

Hamid Hassani and Vincenzo M. Sglavo
Department of Industrial Engineering, University of Trento, Trento, Italy

Zbigniew Hubicki, Monika Wawrzkiewicz, Grzegorz Wójcik, Dorota Kołodyńska and Anna Wołowicz
Department of Inorganic Chemistry, Faculty of Chemistry, Maria Curie-Sklodowska University, Lublin, Poland

Monika Wawrzkiewicz and Zbigniew Hubicki
Maria Curie-Sklodowska University, Faculty of Chemistry, Department of Inorganic Chemistry, Lublin, Poland

Humberto Naoyuki Yoshimura
Center for Engineering, Modelling and Applied Social Sciences, Universidade Federal do ABC, Santo André, SP, Brazil

Paulo Francisco Cesar
School of Dentistry, Universidade de São Paulo, São Paulo, SP, Brazil

Melissa Gurgel Adeodato Vieira and Meuris Gurgel Carlos da Silva
School of Chemical Engineering, University of Campinas, Campinas, SP, Brazil

Ali Talimian and Vincenzo M. Sglavo
Department of Industrial Engineering, University of Trento, Trento, Italy

Maria Angélica Simões Dornellas de Barros and Marcelino Luiz Gimenes
Chemical Engineering Department, State University of Maringá, Maringá, PR, Brazil

Miroslav Kukučka and Nikoleta Kukučka
Envirotech d.o.o., Kikinda, Serbia

Rajendra Sukhadeorao Dongre
Department of Chemistry, RTM Nagpur University, Nagpur, Maharashtra, India

José Ramón Gasca-Tirado
Universidad de Guanajuato, Campus Celaya-Salvatierra, Celaya, Guanajuato, México

Alejandro Manzano-Ramírez
Centro de Investigaciones y Estudios Avanzados del I.P.N., Unidad Querétaro, Querétaro, México

Eric M. RiveraMuñoz and Miguel Apátiga-Castro
Centro de Física Aplicada y Tecnología Avanzada, Universidad Nacional Autónoma de México, Querétaro, México

Rodrigo Velázquez-Castillo and Rufino Nava
División de Investigación y Posgrado, Facultad de Ingeniería, Universidad Autónoma de Querétaro, Querétaro, México

Aarón Rodríguez-López
Universidad Politécnica de Santa Rosa Jáuregui, Santa Rosa Jáuregui, Querétaro, México

Index

www.ingramcontent.com/pod-product-compliance
Lightning Source LLC
LaVergne TN
LVHW062231200926
848484LV00006B/1517